Resources for Teaching Discrete Mathematics

Classroom Projects, History Modules, and Articles

© 2009 by

The Mathematical Association of America (Incorporated)

Library of Congress Control Number 2008941993

ISBN 978-0-88385-184-5

Printed in the United States of America

Current Printing (last digit):

10 9 8 7 6 5 4 3 2 1

Resources for Teaching Discrete Mathematics

Classroom Projects, History Modules, and Articles

Edited by

Brian Hopkins
Saint Peter's College
Jersey City, NJ

Published and Distributed by
The Mathematical Association of America

The MAA Notes Series, started in 1982, addresses a broad range of topics and themes of interest to all who are involved with undergraduate mathematics. The volumes in this series are readable, informative, and useful, and help the mathematical community keep up with developments of importance to mathematics.

MAA Notes

14. Mathematical Writing, by *Donald E. Knuth, Tracy Larrabee, and Paul M. Roberts*.
16. Using Writing to Teach Mathematics, *Andrew Sterrett*, Editor.
17. Priming the Calculus Pump: Innovations and Resources, Committee on Calculus Reform and the First Two Years, a subcomittee of the Committee on the Undergraduate Program in Mathematics, *Thomas W. Tucker*, Editor.
18. Models for Undergraduate Research in Mathematics, *Lester Senechal*, Editor.
19. Visualization in Teaching and Learning Mathematics, Committee on Computers in Mathematics Education, *Steve Cunningham and Walter S. Zimmermann*, Editors.
20. The Laboratory Approach to Teaching Calculus, *L. Carl Leinbach et al.*, Editors.
21. Perspectives on Contemporary Statistics, *David C. Hoaglin and David S. Moore*, Editors.
22. Heeding the Call for Change: Suggestions for Curricular Action, *Lynn A. Steen*, Editor.
24. Symbolic Computation in Undergraduate Mathematics Education, *Zaven A. Karian*, Editor.
25. The Concept of Function: Aspects of Epistemology and Pedagogy, *Guershon Harel and Ed Dubinsky*, Editors.
26. Statistics for the Twenty-First Century, *Florence and Sheldon Gordon*, Editors.
27. Resources for Calculus Collection, Volume 1: Learning by Discovery: A Lab Manual for Calculus, *Anita E. Solow*, Editor.
28. Resources for Calculus Collection, Volume 2: Calculus Problems for a New Century, *Robert Fraga*, Editor.
29. Resources for Calculus Collection, Volume 3: Applications of Calculus, *Philip Straffin*, Editor.
30. Resources for Calculus Collection, Volume 4: Problems for Student Investigation, *Michael B. Jackson and John R. Ramsay*, Editors.
31. Resources for Calculus Collection, Volume 5: Readings for Calculus, *Underwood Dudley*, Editor.
32. Essays in Humanistic Mathematics, *Alvin White*, Editor.
33. Research Issues in Undergraduate Mathematics Learning: Preliminary Analyses and Results, *James J. Kaput and Ed Dubinsky*, Editors.
34. In Eves Circles, *Joby Milo Anthony*, Editor.
35. Youre the Professor, What Next? Ideas and Resources for Preparing College Teachers, The Committee on Preparation for College Teaching, *Bettye Anne Case*, Editor.
36. Preparing for a New Calculus: Conference Proceedings, *Anita E. Solow*, Editor.
37. A Practical Guide to Cooperative Learning in Collegiate Mathematics, *Nancy L. Hagelgans, Barbara E. Reynolds, SDS, Keith Schwingendorf, Draga Vidakovic, Ed Dubinsky, Mazen Shahin, G. Joseph Wimbish, Jr.*
38. Models That Work: Case Studies in Effective Undergraduate Mathematics Programs, *Alan C. Tucker*, Editor.
39. Calculus: The Dynamics of Change, CUPM Subcommittee on Calculus Reform and the First Two Years, *A. Wayne Roberts*, Editor.
40. Vita Mathematica: Historical Research and Integration with Teaching, *Ronald Calinger*, Editor.
41. Geometry Turned On: Dynamic Software in Learning, Teaching, and Research, *James R. King and Doris Schattschneider*, Editors.
42. Resources for Teaching Linear Algebra, *David Carlson, Charles R. Johnson, David C. Lay, A. Duane Porter, Ann E. Watkins, William Watkins*, Editors.
43. Student Assessment in Calculus: A Report of the NSF Working Group on Assessment in Calculus, *Alan Schoenfeld*, Editor.
44. Readings in Cooperative Learning for Undergraduate Mathematics, *Ed Dubinsky, David Mathews, and Barbara E. Reynolds*, Editors.
45. Confronting the Core Curriculum: Considering Change in the Undergraduate Mathematics Major, *John A. Dossey*, Editor.
46. Women in Mathematics: Scaling the Heights, *Deborah Nolan*, Editor.
47. Exemplary Programs in Introductory College Mathematics: Innovative Programs Using Technology, *Susan Lenker*, Editor.
48. Writing in the Teaching and Learning of Mathematics, *John Meier and Thomas Rishel*.
49. Assessment Practices in Undergraduate Mathematics, *Bonnie Gold*, Editor.
50. Revolutions in Differential Equations: Exploring ODEs with Modern Technology, *Michael J. Kallaher*, Editor.

MAA Service Center
P.O. Box 91112
Washington, DC 20090-1112
1-800-331-1MAA   FAX: 1-301-206-9789

Introduction

For some twenty years now, the MAA Notes Series has published secondary materials for undergraduate mathematics courses, such as projects that can be used in teaching calculus. These publications reflect the interests of instructors, providing a means of sharing innovative ideas for teaching calculus, linear algebra, differential equations, statistics, geometry, and abstract algebra. With this book, discrete mathematics joins the list. This collection includes nineteen classroom-tested projects, eleven additional projects based on historical sources, three expository articles considering discrete mathematics topics in more depth, and two articles focused on pedagogy especially related to discrete mathematics.

Why is discrete mathematics only now the subject of such a collection? One possible reason is that, unlike concepts taught in the course itself, discrete mathematics is not well-defined. While there are controversies on how to teach calculus, there is relative unanimity on what topics a calculus course should address. On the other hand, a survey of discrete mathematics courses around the country shows a variety of different topics being covered, goals being sought, and students being served. Does the course cover circuit design and the tools for algorithm analysis? How much logic is covered? Graph theory? Combinatorics? Is this the course where students first learn to write proofs? Are the students mathematics majors, computer science majors, or is the course offered for a general education requirement?

This book does not address those questions. The projects and articles here reflect the wide breadth of topics taught in the diverse discrete mathematics courses offered in universities, colleges, and (increasingly) high schools. I hope that every instructor of discrete mathematics will find projects and articles relevant to the topics of his or her course, and also learn more about other topics, with the realization that those are covered in someone else's course.

The timing of this collection also follows from a number of related activities in the professional organizations. There have recently been workshops and many sessions devoted to discrete mathematics in conferences of the Mathematical Association of America, the Association for Computing Machinery, and the Institute of Electrical and Electronics Engineers. I participated in several of these workshops and sessions, initially with the idea of writing a textbook. But as I discovered the innovative and thoughtful work of so many colleagues, I decided to redirect my creative impulse into putting together this collection. The call went out for classroom-tested projects and articles addressing advanced discrete mathematics topics and teaching issues related to the course. I received submissions from faculty in mathematics departments and computer science departments, from a high school teacher, from new instructors to experts in the field. I am grateful that a wide variety of instructors were willing to provide their work for this volume.

Most of the responses were classroom-tested projects, which vary widely in difficulty and are sometimes distinguished from the advanced articles only by the inclusion of exercises (and solutions!). Some are means of introducing a topic, such as graph theory, strong induction, and motivation for clearly written proofs. Some extend common topics, such as the Towers of Hanoi, the Josephus problem, and Euler's formula. Some discuss applications, such as chemistry, bioinformatics, information storage, and typesetting. Some use technology, including graphing calculators and programming. Some use manipulatives, including integer rods, strings on a pegboard, and pipes from the hardware store. In addition to exercises and solutions in all of these projects, some include open-ended questions and some extension questions suitable for student research. The format for each is a summary, notes to the instructor, references, student worksheets (often mixed with explanatory handouts), and solutions. The classroom projects are ordered topically, following the order of Susanna Epp's *Discrete Mathematics with Applications*.

I am pleased to be able to include eleven of the historical modules developed through New Mexico State University, a continuation of their long-standing program of teaching mathematics from original sources. These too vary from introductory to advanced topics, including combinatorics, set theory, logic, and graph theory. The projects are arranged chronologically by their primary source, allowing readers to follow the historical development of certain discrete mathematics topics. The initial two related projects, though, do not fit into that structure, going from Leibniz to von Neumann in the first, and from Shannon to the abacus in the second. A combined introduction explores how these projects can be used in the classroom, and each article includes exercises in conjunction with the source material, references, and additional notes to the instructor.

The five expository articles examine discrete mathematics content beyond the level of a first course or discuss teaching issues specific to the course. But like many of the projects, some of the articles also defy easy categorization. For example, Shai Simonson's "A Rabbi, Three Sums, and Three Problems" uses fourteenth century mathematics as a springboard for counting problems as an example of the discovery method. To assist the reader in navigating such rich material, the table of contents is annotated with a summary of each project or article and, for the classroom projects, how many 50-minute periods an instructor might dedicate to the activity.

There are many people I want to thank. Bill Marion organized two summer Professional Enhancement Programs on discrete mathematics, where I met Peter Henderson and Susanna Epp; all three have been very helpful and encouraging over the long course of this book's creation. Jerry Lodder, another PREP participant, coordinated work on the historical projects. Michael Jones and Larry Thomas helped with sundry typesetting and graphics issues. Barbara Reynolds, Stephen Maurer, the entire MAA Notes editorial board, and the MAA publications staff have helped immensely in bringing this project to completion. My greatest appreciation goes to the authors of the projects and articles for their patience, creativity, and willingness to share their good work.

In memory of Kenneth P. Bogart, 1943–2005

Contents

II Historical Projects in Discrete Mathematics and Computer Science

III Articles Extending Discrete Mathematics Content

IV Articles on Discrete Mathematics Pedagogy

Part I

Classroom-tested Projects

The Game of "Take Away"

Mark MacLean

Seattle University

Summary

In this project, students play the game "Take Away" and conceive a winning strategy for the game. They then must give a careful written explanation of why their strategies work. I typically use this project in my discrete math class as an introduction to writing proofs. Students often struggle initially with the clarity of their mathematical exposition and with aiming their proofs at an appropriate audience. After completing this activity, we have a class discussion where we critique the students' written explanations.

In most discrete math classes, students' first proofs involve properties of even and odd integers. Since these properties are already familiar to them, students often have trouble discerning what they can and cannot assume that the intended audience knows. Since the game "Take Away" is unfamiliar to most students, I find this project is a more natural starting point from which they begin thinking about the clarity of their own writing.

Alternately (or concurrently), this project could be used as an application of the division algorithm.

Notes for the instructor

Begin class by dividing the students into groups of three or four, and then pass out the worksheet. Allow one 50-minute class period to complete the worksheet, and have the students finish it at home if necessary.

Be sure to walk around the room and observe the groups as they are playing the game. If a group becomes overly frustrated, I might suggest that they try starting with 6 tokens instead of 9, or I suggest that they challenge another group to a game. I usually try to ensure every team knows the winning strategy within 25 minutes. At the end of class I instruct them to continue working at home and to bring a polished written explanation to problem #2 to the next class, as we will critique them. I encourage the students to actually show their explanations to their roommates to see if their writing is clear enough for the roommate to follow the logic.

The next class period, I ask three or four volunteers to write their explanations to question 2 on the board. Over the next 20 minutes, I ask the class to critique these explanations and point out instances where the exposition is unclear, while making some comments myself.

"Take Away" is a variant of the combinatorial game Nim. More information is available in [1].

Bibliography

[1] Berlekamp, Elwyn R., John H. Conway, and Richard K. Guy. *Winning Ways for Your Mathematical Plays*, volume 1, A K Peters, Ltd., 2001.

Worksheet for The Game of "Take Away"

Rules of the game: To play the game, your group will need a pile of nine "tokens." These tokens can be coins, paper clips, small pieces of paper, or any small objects of your choosing. Divide your group into two teams. The two teams will take turns removing tokens from the pile. On each turn, a team must choose to remove either one or two tokens from the pile. The team that removes the last token (or tokens) from the pile wins the game.

Play several rounds of the game, keeping in mind the first problem below.

1. Find a strategy that guarantees your team will win the game. (Note: the existence of such a strategy will depend on whether your team goes first or second. Could there be a guaranteed winning strategy for **both** the first and second team?)

2. In a full paragraph, describe your winning strategy and carefully explain why it works. Pretend that you are explaining the strategy to your roommate, who we'll say knows the rules of the game but has only played a couple of times. Your paragraph must clearly spell out the strategy and convince the roommate that it works.

3. Can you devise a winning strategy if the game starts with 10 tokens? 11? 12? How about n tokens, where n is any positive integer? Describe these winning strategies.

4. In the television series "Survivor: Thailand," the two tribes participated in a competition entitled "Thai 21." In this contest, 21 flags were placed in a circle, and the two tribes took turns removing either 1, 2, or 3 flags from the circle. The team who removed the last flag (or flags) won the game. Is there a winning strategy for Thai 21? Explain.

Solutions

1. There is a winning strategy for the team that goes second. Hence there is no winning strategy for the team that goes first.

2. On each turn, the second team should remove the "opposite" of what the first team removed. That is, if the first team removes 1 token from the pile, the second team should then remove 2 tokens. If the first team removes 2 tokens from the pile, the second team should remove 1. Thus, after each pair of turns, a total of 3 tokens has been removed from the pile, leaving the first team to always choose when the number of tokens in the pile is a multiple of 3. Initially the first team chooses when 9 tokens are in the pile, then they choose when there are 6 tokens, and finally they choose when there are 3 tokens left. In their final turn, the second team takes the remaining tokens and wins the game.

3. If n is congruent to 0 modulo 3: choose to go second, and on each turn always take the opposite of what your opponents just took. If n is congruent to 1 modulo 3: choose to go first, take 1 token, and on each successive turn, always take the opposite of what your opponents just took. If n is congruent to 2 modulo 3: go first, take 2 tokens, then always take the opposite of your opponents. In each strategy you are always leaving your opponent to choose from a multiple of 3. (If students are not familiar with the "mod" terminology, you may simply say that n is equal to 1 plus some multiple of 3, etc.)

4. Here, you want to force your opponent to choose from a multiple of 4. The team that goes first can guarantee themselves a win if they remove 1 flag from the circle. Then, on each successive turn, if their opponents removed m flags, the first team should remove $4 - m$ flags.

Pile Splitting Problem:
Introducing Strong Induction

Bill Marion

Valparaiso University

Summary

In many textbooks in discrete mathematics there are numerous examples for teaching the Weak Form of the Principle of Mathematical Induction, but relatively few elementary problems for applying the Strong Form. What follows is a nice example to draw on when introducing the strong form. It can be presented as a classic puzzle, it has a number of variants and it is inherently recursive.

By introducing the problem (Pile Splitting) as a puzzle, the instructor can engage the students in the process of finding a general solution. She can, then, raise the question as to how they can demonstrate that their conjecture is correct, and, thereby, motivate the need for strong induction. After an induction argument has been presented (the Worksheet includes one such proof), variants of the puzzle can be assigned for the students to work on in class or as a homework assignment.

Notes for the instructor

To give students practice in making conjectures about the solution to the puzzle, they should be asked to solve it themselves. One hands-on approach that works well is to provide each student with a sufficient number of beads or pennies for her to actively play the game enough times with different values of n so as to see a pattern emerge. Those students who correctly conjecture the general solution can assist the others. One way of reaching the conjecture is explained on the Solutions page. During the induction phase a template or "script" for a correct induction argument should be presented to the students (and they should be required to follow the script) so that they become more confident in using the technique and more convinced logically that induction does what we claim it does. The Worksheet includes a complete proof for one puzzle variant in order to give students another example.

In addition to the four variants of the puzzle described in this paper, there are a number of others which can be found in [2]. It should be noted that the author of that article presents a different approach which does not explicitly make use of mathematical induction to prove the general solutions correct.

The recursive nature of the pile splitting problem can lead to a discussion of recursive definitions, recurrence relations, techniques for solving recurrence relations and constructing recursive algorithms to compute the solutions. In the latter case, strong induction comes into play again. It addresses the question: "Does such and such a recursive algorithm, which is designed to compute something, actually do so?"

Bibliography

[1] Rosen, Kenneth. *Discrete Mathematics and Its Applications* 5th ed., McGraw Hill, 2003.

[2] Tanton, James. "A Dozen Questions About: Pile Splitting," *Math Horizons* 12 (2004) 28–31.

Worksheet on the Pile Splitting Problem

Here is a statement of a common version of the pile splitting problem.

> Given n objects in a pile, split the objects into two smaller piles. Continue to split each pile into two smaller piles until there are n piles of size one. At each splitting, compute the product of the size of the two smaller piles. Once there are n piles, sum all the products computed. The result will always be the same no matter how each of the piles is split into two smaller piles. The sum of products is a function of n. Conjecture what the sum is and prove the conjecture correct. [1]

The solution to this pile splitting problem is $(n^2 - n)/2$. Playing with a few examples can provide the necessary insight to come up with the general solution. Applying the strong form of the principle of mathematical induction can demonstrate the correctness of the conjecture, and, *equally as important, that the computation will always produce the same result no matter how the piles are split.*

For the following variant on the standard pile splitting problem, a complete proof of the general solution is provided below.

> We begin with a pile of n objects and proceed to reduce the pile to n piles of size one in the manner described above. Suppose at each splitting the sizes of the two smaller piles are labeled r and s. Now, instead of just computing the product $(r \cdot s)$ of the size of each pair of split piles, we compute the following product: $(r \cdot s) \cdot (r + s)$. And at the end of the process we add all these products. Again, the sum of products turns out to be a function of n: $(n^3 - n)/3$.

Here is the induction argument.

Let $P(n)$ be: for a pile consisting of n stones and split according to the rules above, the sum of all products of the sum and product of each pair of split piles is $(n^3 - n)/3$.

Show $(\forall n \geq 1)\, P(n)$.

Basis Step Show $P(n)$ is true when $n = 1$.

  When $n = 1$, there are no splits; hence the product of sums and products is 0. For the formula, when $n = 1$,

$$\frac{n^3 - n}{3} = \frac{1^3 - 1}{3} = 0.$$

Inductive Step Suppose for any $k > 1$ that $P(1), P(2), \ldots, P(k-1)$ are true. Show this implies that $P(k)$ is true; that is, suppose any pile of j stones where $1 \leq j \leq k - 1$, the sum of all products of the sum and product of each pair of split piles is $(j^3 - j)/3$. Show that this implies the sum of the products of the sums and products is $(k^3 - k)/3$.

  First, divide the pile of k stones into two piles of j and $k - j$ stones. Then, the sum of all the products of sums and products equals the sum of the product of $j + (k - j)$ and $j \cdot (k - j)$ along with all the remaining sums. However, since both j and $k - j$ are between 1 and $k - 1$, the induction hypothesis applies and the sum of the products of sums and products is

$$= (j + (k - j))j(k - j) + \frac{j^3 - j}{3} + \frac{(k - j)^3 - (k - j)}{3}$$

$$= \frac{3(jk^2 - j^2k) + j^3 - j + (k - j)^3 - (k - j)}{3}$$

$$= \frac{3jk^2 - 3j^2k + j^3 - j + k^3 - 3jk^2 + 3j^2k - j^3 - k + j}{3}$$

$$= \frac{k^3 - k}{3}.$$

Thus, since both the **Basis Step** and the **Inductive Step** have been shown to be true, $(\forall n \geq 1)\, P(n)$.

Additional Questions

Consider the following two variations on splitting a pile of n objects. In each case try a few examples, conjecture what the solution should be (again, it turns out to be a function of n) and then use the strong form of the principle of mathematical induction to prove your conjecture. As in the problem statement above, suppose at each splitting the sizes of the two smaller piles are r and s.

1. Split the pile of n objects according to the rules above. At each splitting compute the sum of the reciprocals of the two smaller piles: $(\frac{1}{r} + \frac{1}{s})$. Once there are n piles, multiply all of the sums computed. Again, the result will always be the same no matter how each of the piles is split into two smaller piles.

2. Split the pile of n objects according to the rules above. At each splitting compute the following combinatorial number: $\binom{r+s}{r}$. Once there are n piles, multiply all the combinatorial numbers computed.

Solutions

- For the problem in the worksheet, where at each stage the product $(r \cdot s) \cdot (r + s)$ is computed, it turns out that, for piles of size $1, 2, \ldots, 8$, the sum of all the products is 0, 2, 8, 20, 40, 70, 112 and 168, respectively. Computing the difference of consecutive pairs of integers in the sequence leads to the seven integers 2, 6, 12, 20, 30, 42 and 56 and to the recurrence relation $a_n = a_{n-1} + n(n - 1)$ for $n \geq 2$ with initial condition $a_1 = 0$, the solution to which is $(n^3 - n)/3$.

  At this point it could be said that a solution to the pile splitting problem has been identified and proved correct without using the strong form of induction. That is true. However, it turns out that in most discrete mathematics courses induction is introduced before techniques for solving such linear recurrence relations. If this is the case, after the students have worked on the problem for a while, the instructor might provide some hints or even make a claim as to the general solution and, then, ask the students to use strong induction to prove that claim. (It would be instructive to see if some students can come up with the correct conjecture without reference to recurrence relations and explain their reasoning to the class.)

- For the two puzzles in the Additional Questions Section the solutions are n and $n!$, respectively.

Generalizing Pascal: The Euler Triangles

Sandy Norman and Betty Travis
University of Texas at San Antonio

Summary

This project investigates generalizing Pascal's Triangle to generate the coefficients appearing in the expansion of powers of multinomials of the form $1 + x + x^2 + \cdots + x^k$. Students will initially use a *counting paths* method and then relate the procedure to a generalized triangle, known as an Euler Triangle.

Notes for the instructor

This activity is appropriate for mature Algebra and Precalculus students and can be accomplished in two or three one-hour class periods, depending on the backgrounds of the students. It fits nicely after a unit on binomial expansion. A peg board, with some type of peg or stud, can be used to string paths that can then be counted to generate the coefficients.

Bibliography

[1] A.W.F. Edwards. *Pascal's Arithmetical Triangle*, Johns Hopkins University Press, 2002.

[2] Ronald Graham, Donald Knuth, and Oren Patashnik. *Concrete Mathematics*, Addison Wesley, 1994.

[3] Hilton, Holton, & Pedersen. *Mathematical Reflections*, Springer Verlag, 1997.

[4] Hilton, Holton, & Pedersen. *Mathematical Vistas*, Springer Verlag, 2002.

[5] http://arxiv.org/pdf/math.HO/0505425

[6] http://binomial.csuhayward.edu/Euler.html

[7] mathforum.org/workshops/usi/pascal/pascal_intro.html

[8] mathworld.wolfram.com/PascalsTriangle.html

[9] www.shodor.org/interactivate/activities/pascal1/index.html

I   Generalizing Pascal: The Euler Triangles

Most high school students are familiar with Pascal's Triangle (Fig. 1) as a way to find the coefficients of the binomial $(1 + x)^n$. By adding pairs of coefficients in a given row of the triangle, one can generate the binomial coefficients that appear in the row immediately following.

Recall:

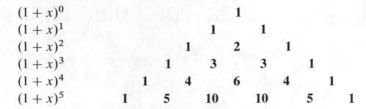

$(1+x)^0$					1				
$(1+x)^1$				1		1			
$(1+x)^2$			1		2		1		
$(1+x)^3$		1		3		3		1	
$(1+x)^4$	1		4		6		4		1
$(1+x)^5$	1	5		10		10		5	1

Figure 1. Pascal's Triangle showing the binomial coefficients up to the 5th power.

The coefficients can then be used to write out the expansion. For example, using the triangle above:

$$(1 + x)^4 = 1 + 4x + 6x^2 + 4x^3 + 1x^4.$$

The coefficients can also be determined by counting the numbers of paths from the apex of the triangle to the entry in question. For example, in the above expansion of $(1 + x)^4$ the coefficient of x^3 is 4. This enumerates the following four paths, each of which must descend at each step to one of the two entries nearest to and below (see Figure 2):

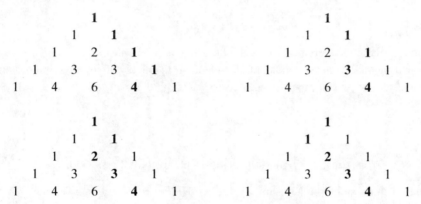

Figure 2. The four paths are indicated by boldface entries.

Is there a similar way to generate the coefficients for $(1 + x + x^2)^n$? For example, if $n = 3$, can we use this method to generate the coefficients for $(1 + x + x^2)^3$?

Students can count the number of paths to a particular entry by drawing out the individual paths on a template[1] of the Euler triangle to be used for trinomials (Figure 3), or by creating the paths with strings (or rubber bands) on a pegboard version of the triangle. (See Photos 1 and 2). In this case, paths must descend at each step to one of the three spaces nearest to and below. For example, there are six paths that can be taken to reach the desired location indicated in Figure 3.

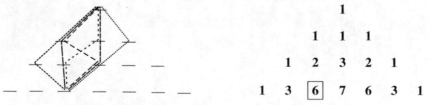

Figure 3. Template for trinomial Euler triangle and completed version.

---

[1] To create the template for the trinomial triangle, we recognize that the number of terms in the expansion of $(1 + x + x^2)^n$ is precisely $2n + 1$. The template, then, will have rows of $1, 3, 5, \ldots$ entries.

Photo 1. Stringing paths on an Euler Triangle.

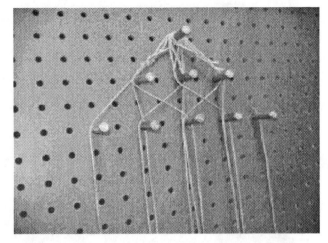

Photo 2. The 1-2-3-2-1 pattern for paths representing the coefficients of $(1 + x + x^2)^2$.

Exercise 1.

Complete the table for $(1 + x + x^2)^n$ for $n = 1, 2,$ and 3 by stringing and counting paths.

$$\underline{}1\underline{}$$

$$\underline{} \quad \underline{} \quad \underline{}$$

$$\underline{} \quad \underline{} \quad \underline{} \quad \underline{} \quad \underline{}$$

$$\underline{} \quad \underline{} \quad \underline{} \quad \underline{} \quad \underline{} \quad \underline{} \quad \underline{}$$

---

2   A recursive method for calculating coefficients

A different — and more convenient — method of obtaining the coefficients for the expansion can be obtained as follows: for each entry other than the initial 1, sum those values that are both adjacent to and above that entry. For this **trinomial** (Figure 4), you add the **three** entries above (if possible), just as in the expansion of a **binomial**, you add the **two** entries above in Pascal's Triangle. In the case of Pascal's triangle, these entries are just the *binomial coefficients*, $\binom{n}{k}$, which represent the coefficients of x^k in the expansion of $(1 + x)^n$. We recall that these binomial coefficients, sometimes read *n choose k*, can be computed using the formula $\binom{n}{k} = \frac{n!}{k!(n-k)!}$. That one can obtain an entry by adding the two previous entries above is a consequence of the recursive identity $\binom{n}{k} = \binom{n-1}{k-1} + \binom{n-1}{k}$.

Euler introduced the notation $\binom{n}{k}_3$ to represent the trinomial coefficients. For example, back in Figure 3, we found the entry represented by $\binom{3}{2}_3$, the coefficient of the x^2 term in $(1 + x + x^2)^3$. More generally, the values $\binom{n}{k}_m$ represent the m-nomial coefficients. These can, like the binomial coefficients, be defined recursively, as follows:

$$\binom{n}{k}_3 = \begin{cases} 0, & n < 0 \text{ or } k < 0 \text{ or } k > 2n, \\[2mm] 1, & n = k = 0, \\[2mm] \binom{n-1}{k-2}_3 + \binom{n-1}{k-1}_3 + \binom{n-1}{k}_3, & n \text{ and } k \text{ not both} = 0. \end{cases}$$

Can you tease out the reason for this rather complicated looking definition? If we examine, the $\binom{4}{3}_3 = 16$ entry from the Euler trinomial triangle in Figure 4, we see that it is equal to $\binom{3}{1}_3 + \binom{3}{2}_3 + \binom{3}{3}_3 = 3 + 6 + 7$. However, if we look at, say, $\binom{3}{5}_3 = 3$, we see that this equals $\binom{2}{3}_3 + \binom{2}{4}_3 + \binom{2}{5}_3 = 1 + 1 + 0$. Even though the triangle itself doesn't have an entry for $\binom{2}{5}_3$, our recursive definition tells us this must be 0, since $k > 2n$ in this instance.

1

1 1 1

1 2 3 **2** **1**

↘ ↓ ↙

1 **3** **6** **7** 6 **3** 1

↘ ↓ ↙

1 4 10 **16** 19 16 10 4 1

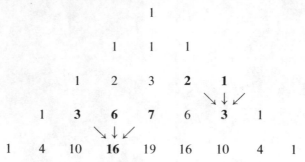

Figure 4. Method for calculating values for entries in the trinomial Euler Triangle.

Exercise 2

- Find the next two rows for the triangle in Figure 4. ______________________

____

- Using your triangle, expand the following:

$(1 + x + x^2)^3 =$ ______________________

$(1 + x + x^2)^6 =$ ______________________

There are a number of interesting observations about the trinomial Euler triangle. Some of these are:

1. The triangle is symmetric, each row being palindromic. Row n follows the pattern 1 $n \cdots n$ 1. (For example, row 3 in Figure 4 above is: 1 3 6 7 6 3 1.)
2. Each new row has two more entries than the previous one.
3. In Pascal's Triangle (for **bi**nomial expansions) the row sum is 2^n. For the **tri**nomial Euler Triangle, what do you note about the sum of the entries in row n?
4. One also notes that the triangular numbers appear along the third diagonal of this Euler triangle. (See Figure 5.)

						1							1
					1	1	1						3
				1	2	3	2	1					9
			1	3	6	7	6	3	1				27
		1	4	10	16	19	16	10	4	1			81
	1	5	15	30	45	51	45	30	15	5	1		243
1	6	21	50	90	126	141	126	90	50	21	6	1	729

Figure 5. Sum of the coefficients in the nth row equals 3^n.

3   Generalizations

Will the processes that we have discussed so far generalize to higher degree polynomial powers? The following exercises will help us discover an answer to this.

Exercise 3. Complete the first three rows of the Euler Triangle for the **quadri**nominal $(1 + x + x^2 + x^3)^n$ by counting paths. Then check your answer by adding the **four** entries above each entry. Verify this algebraically by expanding $(1 + x + x^2 + x^3)^4$.

(Use *DERIVE* or some other Computer Algebra System, if possible.)

Exercise 4. a) What is the sum of the entries in rows 0, 1, 2, 3, and 4 of this triangle? In row n?

b) Using the triangle, write the complete expansion of $(1 + x + x^2 + x^3)^4$.

Of course, we can continue to look at powers of higher and higher degree polynomials — quintinomials, 6-nomials, 7-nomials, etc. Figure 6 gives the first few rows of the quintinomial Euler triangle. An observation we can make that can help determine the shape of the Euler triangle is based on the number of entries in each successive row. For the binomial case (Pascal), each new row had one additional entry. For the trinomial Euler triangle, each new row had two additional entries and for the quadrinomial case, three. We would expect for the quintinomial that the number of entries in each row would increase by four and be determined by adding the nearest five entries from the row above (allowing for blank entries if necessary). For instance, in the center of the triangle in Figure 6, we have $19 = 3 + 4 + 5 + 4 + 3$.

Figure 6. The first few rows of the Euler Triangle for the quintinomial $(1 + x + x^2 + x^3 + x^4)^n$.

Exercise 5.

1. Find the next row for the triangle in Figure 6.

2. What is the sum of the coefficients in each row of the quintinomial Euler triangle (Figure 6)?

3. Create the first few rows of the 6-nomial Euler triangle.

4. What is the sum of the coefficients in each row of the 6-nomial Euler triangle?

5. Generalize how to find the coefficients of any multinomial power $(1 + x + \cdots + x^{m-1})^n$.

6. Show that for an m-nomial Euler triangle, the sum of the entries in row n is just m^n.

---

4   Research Questions

1. For the binomial case, we know that $\binom{n}{k} = \frac{n!}{k!(n-k)!}$. Find a closed form for the trinomial coefficients $\binom{n}{k}_3$.

2. Do the same for the general m-nomial coefficient $\binom{n}{k}_m$.

3. Explain why the $\binom{n}{3}_m$ entry is the nth triangular number ($n \geq 1$).

Solutions

Exercise 1.

$$
\begin{array}{ccccccccccc}
 & & & & & 1 & & & & & \\
 & & & & 1 & & 1 & & 1 & & \\
 & & & 1 & & 2 & & 3 & & 2 & & 1 \\
 & & 1 & & 3 & & 6 & & 7 & & 6 & & 3 & & 1
\end{array}
$$

Exercise 2.

$$1 \quad 5 \quad 15 \quad 30 \quad 45 \quad 51 \quad 45 \quad 30 \quad 15 \quad 5 \quad 1$$

$$1 \quad 6 \quad 21 \quad 50 \quad 90 \quad 126 \quad 141 \quad 126 \quad 90 \quad 50 \quad 21 \quad 6 \quad 1$$

$$1 + 3x + 6x^2 + 7x^3 + 6x^4 + 3x^5 + x^6$$

$$1 + 6x + 21x^2 + 50x^3 + 90x^4 + 126x^5 + 141x^6 + 126x^7 + 90x^8 + 50x^9 + 21x^{10} + 6x^{11} + x^{12}$$

Exercise 3.

$$
\begin{array}{ccccccccccccc}
 & & & & & & 1 & & & & & & \\
 & & & & 1 & & 1 & & 1 & & 1 & & \\
 & & 1 & & 2 & & 3 & & 4 & & 3 & & 2 & & 1
\end{array}
$$

Exercise 4.

a) The sum of the entries for rows 0, 1, 2, 3, 4, and n is respectively 1, 4, 16, 64, 256, and 4^n.

b) $(1 + x + x^2 + x^3)^4 = 1 + 4x + 10x^2 + 20x^3 + 31x^4 + 40x^5 + 44x^6 + 40x^7 + 31x^8 + 20x^9 + 10x^{10} + 4x^{11} + 1x^{12}$.

Exercise 5.

1) 1 5 15 35 70 121 185 255 320 365 381 365 320 255 185 121 70 35 15 5 1

2) The sum of the coefficients on row n is 5^n.

3)

$$
\begin{array}{ccccccccccccccccc}
 & & & & & & & 1 & & & & & & & \\
 & & & & & 1 & & 1 & & 1 & & 1 & & 1 & & 1 & \\
 & & & 1 & & 2 & & 3 & & 4 & & 5 & & 6 & & 5 & & 4 & & 3 & & 2 & & 1 \\
 & 1 & & 3 & & 6 & & 10 & & 15 & & 21 & & 25 & & 27 & & 27 & & 25 & & 21 & & 15 & & 10 & & 6 & & 3 & & 1
\end{array}
$$

4) The sum of the coefficients in row n is 6^n.

5) For the Euler Triangle associated with powers of the m-nomial $1 + x + \cdots + x^{m-1}$, one only needs to add the m entries nearest to and above the entry to be filled. We have more generally

$$\binom{n}{k}_m = \binom{n-1}{k-m+1}_m + \cdots + \binom{n-1}{k-1}_m + \binom{n-1}{k}_m = \sum_{j=0}^{m-1} \binom{n-1}{k-j}_m.$$

6) The sum of the coefficients in the expansion of $(1 + x + \cdots + x^{m-1})^n$ can be obtained by setting $x = 1$ in the expanded form. But this is the same as replacing x by 1 in the unexpanded expression, which gives

$$\underbrace{(1 + 1 + \cdots + 1)}_{m \text{ times}}^n = m^n.$$

Coloring and Counting Rectangles on the Board

Michael A. Jones   and   Mika Munakata

Mathematical Reviews   *Montclair State University*

Summary

We describe the Rectangles on the Board project, an adaptation of an activity for elementary and middle school students that appears in [1]. Students are challenged to determine the coloring of the instructor's 10×10 board, given the restrictions that (1) all 100 squares are colored in one of four colors and (2) the colors form four rectangular regions, one in each color. Our extensions of this project involve counting, symmetry, geometry, and logical reasoning. For example, given the color of some squares, students infer the color of other squares based on geometry. In turn, they use logic and their understanding of this geometry to count the minimal number of squares needed to be revealed so that they can determine all of the squares' colors. Similar reasoning leads them to a best next "guess," when playing the game. Students use combinatorics and symmetry to count the number of ways to color the board.

Notes for the instructor

This project is suitable for mathematics courses at all levels. We have implemented versions of the activity in elementary, middle, and high school classes, as well as in undergraduate mathematics and graduate mathematics education courses. The game-like aspect of the project is engaging and appealing to students. The project can be used to spur all-class discussions, or can be used to promote cooperative learning. Depending on how deeply your class desires to delve into the activity and its extensions, it can take from one to three 50-minute class periods to complete. The worksheet does well to foster communication; hence, we recommend that you have students work in groups and end the project with a whole-class discussion of the solutions. Because many acceptable answers exist for some of the problems, grading many worksheets (e.g., if a large class turns in individual worksheets) can be tedious.

The project extends an activity that appears in Math for Girls and Other Problem Solvers [1]. The book is intended for use in elementary and middle school classrooms to encourage equity and participation. We describe the original activity, called the Multi-Color game, and problems that extend from the activity. The problems related to combinatorics, decision making, proofs, and logic can be used in collegiate discrete mathematics courses. Students are asked not only to provide solutions, but also to explain their reasoning, thereby promoting mathematical communication. We have used versions of this project in courses for pre- and in-service teachers to encourage mathematical reasoning, communication, and cooperative learning. In those classes, we lead discussions about whether or not this activity constitutes mathematics, given many secondary school students' rigid belief that mathematics is about computing solutions and manipulating equations. Middle and elementary school teachers often see possibilities for using the activity to discuss geometry, percents, and fractions.

Before class, the instructor colors a 10×10 board on a piece of paper using four colors, each of which forms exactly one rectangle of integer dimensions (see problem 9 on the worksheet for sample boards). We follow [1] which uses a 10×10 board though any size rectangular board will suffice. Using the matrix coordinates, students "guess" by asking the instructor to reveal the color of one square at a time. The instructor colors the corresponding square on an overhead transparency of a blank 10×10 board. After each square is revealed, students are asked whether the color of that square determines the colors of other squares — called "freebies." After a few guesses, students see that if they use logic and what they know about rectangles, they can conserve guesses by asking for squares that reveal more information about the structure of the board. The object of the game is for the class to determine the coloring of the board in as few guesses as possible. After playing the game with the entire class, have students work on the worksheet in small groups.

For the worksheet, we represent four colors by suits of a deck of cards, ♡, ◇, ♣, and ♠. The board is viewed as a matrix in which individual squares are given in (row, column) position with X denoting 10. The first two questions relate to freebies and other strategy-related considerations that students may have encountered during the game. Questions 3–6 concern the different ways in which the board can be partitioned into four rectangles. We lead students to discover that there are six configuration classes if rotation, reflection, and size of rectangles on the board are disregarded. Students are exposed to this idea in two ways — first by identifying equivalent boards based on geometry and second by noticing that the decisions one makes in partitioning the board force the configuration class. In the last four questions of the worksheet, students are asked to use what they know about strategies and configuration classes to count the total number of ways in which the board can be colored, and to determine the minimal number of guesses necessary to complete a given board.

Bibliography

[1] D. Downie, T. Slesnick, and J. K. Stenmark, *Math for Girls and Other Problem Solvers.* EQUALS, Lawrence Hall of Science, Berkeley, CA, 1981.

Worksheet for Coloring and Counting Rectangles on the Board

1. Some partially filled boards are shown in Figure 1. For each board, answer the following questions:

   i) Based on what has already been revealed, how many "freebies" are there, if any? Identify them and justify your answers.

   ii) What is a good next guess? Why?

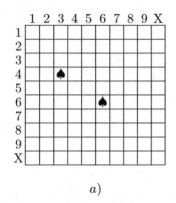
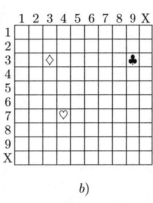
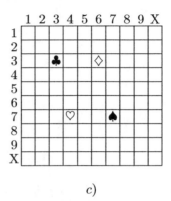

a)   b)   c)

Figure 1

2. The boards below in Figure 2 are partially filled in. Assuming that freebies are correctly determined by the guess, answer the following questions:

   i) Best case scenario: What is the minimum number of additional guesses needed to complete the board? Give examples of which squares you would guess, and what suits those squares would have to be in order for you to complete the board in this best case scenario.

   ii) Worst case scenario: What is the maximum number of additional guesses needed to complete the board? Give a sequence of guesses, along with the suits of the revealed squares, that would result in this worst case scenario.

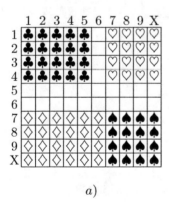
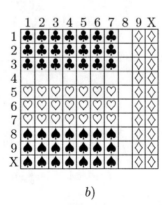
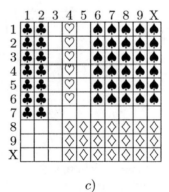

a)   b)   c)

Figure 2

3. For each row of four boards in Figure 3, indicate which board is not like the others. Why?

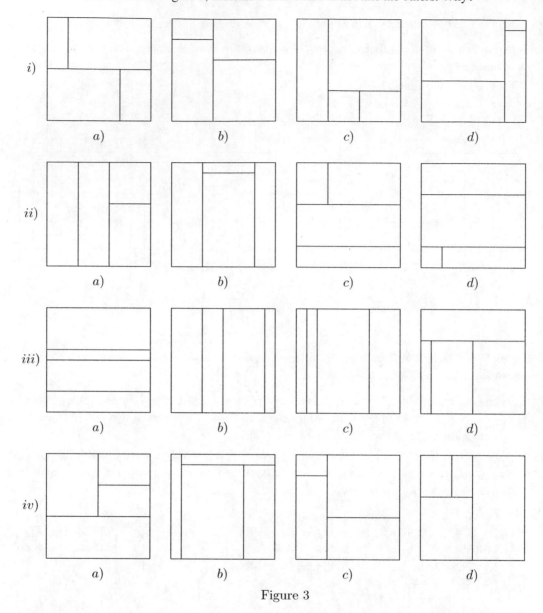

Figure 3

The partitions that belong in the same group for each question in Problem 3 are defined to be in the same *configuration class*. We say that two boards are in the same configuration class if they are equivalent under rotation and reflection, without regard to the size, location, and suit of each rectangle. For example, you should have found that c) in Figure 3i) is not equivalent to the others. You cannot rotate, reflect, or "slide" any of the line segments from a), b), or d) to produce c). We will explore another way to consider configuration classes in the next two problems.

4. Notice that each of the partitions in Problem 3 has at least one vertical or horizontal line segment that forms a border between suits and extends from one edge of the board to the opposite edge. Prove that such a segment always exists for a partition of a square into four rectangles.

5. Partitioning the square into four rectangles can be viewed as an iterative process of three steps such that at each step exactly one rectangle is divided into two. The first step is to divide the 10×10 square into two rectangles with a vertical or horizontal line segment. For example, a horizontal segment has been used to divide a square

into rectangles A and B in Figure 4a). The next step is to divide rectangle A or B into two rectangles using either a horizontal or vertical line. For example, B is divided into two rectangles B and C using a vertical line in Figure 4b). To finish the process of dividing the board into four rectangles, one of the rectangles A, B, or C must be divided into two rectangles. In Figure 4c), we divide rectangle B into rectangles B and D. Note that if A is divided (instead of B), the result is different.

Make a decision tree to indicate all possible ways that a board can be divided into four rectangles. Realize that certain decisions are equivalent. A horizontal first line and a vertical first line are indistinguishable because the resulting boards are equivalent under rotation. Partitions are distinguished from one another by the relative positions of the rectangles to one another, but not by the size of the rectangles.

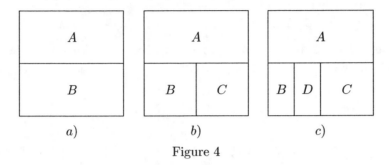

Figure 4

6. The terminal nodes of the decision tree from problem 5 (here, the nodes resulting from three decisions) also represent configuration classes. How many different configuration classes were produced from your decision tree?

7. For each of the configuration classes you found in problem 6, find the minimal number of guesses required to complete the board. Keep in mind that within a single configuration class, different numbers of minimal guesses may be required depending on the positions of the borders. You can view this as moving the boundaries while staying within the class, but changing the minimal number of guesses. For example, Figure 5 shows boards from the same configuration class that require different minimal numbers of guesses.

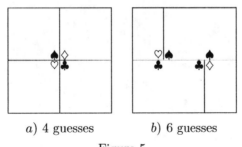

a) 4 guesses   b) 6 guesses

Figure 5

For each of the other configuration classes find all possible numbers of minimal guesses by drawing representative boards.

8. How many different colorings of the board are possible? Hint: Use the configuration classes to guide your work. Realize that two boards in the same configuration class that are equivalent under rotation produce different colorings. Further, positions of the borders and the distribution of the suits produce different colorings too.

9. For the completed boards in Figure 6, what is the minimum number of squares that can remain unknown so that more than one board is possible? Give examples of which squares would have to remain unknown in order to achieve this minimal number.

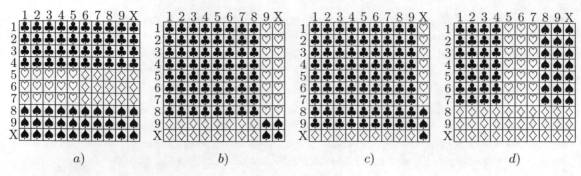

Figure 6

10. Determine the values of n such that knowing a specific n squares still does not complete the board, but knowing an $(n + 1)$st square allows you to complete the board. Use sample boards to explain.

Solutions

1.   *i*) Solutions for Figures 1*a*), 1*b*), and 1*c*) appear below.

      1*a*) There are 10 freebies. They are in the rectangle with vertices at (6,3), (6,6), (4,6), and (4,3).

      1*b*) No freebies.

      1*c*) There are 28 freebies. There are 8 ♣ in the rectangle defined by vertices (3,1), (3,3), (1,3), and (1,1). There are 15 ♠ freebies in the rectangle defined by vertices (X,7), (X,X), (7,X), and (7,7). There are 3 ♡ freebies at (8,4), (9,4), and (X,4) and 2 ◇ freebies at (1,6) and (2,6).

   ii) Answers vary. A good guess has the potential to reveal many freebies. For example, in Figure 1*b*), a guess in the rectangle with vertices (X,5), (X,X), (4,X), and (4,5) could reveal a ♠ and lead to many freebies, as the board would have the same structure as Figure 1*c*). Also, students might look for borders between suits or guess squares that have more possibilities for their suits. For the latter, consider the board in Figure 1*b*): squares in the rectangle given by vertices (6,4), (6,8), (1,8), and (1,4) can be any of the four suits.

2.   *i*) Answers vary. Examples of acceptable solutions are given for each figure.

      2*a*) 1 guess: If (6,6) is ♡, then we're done.

      2*b*) 1 guess: If (4,8) is ♣, then we're done.

      2*c*) 2 guesses: If (8,3) is ♣, and (7,5) is ♠, then we're done.

   ii) Answers vary. Examples of acceptable solutions are given for each figure.

      2*a*) 5 guesses; 1 guess to reveal the vertical border between ♡ and ♣ and a maximum of 2 each to determine the horizontal borders between ♣ and ◇ and between ♡ and ♠. For example, if (1,6) is ♣; (6,1) is ◇; (5,1) is ♣; (6,X) is ♠; and (5,X) is ♡, then the board is completed.

      2*b*) 2 guesses; 1 guess to reveal the vertical border between ◇ and the other suits and 1 guess to reveal the horizontal border between ♣ and ♡. For example, if (X,8) is ◇, and (4,1) is ♣, then the board is completed.

      2*c*) 4 guesses; 1 guess to reveal the horizontal border between ◇ and the 2 suits ♡ and ♠ above it, 1 guess to reveal the vertical border between ♡ and ♠, 1 guess to reveal the vertical border between ♣ and ♡, and 1 guess to reveal the border between ♣ and ◇. For example, if (7,X) is ♠; (1,5) is ♠; (1,3) is ♡; (X,1) is ◇, then the board is completed.

3. 3*i*) *c*; 3*ii*) *b*; 3*iii*) *d*; 3*iv*) *c*.

4. Consider the four corners of the board. If any adjacent pair is the same color, then one of the four rectangles would extend from one edge of the board to the opposite edge. This rectangle creates either a vertical or horizontal border that traverses the board.

We find it useful to solve this problem by viewing the board as a unit square, partitioned into four rectangles. If the four corners of the board are colored distinctly, then each colored rectangular region has a single vertex in the interior of the square; the other vertices that define the rectangle are on the boundary of the square. Denote the four rectangular regions as A, B, C, and D such that these contain the vertices of the unit square $(0, 1)$, $(1, 1)$, $(1, 0)$, and $(0, 0)$, respectively, and their interior vertices are denoted a, b, c, and d, respectively. Let $a = (x_a, y_a)$, $b = (x_b, y_b)$, $c = (x_c, y_c)$, and $d = (x_d, y_d)$.

Rectangles A and B must share a vertical border which implies that $x_a = x_b$. Similarly, C and D share a vertical border and $x_c = x_d$. Because A and D share a horizontal border, then $y_a = y_d$. Likewise, B and C share a horizontal border such that $y_b = y_c$. There are nine ways in which the x and y values of the vertices can be ordered:

1) $y_a < y_b$

       a) $x_a < x_c$ *b*) $x_a = x_c$ *c*) $x_a > x_c$

2) $y_a = y_b$

       a) $x_a < x_c$ *b*) $x_a = x_c$ *c*) $x_a > x_c$

3) $y_a > y_b$

       a) $x_a < x_c$ *b*) $x_a = x_c$ *c*) $x_a > x_c$

Cases 1*b*), 2*b*) and 3*b*) all have $x_a = x_c$. Thus, these colorings have a vertical border that traverses the board. Similarly, cases 2*a*), 2*b*) and 2*c*) all have $y_a = y_b$; these colorings have a horizontal border that traverses the board.

In case 1*a*), $y_a < y_b$ and $x_a < x_c$ as pictured in Figure 7*a*). The vertices a, b, c and d form a rectangle that is not colored. Hence, this case cannot occur. Due to symmetry, case 3*c*) cannot occur for the same reason.

In case 1*c*), $y_a < y_b$ and $x_a > x_c$ as pictured in Figure 7*b*). The vertices c, b, a and d form a rectangle that is colored twice (as part of rectangles A and C). So, this case cannot occur. Due to symmetry, case 3*a*) cannot occur for the same reason.

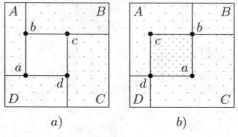

Figure 7

5. The decision tree appears in Figure 8. Because an initial horizontal or vertical division of the square into two rectangles is rotationally equivalent, we identify the first action by the notation $h \sim v$. From solution 4, there exists at least one line segment that extends from one edge of the board to the opposite edge. There can be 3, 2, or 1 such parallel line segments. For 3 parallel line segments, there is only one configuration, *a*). For 2 such segments, the two possibilities are given by *b*) and *c*). For 1, all configurations are accounted for by *d*), *e*), and *f*).

   Notice that the same sequential division of the tree into rectangles does not yield all ways that a square can be divided into five rectangles. As an example, the division of the 10×10 square into rectangles in Figure 7*a*) cannot be generated by extending the decision tree approach. This follows because no border traverses the entire length of the 10×10 board while the first line drawn in the decision tree approach traverses the entire board.

6. There are 6 configuration classes. The 2 boards denoted by *c*) in Figure 8 are equivalent. The other 5 terminal nodes in the decision tree in Figure 8 belong to unique configuration classes.

7. Figure 9 consists of boards for each configuration class that yield all of the possible minimal numbers of guesses. We consider the boards from the terminal nodes of the decision tree in left-to-right order. Terminal node *e*) is described as part of the problem statement and is shown in Figure 5.

8. Using the terminal nodes of the tree in Figure 8, we count how many ways there are to determine the suits of each configuration class. For node *a*), of the 9 possible horizontal borders in the 10×10 board, 3 such borders

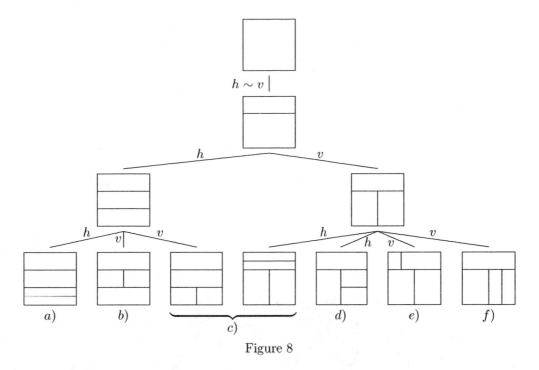

Figure 8

define the 4 regions. There are $\binom{9}{3}$ ways to choose these horizontal borders. Suits in each of these boards can be allocated in 4! ways. And, because horizontal and vertical borders arc in the same configuration class under rotation, the total number of boards in this configuration class is $2 \cdot 4! \cdot \binom{9}{3}$.

For node b), there are $\binom{9}{2}$ ways to select the horizontal borders. For each of these pairs of horizontal borders, there are 9 ways to choose the vertical border making the cross-bar of the H. Suits for the 4 regions can be allocated in 4! ways. Once again, the initial borders could have been vertical, instead of horizontal. Hence, there are $2 \cdot 4! \cdot 9 \cdot \binom{9}{2}$. For nodes c), the only difference is that the vertical border is in the top or bottom region. Therefore, there are twice as many boards for configuration class c as for node b), or $2 \cdot \left[2 \cdot 4! \cdot 9 \cdot \binom{9}{2} \right]$ completed boards.

For node d), there are $\binom{9}{2}$ ways to select the horizontal borders. The vertical border determines how far across from the right side the lower horizontal border goes. There are 9 ways to select the vertical border that extends from the upper horizontal border to the bottom of the board. Because the lower horizontal border could extend from either the left or the right edge to the vertical border, choosing left or right doubles the number of boards. Also, there are 2 choices for which horizontal border traverses the board, either the upper or the lower, doubling again the number of boards. Because the initial borders could have been vertical and considering the 4! ways to allocate suits to the 4 regions, there are $2 \cdot 2 \cdot \left[2 \cdot 4! \cdot 9 \cdot \binom{9}{2} \right]$ completed boards.

For node e), it is possible for both the horizontal and vertical borders to traverse the board, making a '+,' as in Figure 5a). To prevent double counting, we first consider those that do not make a '+' sign. There are 9 choices for the horizontal border that traverses the board. The top region's vertical border can be selected in 9 ways, while the bottom region's vertical border can be selected in 8 ways, to prevent the '+'. Suits can be allocated to the four regions in 4! ways. With the cases in which the vertical border is the only one that traverses the board, there are $2 \cdot [4! \cdot 9 \cdot 9 \cdot 8]$ ways to complete the board in this configuration class without '+' cases. The '+' cases can be completed in $4! \cdot 9 \cdot 9$ ways because the horizontal border can be chosen in 9 ways, the vertical border can be chosen in 9 ways, and the four suits can be allocated to the four regions in 4! ways. Hence, there are $2 \cdot [4! \cdot 9 \cdot 9 \cdot 8] + 4! \cdot 9 \cdot 9$ distinct boards in this configuration class.

For node f), there are 9 ways to choose the horizontal border, $\binom{9}{2}$ ways to choose the two vertical borders, and the vertical borders could be above or below the horizontal border. The four suits can be allocated to the four

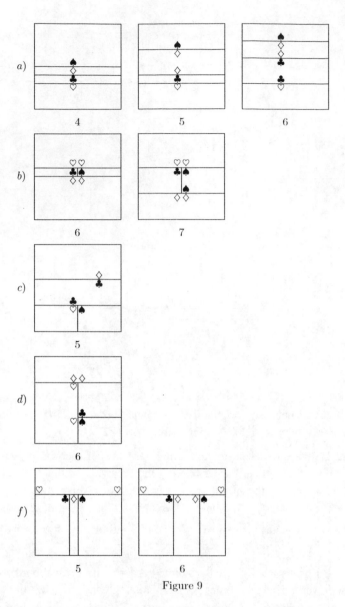

Figure 9

regions in 4! ways. Because the initial horizontal border could have been vertical, we multiply our result by 2. Hence, there are $2 \cdot 4! \cdot 2 \cdot \binom{9}{2} \cdot 9$ completed boards in this configuration class.

Adding the total number of boards for each configuration class there are

$$4! \left[17 \cdot 9^2 + 2 \cdot 9^2 \cdot \binom{9}{2} + 2 \cdot \binom{9}{3} \right]$$

or 177048 possible boards.

9. Answers vary. Examples of acceptable solutions are given.

   a) 3 squares: (7,6), (6,6), and (5,6).

   b) 2 squares: (8,9) and (8,X).

   c) 1 square: (8,X).

   d) 7 squares: (1,5), (2,5), (3,5), (4,5), (5,5), (6,5), and (7,5).

10. All values of n from 3 to 99. For $n = 3$, the board in Figure 5a) is an example where knowing three squares (♠, ♡, and ◇) isn't enough to complete the board, but knowing a specific 4th square (♣) does complete the board. For $n = 99$, knowing the suits of all squares except for (8,X) in Figure 6c) is not enough to complete the board because (8,X) could be ♡ or ♠. Other values can be constructed by generalizing the boards in Figure 6.

Fun and Games with Squares and Planes

Maureen T. Carroll and Steven T. Dougherty
University of Scranton

Summary

This project is intended to introduce students to the concepts of mutually orthogonal Latin squares and their relationship to finite affine planes. These topics are introduced in the first section. After describing how tic-tac-toe is played on an affine plane, the second section explores player strategies. Playing the game will help students understand the combinatorial and geometric notions described, and build geometric intuition for these objects.

Notes for the instructor

Students must play the game in order to understand the strategy arguments. You can have them play against each other in class or turn in their game sheets as an exercise. While the last exercise may be an exercise in frustration, it is important for students to repeatedly play the game.

For a class project, have the members of your class play a "Tic-tac-toe on the affine plane of order 4" tournament. Our Mathematics Club holds an annual tic-tac-toe tournament at the University of Scranton with prizes for the top finishers. We have best-of-three matches to decide the winner of each random pairing, with a toss of a coin deciding who makes the first move. While perfect play will result in a win for the first player, you can rely on your students to make mistakes!

As an additional project, ask your students to create a new game to play on an affine plane. For example, how about a game where the player who claims the last unclaimed point on a line loses? As another example, what happens on π_n if a player needs only $n - 1$ of their marks on any line to win? Some of these games may be easy to analyze strategically while others could be extremely difficult.

There are other examples of generalizations of tic-tac-toe. Several examples of these are given in the reading list below. For those who wish to learn about the weight function techniques we used to prove our game theoretic results, see our paper listed below.

Bibliography

[1] Carroll, M. T. and S. Dougherty. "Tic-Tac-Toe on a Finite Plane," *Mathematics Magazine* 77 (2004) 260–274.

[2] Euler, L. "Recherches sur une nouvelle espèce des quarrés magiques," *Leonardi Euleri Opera Omnia Ser. I*, Vol 7, 291–392, Tuebner, Berlin-Leipzig, 1923.

[3] Golomb, S. and A. Hales. "Hypercube Tic-Tac-Toe," *More Games of No Chance*, Richard J. Nowakowski (editor), Mathematical Sciences Research Institute Publications No. 42, Cambridge University Press, 2002, 167–182.

[4] Mullen, G. "A candidate for the 'Next Fermat Problem'," *The Mathematical Intelligencer* 17 (1995) 18–22.

[5] Patashnik, O. "Qubic $4 \times 4 \times 4$ Tic-Tac-Toe," *Mathematics Magazine* 53 (1980) 202–216.

[6] Paul, J. L. "Tic-Tac-Toe in *n*-Dimensions," *Mathematics Magazine* 51 (1978) 45–49.

[7] Tarry, G. "Le problème des 36 officers," *Compte Rendu Ass. Franc. Pour l'avacement des Sciences*, Vol 2, 170–203, 1901.

[8] Weeks, J. "Torus and Klein Bottle Games," available at `http://www.geometrygames.org`.

I Worksheet on Latin squares

In 1782, the great Leonhard Euler posed the following question. Can 36 officers, with six different ranks from six different regiments, be arranged in a 6 by 6 square such that each row and column has each rank and regiment exactly once? With this deceptively easy question Euler started one of the most widely studied and productive branches of discrete mathematics.

As often happens in mathematics, the ease with which a question is posed can be adversely proportional to the difficulty in finding a solution. A solution to Euler's question remained unpublished until 1901 when a French colonel named Tarry computed (by hand) all possible arrangements. It took Tarry one year to do what would now take a computer under an hour, eventually finding that there was no such arrangement. This is exactly what Euler had predicted, but could not prove.

Euler approached the problem in the following way. He used Latin letters (A,B,C,D, ...) to represent the ranks and Greek letters $(\alpha, \beta, \gamma, \ldots, \ldots)$ to represent the regiments. So each officer now had a unique representation consisting of one Latin letter and one Greek letter, an ordered pair. Euler then split the problem into arranging the regiments and arranging the ranks. To arrange either you simply had to make a square in which each letter appeared exactly once in each row and each column. This object has come to be known as a Latin square regardless of the alphabet employed. For example here are two 4 by 4 Latin squares.

$$\begin{pmatrix} A & B & C & D \\ B & A & D & C \\ C & D & A & B \\ D & C & B & A \end{pmatrix} \qquad \begin{pmatrix} \alpha & \beta & \gamma & \delta \\ \delta & \gamma & \beta & \alpha \\ \beta & \alpha & \delta & \gamma \\ \gamma & \delta & \alpha & \beta \end{pmatrix} \tag{1}$$

If Euler had posed a 16 officer problem, a solution could be constructed by overlapping the two Latin squares above so that each ordered pair of Latin and Greek letters appears exactly once. We say that the two Latin squares are *orthogonal* if this occurs. Overlapping the two squares above produces the Graeco-Latin square:

$$\begin{pmatrix} A\alpha & B\beta & C\gamma & D\delta \\ B\delta & A\gamma & D\beta & C\alpha \\ C\beta & D\alpha & A\delta & B\gamma \\ D\gamma & C\delta & B\alpha & A\beta \end{pmatrix} \tag{2}$$

Exercise 1.1. Verify that each letter appears exactly once in each row and column of both squares of (1), and that each ordered pair appears exactly once in (2).

This Graeco-Latin square proves that it is possible to arrange 16 officers in a 4 by 4 square such that each of the ranks and regiments appears exactly once in each row and column. The ease of this construction makes the impossibility of the 36 officer problem even stranger. To further add to the mysterious nature of this problem, it *is* known that n^2 officers can be arranged in such an n by n square for all n except $n = 2$ and $n = 6$.

Interestingly, we can expand the structure in (2) by using the the natural numbers, $0, 1, 2, 3, \ldots$, as our alphabet. This is the alphabet that is typically used for all Latin squares since it allows for the use of the algebra of the natural numbers in constructions. Consider the following Latin square:

$$\begin{pmatrix} 0 & 1 & 2 & 3 \\ 2 & 3 & 0 & 1 \\ 3 & 2 & 1 & 0 \\ 1 & 0 & 3 & 2 \end{pmatrix} \tag{3}$$

We can overlap this 4 by 4 square with each of the others in (1) as shown below.

$$\begin{pmatrix} 0A & 1B & 2C & 3D \\ 2B & 3A & 0D & 1C \\ 3C & 2D & 1A & 0B \\ 1D & 0C & 3B & 2A \end{pmatrix} \qquad \begin{pmatrix} 0\alpha & 1\beta & 2\gamma & 3\delta \\ 2\delta & 3\gamma & 0\beta & 1\alpha \\ 3\beta & 2\alpha & 1\delta & 0\gamma \\ 1\gamma & 0\delta & 3\alpha & 2\beta \end{pmatrix} \tag{4}$$

Exercise 1.2. Verify that each ordered pair appears exactly once in both squares of (4).

In (2) and (4) we have shown that the Latin squares given in (1) and (3) are Mutually Orthogonal Latin Squares (MOLS) of order 4. This means that there are three 4 by 4 squares, each pairing of which is orthogonal. Here is a way to represent these three MOLS using only the first four natural numbers as our alphabet.

$$\begin{pmatrix} 0 & 1 & 2 & 3 \\ 1 & 0 & 3 & 2 \\ 2 & 3 & 0 & 1 \\ 3 & 2 & 1 & 0 \end{pmatrix} \qquad \begin{pmatrix} 0 & 1 & 2 & 3 \\ 2 & 3 & 0 & 1 \\ 3 & 2 & 1 & 0 \\ 1 & 0 & 3 & 2 \end{pmatrix} \qquad \begin{pmatrix} 0 & 1 & 2 & 3 \\ 3 & 2 & 1 & 0 \\ 1 & 0 & 3 & 2 \\ 2 & 3 & 0 & 1 \end{pmatrix} \tag{5}$$

Exercise 1.3. Construct another Latin square of order 4, different from those in (5) but using the same alphabet.

For the following exercise, consider the following Latin squares of order 4.

$$P = \begin{pmatrix} 0 & 1 & 2 & 3 \\ 3 & 0 & 1 & 2 \\ 2 & 3 & 0 & 1 \\ 1 & 2 & 3 & 0 \end{pmatrix} \qquad Q = \begin{pmatrix} 0 & 1 & 2 & 3 \\ 2 & 3 & 0 & 1 \\ 1 & 2 & 3 & 0 \\ 3 & 0 & 1 & 2 \end{pmatrix} \tag{6}$$

Exercise 1.4.

(a) Are P and Q orthogonal to each other?

(b) Is P orthogonal to the Latin squares in (5)? Is Q?

(c) Using the same alphabet, can you find another Latin square, R, for which P, Q and R are MOLS?

At this point you may have conjectured that there are at most three MOLS of order 4. We shall now show that the largest number of MOLS of order n is $n - 1$. Suppose we have a set of MOLS of order n. We can assume that the first row of each is arranged in ascending order, $0, 1, 2, 3, 4, \ldots, n - 1$, since if it were not we could simply rename the elements so that it fits this arrangement. Consider the elements in the first column of the second row in these squares. These elements must be distinct and different from 0. Why? They must be distinct since the ordered pairs $(0, 0), (1, 1), \ldots, (n - 1, n - 1)$ are produced in the first row when overlapping any two of these squares. They must be different from 0 since there can only be one 0 in the first column. Therefore we only have $n - 1$ possibilities, and we can have at most $n - 1$ MOLS of order n.

If we have a set of $n - 1$ MOLS of order n then we say that it is a complete set of MOLS of order n. Here is the principal question which has generated over 200 years of research and remains largely unsolved.

Fundamental Question: For which n are there complete sets of MOLS?

This is another example of a question which is easily stated, can be explained to almost anyone, and has withstood the attempts of a great number of mathematicians for hundreds of years. It is precisely because of this that Gary Mullen, a well-known mathematician in this field, recently suggested it as the next Fermat problem. As we will see, this question is equivalent to one of the most important questions of finite geometry, namely when do projective and affine planes exist.

There are some simple constructions for a complete set of MOLS of prime order which we shall now describe. First we recall some basic definitions of modular arithmetic. We say that $a \equiv b \pmod{n}$ if $a - b$ is a multiple of n. We can take as the representatives of all natural numbers modulo n the set $\{0, 1, 2, \ldots, n - 1\}$, and use this set as our alphabet. When we add or multiply a and b we take the representative from this set that is equivalent to the sum or product modulo n. For example, $3 + 5 \equiv 1 \pmod{7}$ and $3(4) \equiv 5 \pmod{7}$. Using these operations, we can construct tables not unlike elementary school multiplication tables, but utilizing modular arithmetic. As an example, addition and multiplication tables for modulo 3 arithmetic are shown below. Notice that each cell in the grid is the result of its row heading and column heading under the specified operation modulo 3.

+	0	1	2		*	0	1	2
0	0	1	2		0	0	0	0
1	1	2	0		1	0	1	2
2	2	0	1		2	0	2	1

Exercise 1.5. Construct the addition and multiplication tables for modulo 5 arithmetic.

We shall now attempt to construct $n - 1$ MOLS, $L^1, L^2, \ldots, L^{n-1}$, of order n using the set $\{0, 1, 2, \ldots, n - 1\}$ as our alphabet. We let L_{ij}^k denote the entry in the i-th row and j-th column of the Latin square, or matrix, L^k.

Orthogonal Construction: Define $L_{ij}^k = ki + j \pmod{n}$, where i, j and k are members of our alphabet set.
   For example, when $n = 3$ we have

$$L^1 = \begin{pmatrix} 0 & 1 & 2 \\ 1 & 2 & 0 \\ 2 & 0 & 1 \end{pmatrix} \quad \text{and} \quad L^2 = \begin{pmatrix} 0 & 1 & 2 \\ 2 & 0 & 1 \\ 1 & 2 & 0 \end{pmatrix}. \tag{7}$$

Exercise 1.6. Verify that L^1 and L^2 are orthogonal.

Notice that this construction technique has produced the largest possible set of MOLS of order 3 since we know there can be at most two. Will this construction technique work for all possible values of n?

Exercise 1.7. Construct the L^k's when $n = 4, 5$ and 6:
(a) Use the Orthogonal Construction technique to form L^1, L^2, and L^3 when $n = 4$.
(b) Use the construction to form L^1, L^2, L^3 and L^4 when $n = 5$.
(c) Use the construction to form L^1, L^2, L^3, L^4 and L^5 when $n = 6$.
(d) Did the construction yield $n - 1$ MOLS of order n in these three cases? If not, do you have a conjecture as to when this construction will yield $n - 1$ MOLS?

If you conjectured that this construction technique only works for prime n then you are correct. (You will be asked to supply a proof for this in the last project.) A similar construction technique is used to generate the three MOLS of order 4 but the addition and multiplication operations are defined in a different way. Specifically, it is not modulo 4 arithmetic but the arithmetic of the finite field of order 4.

The relationship between MOLS and finite geometry was hinted at in reference to the Fundamental Question. Let's look at an example to explore the nature of this connection. Arrange nine points in a square and connect each set of three horizontal points and each set of three vertical points, using a different colored marking device to connect each set. For the next part, think of your grid of points as corresponding to L^1 in (7), and connect each set of three points that have a common symbol, using a different color for each set. Do the same with L^2. The result should be similar to the following diagram, but much more colorful.

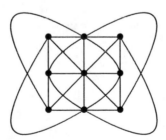

Each connection of three points is called a line. How many lines are there? Your answer should be the same as the number of colors used, namely twelve. Notice that any two lines either meet once or are parallel (no points in common). This object is known as an affine plane of order 3. In the following four exercises you will construct MOLS for higher order affine planes. These constructions are needed to proceed to the games in the second part of the project.

Exercise 1.8. Construct the affine plane of order 4: Put 16 points in a square and draw a line for each set of four vertical and horizontal points. Next, using the three MOLS of order 4 given in (5), connect each set of four points with the corresponding common symbols. Use a different color for each line to make the connections more visible.

Exercise 1.9. Use the 4 MOLS of order 5 from Exercise 1.7 to draw the corresponding affine plane of order 5 using the method outlined in Exercise 1.8.

Exercise 1.10. Use the Orthogonal Construction technique to construct 6 MOLS of order 7 and then use them to draw the corresponding affine plane of order 7 as in Exercise 1.8.

Exercise 1.11. Prove the following result. The Orthogonal Construction technique produces $p - 1$ MOLS of order p if and only if p is a prime. Here is an outline of one way to proceed: First show that L^k is a Latin square, that is, show that each element appears once in each row and column. Then prove that if p is a prime then L^k and L^j are orthogonal by showing that each ordered pair is represented exactly once. The following fact will be useful. If p is prime and $ax \equiv b \pmod{p}$ then there is a unique x in $\{0, 1, \ldots, p - 1\}$ satisfying the equation. Next, show it will not work when p is not a prime. An easy way to do this is to show that the construction will not produce Latin squares in all cases.

Epilogue

Now that you have developed an understanding for the Fundamental Question, you may be wondering whether there has been any progress towards a solution. Well, it is known that a complete set of MOLS exists for all orders p^k where p is prime and k is a positive integer. It is not known if there exists a complete set of MOLS for any other value. The problem has proven to be unbelievably difficult. For example, it has been shown that there is no complete set of MOLS of order 10. It took numerous theoretical papers and one year of computation on a supercomputer to settle this particular case. It did not determine the largest number of MOLS of order 10, only that there are not 9 of them. The difficulty encountered with the $n = 10$ case is what is known as combinatorial explosion. While it is easy to settle the 6 by 6 case with the aid of a computer, it is impossible for a computer to completely settle the 10 by 10 case since the number of Latin squares increases too quickly.

2 Worksheet on Games

The great power of describing mathematics in geometric terms is that it allows us to apply the intuition we have developed since we were children to very complicated problems. For example, understanding the possible simultaneous solutions to equations of the form $(x - h)^2 + (y - k)^2 = r^2$ and $ax + by + c = 0$ is greatly simplified upon realizing that these equations may be described geometrically as a circle and a line, respectively. Likewise, while it is possible to study complete sets of MOLS simply by their definition, our understanding will be greatly enhanced by viewing them as a geometry. Specifically, they can be seen as an affine plane which satisfies axioms similar to those we know well from the Cartesian plane. Even though we will tinker slightly with the definitions of point, line and plane in this new context, we can still apply a good deal of the intuition that we have developed for the Cartesian plane. In this section, we shall describe a very effective way to understand the geometry described in the previous section by using an unlikely learning tool, tic-tac-toe. By playing this game, we will develop geometric intuition which will aid in the understanding of MOLS and affine planes.

Tic-tac-toe and variations on its theme are abundant, appearing in children's books, journals, game theory textbooks and on the Internet. The rules for tic-tac-toe are simple enough for a young child to learn. A player must mark all of the positions on a line in order to win the game. While the standard version of tic-tac-toe is played on a 3 by 3 grid, it can easily be expanded to any n by n grid, with the usual horizontal, vertical and diagonal lines as the winning lines. We could take this generalization a bit further by playing on an $n \times n \times n$ cube or a hypercube, that is, a cube in n dimensions. You can challenge the computer to a game of tic-tac-toe played on a torus or a Klein bottle on the Internet at www.geometrygames.org. We will use a different game board, namely the affine plane that was introduced in the previous section. That is, we shall describe the game played on a complete set of MOLS.

The diagrams created in the previous section will be used to develop intuition for affine planes. As a start, let's look at the representation of the affine plane of order 3 constructed in the previous section. It may be helpful to use your colored diagram as a reference. Labels have been inserted so that we may refer to individual points.

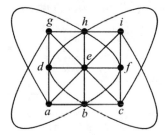

This plane has 9 points and 12 lines, each of these lines consists of three points and, in your representation, each has a different color. As you may have speculated, this is called the affine plane *of order 3* since each line has exactly three points, and it is denoted as π_3. Referring to each line as a set of three points, the lines are:

$$\{a, b, c\}, \{d, e, f\}, \{g, h, i\},$$
$$\{a, d, g\}, \{b, e, h\}, \{c, f, i\},$$
$$\{a, e, i\}, \{c, e, g\}, \{a, h, f\},$$
$$\{g, b, f\}, \{i, b, d\}, \{c, h, d\}.$$

These lines appear to be quite different from those found in the Cartesian plane but we shall see that they satisfy similar axioms. Lines in the Cartesian plane are continuous, contain infinitely many points, and their representation is described as straight. Lines on this affine plane are discrete, contain a finite number of points, and their representation cannot necessarily be drawn with a straightedge. It is an example of a finite plane. Though we have "connected the dots" in our representation, there are no points in between which belong to the plane. When we connect the dots we are merely representing the simple fact that the connected points form a line. Therefore, our representation of any line in a finite affine plane does not need to be straight.

The concept of parallel lines remains unchanged in this context. They are simply lines having no points in common, for example $\{g, b, f\}$ and $\{c, h, d\}$. Additionally, we say that a line is parallel to itself in the same way we do on the Cartesian plane. The lines $\{g, b, f\}$ and $\{i, b, d\}$ are not parallel since they meet at the point b, in other words b is a point on both lines. Notice that an intersection of lines in the representative diagram does not necessarily indicate a point common to both lines. For example, the lines $\{i, b, d\}$ and $\{c, h, d\}$ appear to meet in the diagram, but upon further inspection you will notice that they do not have a point in common.

Exercise 2.1. Questions about the affine plane of order 3, π_3:
(i) List the lines which are parallel to $\{a, b, c\}$.
(ii) List the lines which are parallel to $\{a, d, g\}$.
(iii) List the lines which are parallel to $\{g, b, f\}$.
(iv) List the lines which are parallel to $\{c, e, g\}$.

The answers to parts (i), (ii), (iii) and (iv) are called parallel classes. The affine plane of order 3 has 4 parallel classes of 3 lines each since it has 12 lines in total.

Exercise 2.2. Questions about the affine plane of order 4, π_4, constructed in Exercise 1.8 in the previous section: First, label the 16 points in the same manner as in π_3, starting with the letter a in the lower left corner and proceeding accordingly.
(i) List the lines which are parallel to $\{a, b, c, d\}$.
(ii) List the lines which are parallel to $\{a, e, i, m\}$.
(iii) List the lines which are parallel to $\{d, g, j, m\}$.
(iv) List the lines which are parallel to $\{a, g, l, n\}$.
(v) List the lines which are parallel to $\{a, h, j, o\}$.

Exercise 2.3. One of the primary techniques of mathematics is generalizing to all cases from a few known cases. In this exercise, use your answers from the previous two exercises to finish filling in the first two columns. Using similar techniques, consult your solutions to Exercises 1.9 and 1.10 to fill in the third and fourth columns. Try to recognize the patterns emerging in order to formulate some conjectures about the general case, π_n.

Number of ⇓ in ⇒	π_3	π_4	π_5	π_7	π_n
points	9	16			
lines	12				
points on each line	3	4			
lines through any point					
points where two non-parallel lines meet					
lines parallel to any given line [Note: Every line is parallel to itself.]	3				
parallel classes	4				

An example of the method for constructing these finite affine planes from their MOLS, specifically for π_3, was given after Exercise 1.7, but the reason for this method was not. It is clear from these exercises that the underlying structure has a great deal of symmetry. We cannot take any finite set of points, connect them as we wish, and call it an affine plane. As is always the case in mathematics, there are rules to be followed. Specifically, an affine plane is a nonempty set of points, P, and a nonempty collection of subsets of points (called lines) which satisfy the following three axioms:

(1) through any two distinct points there exists a unique line;

(2) if p is a point, ℓ is a line, and p is not on line ℓ, then there exists a unique line, m, that passes through p and is parallel to ℓ;

(3) there are at least two points on each line, and there are at least two lines.

Let's verify one instance of the second axiom on the affine plane of order 4. The point f is not on the line $\{a, g, l, n\}$. There are five lines through f, namely $\{e, f, g, h\}$, $\{b, f, j, n\}$, $\{a, f, k, p\}$, $\{d, f, i, o\}$ and $\{c, f, l, n\}$, of which only $\{d, f, i, o\}$ is parallel to $\{a, g, l, n\}$.

Exercise 2.4. Verify that your representation of π_3 satisfies these axioms.

When these rules are followed we can create a finite or an infinite affine plane. The Cartesian plane, with points and lines defined as usual, is the example we typically envision when reading this definition. It is an example of an infinite affine plane. Since we intend to play tic-tac-toe on finite planes, we will focus on these. Every line on a finite affine plane has the same number of points, and as you have surely noticed, this number is called the order of the plane. In an affine plane of order n there are n^2 points. If you are thinking ahead towards playing the game, you may notice that n^2 is the perfect number of points needed for an $n \times n$ grid.

Since we may not always have a representative diagram as a reference, how can we easily identify the lines through any given point without the use of one? Well, the horizontal and vertical lines should be clearly visible. As for the others, you may recall that in the previous section we constructed the representative diagrams by using the MOLS. So, even if we did not have a diagram we could still easily find the other lines by consulting the MOLS. In each Latin square, the points corresponding to identical symbols form a line. So, we can find one line per Latin square through any given point. On the affine plane of order 4, for example, through any given point we have one horizontal line, one vertical line, and one line from each of the three MOLS of order 4, for a total of five lines. For example, consulting both the MOLS in (5) and your labeled representation of π_4 from Exercise 2.2 we can easily see the five lines through the point labeled "a" on π_4:

$\{a, b, c, d\}$	horizontal line
$\{a, e, i, m\}$	vertical line
$\{a, f, k, p\}$	line determined by the symbol 3 in first Latin square
$\{a, g, l, n\}$	line determined by the symbol 1 in second Latin square
$\{a, h, j, o\}$	line determined by the symbol 2 in third Latin square

Exercise 2.5.

(a) Using the labeled representation of π_3, identify the two lines determined by the symbol 1 in the associated MOLS L^1 and L^2 which were given in (7) in the previous section.

(b) Using your labeled representation of π_4 from Exercise 2.2, identify the three lines determined by the symbol 2 in the associated MOLS given in (5).

(c) Label the points of your representation of π_5 from Exercise 1.9 by starting with the letter a in the lower left corner and proceeding as we have on the other planes. Identify the four lines determined by the symbol 3 in the associated MOLS found in Exercise 1.7(b).

Now that you have an understanding for our new game board, let's play tic-tac-toe on these planes. When we play on what looks to be a standard 3×3 board, we will actually be playing the game on the affine plane of order 3. If we were playing standard 3×3 tic-tac-toe, there would be 8 ways to win, namely the 3 horizontal, 3 vertical, and 2 diagonal lines. Since we are playing on the affine plane of order 3 there are 12 ways to win, namely the 12 lines of the plane. The old ways to win are still there, but we've added four new winning lines which are given below.

We will assume that X always makes the first move and we will use subscripts to denote the order of play. So, in the example given below, X_2 is the second move of player X, and O_1 is the first move of player O. Here is a sample game where player X has won. Can you identify the winning line?

X_1	X_3	O_3
	O_1	X_2
X_4	O_2	

Exercise 2.6. Play several games of tic-tac-toe on the affine plane of order 3. Be sure to play some games as X and others as O if you have an actual opponent.

Did you notice one of the ways this game differs from standard 3×3 tic-tac-toe is the placement of the first move? In standard 3×3 tic-tac-toe the center point sits on 4 lines, the corner points sit on 3 lines, and all other points sit on 2 lines. This difference places greater importance on the first move of the game. On any affine plane, the first player's move could just as well be made with his eyes closed since every point sits on the same number of lines. In other words, there is no advantage to be gained by claiming a particular first point on the plane.

In the standard 3×3 version of tic-tac-toe, we usually learn as young children that the second player can always force a draw as long as they make the correct choices. Determining the outcome of play assuming that players make the best possible choices is a basic question in game theory. Let's start small and see what conclusions we can make about playing on the smallest affine plane, that of order 2. The diagram for this plane of 4 points and 6 lines is given below. (There is only one Latin square associated with this plane. What is it?)

Notice that once X_1 and O_1 are chosen, player X will win on his second move since there exists a line between any two points on an affine plane. In this situation we say that X has a winning strategy. This was too easy. Let's consider the affine plane of order 3.

Try to play tic-tac-toe on π_3 until you can determine the outcome of play on this plane. You will find that player X always wins. By the reasoning given above, X_1 and O_1 may be assumed to be chosen arbitrarily. Choose X_2 as any point not on the line containing X_1 and O_1. Since there exists a line between any two points on an affine plane, O_2 must be placed on the line containing X_1 and X_2 (otherwise X wins on the next move). Likewise, X_3 must block the line containing O_1 and O_2. Player O must now block either the line containing X_1 and X_3 or the line containing X_2 and X_3, (it is a simple matter to see that O does not already have these lines blocked). Now, X_4 completes the line that O_3 did not block, and X wins the game. We have shown that X has a winning strategy by providing the method by which X can win any game on the affine plane of order 3. Now that we've provided a winning strategy on this plane, let's move on to the plane of order 4.

Exercise 2.7. Play tic-tac-toe on π_4 (use either the diagram created in Exercise 1.8 or the MOLS given in (5)), and try to determine the outcome of play on this plane.

As soon as we venture beyond the first two affine planes we find that the complexity of the game increases dramatically. This jump in difficulty when moving from one case to the next is not uncommon in mathematics. Indeed, we saw this very phenomenon in the attempt to find a complete set of MOLS of order 10 in the previous section. As for tic-tac-toe on the affine plane of order 4, the additional points and lines generate a far greater number of possible moves for each player. If you found that you cannot provide an easy move-by-move analysis as we did for the previous two planes then you are not alone. Perhaps we do not possess the computational fortitude that Tarry showed when he solved the 36 officer problem by hand, or perhaps we are merely availing ourselves of our computational technology, but in either case we relied on a computer to tell us that player X has a winning strategy by exhausting all possible outcomes. We do not, however, know what this winning strategy is in general. This means that we cannot describe an easily applied winning algorithm (for a person to follow) which prescribes each move. We can, however, teach a computer to play so that it never loses, but a vast number of tree searches are required at each move, placing this algorithm far beyond the capabilities of a human player.

Epilogue

What about the affine planes of order 5 and order 7 that you constructed in the previous section? We determined that player O can always force a draw on these planes just as the second player can in the standard 3×3 tic-tac-toe you played as a child. In fact, player O can force a draw on all affine planes of order greater than 4. If you presume that since a move-by-move analysis is not possible for the affine plane of order 4 then it surely must not be possible for higher ordered planes, then you are correct. We did not, however, have to rely on a computer to determine the outcome of play on these higher ordered planes. We were able to prove this result by using a technique that goes beyond the scope of this project, namely weight functions.

Solutions

Solution 3 Numerous examples can be given, for example:
$$\begin{pmatrix} 0 & 1 & 2 & 3 \\ 3 & 0 & 1 & 2 \\ 2 & 3 & 0 & 1 \\ 1 & 2 & 3 & 0 \end{pmatrix}$$

Solution 4 (a) No, for example the pair $(0, 3)$ appears twice but $(0, 1)$ does not appear. (b) Neither P nor Q are orthogonal. (c) No.

Solution 5

$$\begin{pmatrix} + & 0 & 1 & 2 & 3 & 4 \\ \hline 0 & 0 & 1 & 2 & 3 & 4 \\ 1 & 1 & 2 & 3 & 4 & 0 \\ 2 & 2 & 3 & 4 & 0 & 1 \\ 3 & 3 & 4 & 0 & 1 & 2 \\ 4 & 4 & 0 & 1 & 2 & 3 \end{pmatrix} \qquad \begin{pmatrix} * & 0 & 1 & 2 & 3 & 4 \\ \hline 0 & 0 & 0 & 0 & 0 & 0 \\ 1 & 0 & 1 & 2 & 3 & 4 \\ 2 & 0 & 2 & 4 & 1 & 3 \\ 3 & 0 & 3 & 1 & 4 & 2 \\ 4 & 0 & 4 & 3 & 2 & 1 \end{pmatrix}$$

Solution 1.7

(a)

$$L^1 = \begin{pmatrix} 0 & 1 & 2 & 3 \\ 1 & 2 & 3 & 0 \\ 2 & 3 & 0 & 1 \\ 3 & 0 & 1 & 2 \end{pmatrix} \quad L^2 = \begin{pmatrix} 0 & 1 & 2 & 3 \\ 2 & 3 & 0 & 1 \\ 0 & 1 & 2 & 3 \\ 2 & 3 & 0 & 1 \end{pmatrix} \quad L^3 = \begin{pmatrix} 0 & 1 & 2 & 3 \\ 3 & 0 & 1 & 2 \\ 2 & 3 & 0 & 1 \\ 1 & 2 & 3 & 0 \end{pmatrix}$$

(b)

$$L^1 = \begin{pmatrix} 0 & 1 & 2 & 3 & 4 \\ 1 & 2 & 3 & 4 & 0 \\ 2 & 3 & 4 & 0 & 1 \\ 3 & 4 & 0 & 1 & 2 \\ 4 & 0 & 1 & 2 & 3 \end{pmatrix} \quad L^2 = \begin{pmatrix} 0 & 1 & 2 & 3 & 4 \\ 2 & 3 & 4 & 0 & 1 \\ 4 & 0 & 1 & 2 & 3 \\ 1 & 2 & 3 & 4 & 0 \\ 3 & 4 & 0 & 1 & 2 \end{pmatrix} \quad L^3 = \begin{pmatrix} 0 & 1 & 2 & 3 & 4 \\ 3 & 4 & 0 & 1 & 2 \\ 1 & 2 & 3 & 4 & 0 \\ 4 & 0 & 1 & 2 & 3 \\ 2 & 3 & 4 & 0 & 1 \end{pmatrix} \quad L^4 = \begin{pmatrix} 0 & 1 & 2 & 3 & 4 \\ 4 & 0 & 1 & 2 & 3 \\ 3 & 4 & 0 & 1 & 2 \\ 2 & 3 & 4 & 0 & 1 \\ 1 & 2 & 3 & 4 & 0 \end{pmatrix}$$

(c)

$$L^1 = \begin{pmatrix} 0 & 1 & 2 & 3 & 4 & 5 \\ 1 & 2 & 3 & 4 & 5 & 0 \\ 2 & 3 & 4 & 5 & 0 & 1 \\ 3 & 4 & 5 & 0 & 1 & 2 \\ 4 & 5 & 0 & 1 & 2 & 3 \\ 5 & 0 & 1 & 2 & 3 & 4 \end{pmatrix} \quad L^2 = \begin{pmatrix} 0 & 1 & 2 & 3 & 4 & 5 \\ 2 & 3 & 4 & 5 & 0 & 1 \\ 4 & 5 & 0 & 1 & 2 & 3 \\ 0 & 1 & 2 & 3 & 4 & 5 \\ 2 & 3 & 4 & 5 & 0 & 1 \\ 4 & 5 & 0 & 1 & 2 & 3 \end{pmatrix} \quad L^3 = \begin{pmatrix} 0 & 1 & 2 & 3 & 4 & 5 \\ 3 & 4 & 5 & 0 & 1 & 2 \\ 0 & 1 & 2 & 3 & 4 & 5 \\ 3 & 4 & 5 & 0 & 1 & 2 \\ 0 & 1 & 2 & 3 & 4 & 5 \\ 3 & 4 & 5 & 0 & 1 & 2 \end{pmatrix}$$

$$L^4 = \begin{pmatrix} 0 & 1 & 2 & 3 & 4 & 5 \\ 4 & 5 & 0 & 1 & 2 & 3 \\ 2 & 3 & 4 & 5 & 0 & 1 \\ 0 & 1 & 2 & 3 & 4 & 5 \\ 4 & 5 & 0 & 1 & 2 & 3 \\ 2 & 3 & 4 & 5 & 0 & 1 \end{pmatrix} \quad L^5 = \begin{pmatrix} 0 & 1 & 2 & 3 & 4 & 5 \\ 5 & 0 & 1 & 2 & 3 & 4 \\ 4 & 5 & 0 & 1 & 2 & 3 \\ 3 & 4 & 5 & 0 & 1 & 2 \\ 2 & 3 & 4 & 5 & 0 & 1 \\ 1 & 2 & 3 & 4 & 5 & 0 \end{pmatrix}$$

(d) For $n = 4$ and $n = 6$ the construction did not even produce Latin squares for L^i when i was not relatively prime to n. For $n = 5$ the construction did work. The correct conjecture is that it will work if and only if n is prime.

Solution 1.8   Simply connect the points corresponding to each number in the given Latin squares.

Solution 1.9   Simply connect the points corresponding to each number in the given Latin squares.

Solution 1.10

$$L^1 = \begin{pmatrix} 0 & 1 & 2 & 3 & 4 & 5 & 6 \\ 1 & 2 & 3 & 4 & 5 & 6 & 0 \\ 2 & 3 & 4 & 5 & 6 & 0 & 1 \\ 3 & 4 & 5 & 6 & 0 & 1 & 2 \\ 4 & 5 & 6 & 0 & 1 & 2 & 3 \\ 5 & 6 & 0 & 1 & 2 & 3 & 4 \\ 6 & 0 & 1 & 2 & 3 & 4 & 5 \end{pmatrix} \quad L^2 = \begin{pmatrix} 0 & 1 & 2 & 3 & 4 & 5 & 6 \\ 2 & 3 & 4 & 5 & 6 & 0 & 1 \\ 4 & 5 & 6 & 0 & 1 & 2 & 3 \\ 6 & 0 & 1 & 2 & 3 & 4 & 5 \\ 1 & 2 & 3 & 4 & 5 & 6 & 0 \\ 3 & 4 & 5 & 6 & 0 & 1 & 2 \\ 5 & 6 & 0 & 1 & 2 & 3 & 4 \end{pmatrix} \quad L^3 = \begin{pmatrix} 0 & 1 & 2 & 3 & 4 & 5 & 6 \\ 3 & 4 & 5 & 6 & 0 & 1 & 2 \\ 6 & 0 & 1 & 2 & 3 & 4 & 5 \\ 2 & 3 & 4 & 5 & 6 & 0 & 1 \\ 5 & 6 & 0 & 1 & 2 & 3 & 4 \\ 1 & 2 & 3 & 4 & 5 & 6 & 0 \\ 4 & 5 & 6 & 0 & 1 & 2 & 3 \end{pmatrix}$$

$$L^4 = \begin{pmatrix} 0 & 1 & 2 & 3 & 4 & 5 & 6 \\ 4 & 5 & 6 & 0 & 1 & 2 & 3 \\ 1 & 2 & 3 & 4 & 5 & 6 & 0 \\ 5 & 6 & 0 & 1 & 2 & 3 & 4 \\ 2 & 3 & 4 & 5 & 6 & 0 & 1 \\ 6 & 0 & 1 & 2 & 3 & 4 & 5 \\ 3 & 4 & 5 & 6 & 0 & 1 & 2 \end{pmatrix} \quad L^5 = \begin{pmatrix} 0 & 1 & 2 & 3 & 4 & 5 & 6 \\ 5 & 6 & 0 & 1 & 2 & 3 & 4 \\ 3 & 4 & 5 & 6 & 0 & 1 & 2 \\ 1 & 2 & 3 & 4 & 5 & 6 & 0 \\ 6 & 0 & 1 & 2 & 3 & 4 & 5 \\ 4 & 5 & 6 & 0 & 1 & 2 & 3 \\ 2 & 3 & 4 & 5 & 6 & 0 & 1 \end{pmatrix} \quad L^6 = \begin{pmatrix} 0 & 1 & 2 & 3 & 4 & 5 & 6 \\ 6 & 0 & 1 & 2 & 3 & 4 & 5 \\ 5 & 6 & 0 & 1 & 2 & 3 & 4 \\ 4 & 5 & 6 & 0 & 1 & 2 & 3 \\ 3 & 4 & 5 & 6 & 0 & 1 & 2 \\ 2 & 3 & 4 & 5 & 6 & 0 & 1 \\ 1 & 2 & 3 & 4 & 5 & 6 & 0 \end{pmatrix}$$

Solution 1.11   The Orthogonal Construction technique produces $p-1$ MOLS of order p if and only if p is a prime.
Proof. Assume p is a prime. If $L^k_{i,j} = L^k_{i,j'}$ then $ki + j = ki + j'$ which implies $j = j'$. Thus no element appears more than once in a row. If $L^k_{i,j} = L^k_{i',j}$ then $ki + j = ki' + j$ implies $ki = ki'$ and then $k(i - i') = 0$. There is a unique solution to $kx = 0 \pmod{p}$, namely 0 and so $i = i'$ and no element appears more than once in a column and L^k is a Latin square.

Next we shall show that L^k and $L^{k'}$ are orthogonal. Assume $(L^k_{i,j}, L^{k'}_{i,j}) = (L^k_{i',j'}, L^{k'}_{i',j'})$ then $ki + j = ki' + j'$ and $k'i + j = k'i' + j'$. This gives that $k(i - i') = j' - j$ and $k'(i - i') = j' - j$. However, there is a unique solution to $(i - i')x = (j' - j)$ and so $k = k'$. Hence if $k \neq k'$ the same ordered pair cannot appear more than once.

If p is not a prime then $p = ab$ for $1 < a, b < p$. Consider L^a: in the row corresponding to b we have $L^a_{b,j} = ab + j = j$ and in the 0th row we have $L^a_{a0+j} = j$ so that L^a is not a Latin square.   □

Solution 2.1
(i) $\{a,b,c\}, \{d,e,f\}, \{g,h,i\}$
(ii) $\{a,d,g\}, \{g,e,h\}, \{c,f,i\}$
(iii) $\{g,b,f\}, \{h,d,c\}, \{i,e,a\}$
(iv) $\{c,e,g\}, \{a,f,h\}, \{b,d,i\}$

Solution 2.2
(i) $\{a,b,c,d\}, \{e,f,g,h\}, \{i,j,k,l\}, \{m,n,o,p\}$
(ii) $\{a,e,i,m\}, \{b,f,j,n\}, \{c,g,k,o\}, \{d,h,l,p\}$
(iii) $\{d,g,j,m\}, \{c,h,i,n\}, \{b,e,l,o\}, \{a,f,k,p\}$
(iv) $\{a,g,l,n\}, \{b,h,k,m\}, \{d,f,i,o\}, \{c,e,j,p\}$
(v) $\{a,h,j,o\}, \{c,f,l,m\}, \{d,e,k,n\}, \{b,g,i,p\}$

Solution 2.3

Number of $\Downarrow$ in $\Rightarrow$	π_3	π_4	π_5	π_7	π_n
points	9	16	25	49	n^2
lines	12	20	30	58	$n^2 + n$
points on each line	3	4	5	7	n
lines through any point	4	5	6	8	$n + 1$
points where two non-parallel lines meet	1	1	1	1	1
lines parallel to any given line [Note: Every line is parallel to itself.]	3	4	5	7	n
parallel classes	4	5	6	8	$n + 1$

Solution 2.5

   (a) $\{a, f, h\}, \{h, d, c\}$

   (b) $\{o, l, e, b\}, \{o, i, f, d\}, \{0, j, h, a\}$

   (c) $\{x, r, l, f, e\}, \{x, q, o, h, a\}, \{x, p, m, j, b\}, \{x, t, k, g, c\}$

Solution 2.6   You should find that X has a winning strategy.

Solution 2.7   Player X has a winning strategy but, as the next paragraph in the worksheet notes, this strategy is not easy to find. After playing the game extensively by hand, you may appreciate the online version at

   `http://academic.scranton.edu/faculty/carrollm1/tictactoe/tictactoea4.html`

Exploring Recursion with the Josephus Problem
(Or how to play "One Potato, Two Potato" for keeps)

Douglas E. Ensley and James E. Hamblin
Shippensburg University

Summary

The Josephus problem is addressed in many discrete mathematics textbooks as an exercise in recursive modeling, with some books (e.g., [1] and [3]) even using it within the first few pages as an introductory problem to intrigue students. Since most students are familiar with the use of simple rhymes (like Eeny-meeny-miney-moe) for decision-making on the playground, they are comfortable with the physical process involved in this problem. For students who may wish to pursue this topic independently, [4] and [5] provide nice surveys and bibliographies, and the website [2] provides web-based tools for exploring the problem directly. The activities presented here are intended to be completed by students in a single class period early in the semester. We find that an opening student-centered problem can get the class involved and set a good tone for the semester. Moreover, we find that many issues arising from this particular problem can be built upon throughout the course. The next section provides some suggestions for connections to other parts of the course.

Notes for the instructor

The Josephus problem can be explored through role playing or through carefully constructed pencil and paper activities, depending on the amount of time one wishes to devote to it. We list below some of the things we discuss just before the activity as well as some of the contexts in which we have students revisit the problem later on.

- A good preliminary discussion on recursion can be initiated with the following problems.

  a. Pose the question, "What is $1 + 2 + 3 + \cdots + 19 + 20$?" This provides a good opportunity to share the creative idea of regrouping in order to sum 10 copies of 21 for a total of 210.

  b. Followup with the question, "What is $1 + 2 + 3 + \cdots + 20 + 21$?" Some students will try the regrouping trick, but at least one should point out that you can simply add 21 to the previous answer.

  c. This idea of using a "similar but simpler" problem that has been solved previously is the very essence of recursive thinking.

- The activities presented here have been written to be completed with paper and pencil, but with the investment of more time one can have students act out the roles. This is a good ice-breaking activity early in the semester, but it does take more time. Through role playing, students will discover for themselves issues like "We need to remember who was first," and "We need a system for describing who is the last one left."

- There will be several opportunities later in a discrete mathematics course when one can reprise the Josephus game as a source for exercises and motivational examples. Computer science courses often use this problem as an exercise in recursive programming or in maintaining circular linked lists. Hence, with some cooperation from a friendly computer science instructor, this problem can prove useful in more than one context.

a. (Mathematical induction) In the Josephus problem with skip number 2, prove that for all integers $n \geq 0$, if the game starts with 2^n players, then the person in position 1 will be the last person left. (This uses induction with the induction step involving the one pass all the way around the circle for the first time in which the even numbered people are eliminated.)

b. (Follow up) In the Josephus problem with skip number 2, if $0 \leq k < 2^n$ and the game starts with $2^n + k$ players, then the person in position $2k + 1$ will be the last person left. This is a non-inductive argument consisting of removing the first k (even numbered) people and then applying (a) to the remaining circle of size 2^n.

c. (Binary representation of numbers) Define the cyclic left shift of a binary numeral b as the number obtained from shifting the leading (i.e., leftmost) 1 bit to the rightmost end of the numeral. For example, the cyclic left shift of the binary numeral 1001101 is the numeral 0011011, which is the same as 11011. Show that if $0 \leq k < 2^n$, then the cyclic left shift of the binary representation of $2^n + k$ is the binary representation of $2k + 1$. Hence, the cyclic left shift of a number m gives the last person left in the m person Josephus game with skip number 2. This gives an "application flavor" to the study of binary numbers that may make them more intriguing.

d. (Modular arithmetic) When introducing modular arithmetic, an analogy can be made to the Josephus problem in which the original circle of people are numbered 0 through $n - 1$. In particular, the patterns within the tables of "last person left" all have the relationship "add k" but with the provision that the addition "wraps around the circle" to refer to the actual people.

Bibliography

[1] Ensley, D. E. and J. W. Crawley. *Discrete Mathematics: Mathematical Reasoning and Proof with Puzzles, Patterns and Games*, New York: John Wiley, 2006.

[2] Ensley, D. E. and J. W. Crawley. Companion website for *Discrete Mathematics*, at
    `http://webspace.ship.edu/~deensley/DiscreteMath/`

[3] Graham, R., D. Knuth and O. Patashnik. *Concrete Mathematics*, Reading, MA: Addison-Wesley, 1994.

[4] Herstein, I. N. and I. Kaplansky. *Matters Mathematical*, New York: Chelsea Publishing Company, 1974.

[5] Schumer, P. D. "The Josephus problem: Once more around," *Mathematics Magazine* 75 (2002) 12–17.

Worksheet on Exploring Recursion with the Josephus Problem
(Or how to play "One Potato, Two Potato" for keeps)

Introduction

Ancient mathematics problems that still hold their own are always fun to play with. A particularly good one, which happens to be named for a first century historian, has its origins in the Jewish - Roman war. The historian Flavius Josephus was apparently trapped by the Romans in a cave with 40 fellow Jewish rebels. As good soldiers they decided on suicide rather than capture, so they formed a circle and agreed that every third person would be killed until no one was left.

Josephus and a friend were more keen on being captured than their colleagues, so they quickly found the spots to stand to ensure they were the two remaining at the end of the grisly proceedings. Hence, the mathematically inept suffered an untimely demise while Josephus and his friend lived to tell the tale.

This morbid story doesn't seem like much of a game or puzzle, but it has the same basic structure (with terminal consequences) as the age old way of choosing someone from a group: the "one potato, two potato" algorithm. We will spend some time in class today playing this type of game and analyzing our results.

Analyzing the Josephus Problem

In general, when we play the "Josephus game," there will be a certain number of people standing in a circle, and a "skip number" that tells us how many people to count before removing someone from the circle. In the classical example described above, the number of people is 41 and the skip number is 3.

Let's look at a simpler example. This time, there will be only six people in the circle, but we will keep the skip number at 3. We'll continue to play until there is only one person remaining. Let's say the people, named Ann, Beth, Chris, Dave, Emma, and Fred, are arranged as shown in Figure 1.

Figure 1. Six people in a circle.

In this case, we decide to start counting with Ann. We count Ann and Beth, and when we get to the third person, Chris, he is removed from the circle. With Chris gone, we continue counting with Dave. We count Dave and Emma, and when we get to Fred, he is eliminated from the circle. Now there are four people left in the circle: Ann, Beth, Dave, and Emma, and the counting continues with Ann. We count Ann and Beth, and then Dave is eliminated. The current situation is displayed in Figure 2.

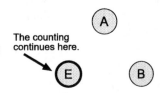

Figure 2. Three people remain.

We next count Emma and Ann, and remove Beth, and the counting once again continues with Emma. We count Emma, Ann, and then Emma is removed, so Ann is the person who is left standing at the end.

An important thing to notice about this process is that we need to know which person to start the counting with at each step, including the first step. If we remove a couple of people and then go on a coffee break, we might come back and forget who to resume the counting with.

For discussion: Can you think of a way that we could remember which person we need to start the counting with at each step?

One solution is for us to put a funny hat on the person we need to start the counting with at each step. In our diagrams, we will put a thick circle around the "starting person."

Let's try the game again, this time with seven people (named A, B, C, D, E, F, and G) and removing every fifth person. Recall that we say that the "skip number" is equal to 5. Figure 3 shows diagrams illustrating how such a game progresses. Note that the players are removed in the order E, C, B, D, G, A, and person F is the last one standing.

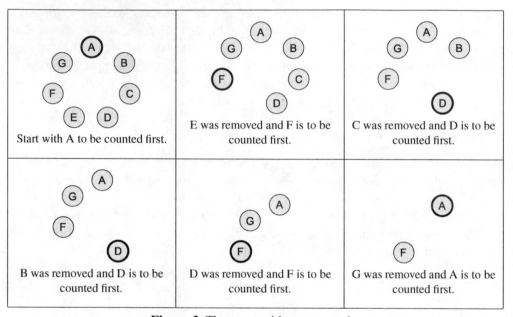

Figure 3. The game with seven people.

Exercise 1. On your own, play the Josephus game with n players and a skip number of k for each of the following values. Determine who is the last person standing.

   a. $n = 6, k = 2$ b. $n = 10, k = 3$ c. $n = 11, k = 3$

Changing the Starting Player

What happens if we decide to keep the values of n and k the same, but change the person we start the game with? How does this affect the outcome? Let's go back to the example with seven people and a skip number of 5. Let's say the people are named Terry, Ursula, Vivian, Walter, Xander, Yolanda, and Zack, and we want to start the game with Walter as Figure 4 shows.

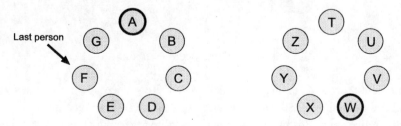

Figure 4. What's in a name?

For discussion: In the game shown on the left in Figure 4, F is the last person standing. Who will be the last person standing in the game shown on the right? Can you figure it out without playing the entire game again?

If you said that Ursula would be the last person standing, you are correct! When we have seven people and a skip number of 5, the last person standing is the sixth one around the circle from the starting player. (Here we count the starting person as the first player around the circle.) In mathematical notation, we will write this as $J(7, 5) = 6$.

The $J(n, k)$ notation is very handy for describing the last person left in the Josephus game that starts with n people in the circle and eliminates every k^{th} one. For example, the result of our first example can be described by simply writing $J(6, 3) = 1$.

Exercise 2. Go back to the three games you played in Exercise 1. Using the mathematical notation we have defined, find the value of $J(n, k)$ in each of the following cases.

   a. $J(6, 2)$ b. $J(10, 3)$ c. $J(11, 3)$

Recursion: Using What Came Before

This idea of changing the starting player can be very helpful for finding patterns in the Josephus problem. Consider the game with eight people and a skip number of 5, as shown in Figure 5. After the first step of this game, E is eliminated

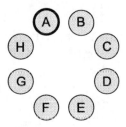

Figure 5. Beginning the game with eight people.

from the circle, and we have the situation in Figure 6.

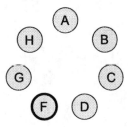

Figure 6. After person E is eliminated.

Now what? Well, we continue to play the game as before, or we might notice that we have seen this situation before. This is a game with **seven** players and a skip number of 5. We already determined that the last person standing in this game is the sixth person around the circle from the starting player. In this case, that means that C is the last person standing.

For discussion: Finish playing the game to verify that C is the last one standing.

Here is another example of this idea. In Exercise 2(c), you determined that $J(11, 3) = 7$. That is, in a game with eleven people and a skip number of 3, the seventh person around the circle from the starting person will be the last one standing. How does this help us determine the value of $J(12, 3)$? Consider the first step of the game with twelve people and a skip number of 3. The first person eliminated is person 3, and person 4 becomes the new starting player.

Now there are eleven people remaining, and we know that the last one standing will be the seventh person around the circle starting with person 4. This is person 10. You can verify on your own that $J(12, 3) = 10$.

Finding the Pattern

There is a pattern to how the position of the last survivor changes as we change the number of people initially standing in the circle. To see this pattern, we need to experiment and compute the answer for many different examples. In the table in Exercise 3, the top row shows the number of people in the circle, and the bottom row shows the position of the last person standing when the skip number is 3. The values we have already determined are filled in for you.

Exercise 3. Fill in the rest of the table, either by playing each game or by appealing to the "using-what-came-before" strategy.

n	3	4	5	6	7	8	9	10	11	12	13	14
$J(n, 3)$				1					7	10		

a. What pattern do you notice in the table?

b. Can you explain in terms of the "using-what-came-before" strategy why this pattern holds?

c. On your own, make a similar table but change the skip number to 4. Can you predict what pattern you will see?

An easier variation

A game that's a little better suited for detailed analysis is the variation where every second person is eliminated — that is, the skip number is 2. The game will officially be played with people named $1, 2, \ldots, n$ in a circle (with the numbers going clockwise). We go around the circle clockwise getting rid of every second person (Person 2 is the first to go) until no one is left. For example, if we start with four people, then the people are eliminated in the order $2, 4, 3, 1$, so person 1 is the last survivor.

We will let $J(n)$ denote the last survivor in the game which starts with n people and has a skip number of 2. (That is, we use $J(n)$ instead of $J(n, 2)$.)

Exercise 4. Fill in the rest of the table, either by playing each game or by appealing to the "using-what-came-before" strategy.

n	1	2	3	4	5	6	7	8	9	10	11	12	13	14	15	16
$J(n)$			1													

n	17	18	19	20	21	22	23	24	25	26	27	28	29	30
$J(n)$														

a. How is the value of $J(n)$ related to the value of $J(n - 1)$?

b. What will be the next value of n for which $J(n) = 1$?

c. How would you describe a formula for $J(n)$ that would allow someone to quickly figure out the last place in line given any n?

Josephus and his buddy

In the original story, Josephus actually escapes with a friend, so in reality he had to know the positions of the last two survivors of this macabre game. To keep it simple, let's still use the game with skip number 2, but now we will use $F(n)$ to denote the required position of the friend in the Josephus game starting with n people.

Exercise 5. Play the Josephus game (with every second person eliminated, as above) for various n and record the numbers $J(n)$ and $F(n)$ of the last person alive and of the next-to-the-last person alive, respectively. Find more values than in the table below if you think it is helpful to do so. Remember to try to use things you already know as you tackle larger and larger values of n.

n	12	13	14	15	16	17	18	19	20	21	22	23	24
$J(n)$													
$F(n)$													

a. How is the value of $F(n)$ related to the value of $F(n-1)$?

b. What will be the next value of n for which $F(n) = 1$?

c. Is there a direct relationship between $J(n)$ and $F(n)$?

Further questions for exploration

The following problems, as well as the ones above, can be explored with the applet found under Section 1.1 on the website

http://webspace.ship.edu/~deensley/DiscreteMath/flash/

Exercise 6. Fill in the following table using the "One potato, two potato" game on n people, starting the first "one potato" on person 1. For those not familiar with this method of choosing a person on the playground, this is simply the Josephus problem with every **eighth** person eliminated. That is, in the table below we use $P(n)$ to mean the same thing as $J(n, 8)$ from the previous discussion.

n	10	11	12	13	14	15	16	17	18	19	20	21	22	23	24
$P(n)$															

a. If the _____ students in this class stand in a circle in alphabetical order and do "one potato, two potato", who will be the last person left?

b. Suppose in the game with 6 people, Josephus is person 1 but before the game starts, the Roman leader says, "Hey Joey, *you* pick the skip number." What should he say so that he is the last person left?

c. Is it possible for Josephus to always come up with a response to the previous question no matter how many people are originally in the circle?

Solutions

Exercise 1. We will use the conventions of labeling the people A, B, C, etc. clockwise around the circle and starting our count with person A.

  a. For $n = 6$ and $k = 2$, the last person left is E.

  b. For $n = 10$ and $k = 3$, the last person left is D.

  c. For $n = 11$ and $k = 3$, the last person left is G.

Exercise 2.

  a. $J(6, 2) = 5$.

  b. $J(10, 3) = 4$.

  c. $J(11, 3) = 7$.

Exercise 3. Here is the completed table:

n	3	4	5	6	7	8	9	10	11	12	13	14
$J(n, 3)$	2	1	4	1	4	7	1	4	7	10	13	2

  a. For all $n \geq 2$, person $J(n, 3)$ is three more around (clockwise) the original circle from person $J(n - 1, 3)$.

  b. If the kth person around the circle of $n - 1$ people is the last one remaining, then in the game that starts with n people, after one person is eliminated the first person in the remaining circle of $n - 1$ is person 4. The kth person in *this* circle, is the $(k + 3)^{th}$ person in the original circle.

  c. For all $n \geq 2$, person $J(n, 4)$ is four more around (clockwise) the original circle from person $J(n - 1, 4)$.

Exercise 4. Here is the completed table:

n	1	2	3	4	5	6	7	8	9	10	11	12	13	14	15	16
$J(n)$	1	1	3	1	3	5	7	1	3	5	7	9	11	13	15	1

n	17	18	19	20	21	22	23	24	25	26	27	28	29	30
$J(n)$	3	5	7	9	11	13	15	17	19	21	23	25	27	29

  a. For all $n \geq 2$, person $J(n)$ is two more around (clockwise) the original circle from person $J(n - 1)$.

  b. The next value of n for which $J(n) = 1$ will be $n = 32$. It appears that $J(n) = 1$ if and only if n is a power of 2.

  c. Given n people originally, let m be the smallest power of 2 less than or equal to n. Eliminate people $2, 4, \ldots, 2(n - m)$. This leaves the game with m people, the first of whom is person $2(n - m) + 1$. According to the observation in part (b) of this exercise, this person will be the last person left at the end of the entire process.

Exercise 5. Here is the completed table:

n	12	13	14	15	16	17	18	19	20	21	22	23	24
$J(n)$	9	11	13	15	1	3	5	7	9	11	13	15	17
$F(n)$	1	3	5	7	9	11	13	15	17	19	21	23	1

  a. For all $n \geq 2$, person $F(n)$ is two more around (clockwise) the original circle from person $F(n - 1)$.

  b. The next value of n for which $F(n) = 1$ will be $n = 48$. It appears that $F(n) = 1$ if and only if $n = 3 \cdot 2^k$ for some value of $k \geq 0$.

  c. $J(n) - F(n) = 2^k$ when the integer k can be chosen so that $3 \cdot 2^{k-1} \leq n < 2^{k+1}$, and $F(n) - J(n) = 2^k$ when the integer k can be chosen so that $2^{k+1} \leq n < 3 \cdot 2^k$.

Exercise 6. Here is the completed table:

n	10	11	12	13	14	15	16	17	18	19	20	21	22	23	24
$P(n)$	1	9	5	13	7	15	7	15	5	13	1	9	17	2	10

a. This answer will depend on the number of people in your class. Suppose there are 32 people in your class. Using the pattern of "adding 8" relative to the number in the circle, we find that $J(32, 8) = 17$.

b. Using a skip number of 60 will work for sure (see the next answer), but the smallest number that will work is $k = 3$.

c. For the game with n people, using k that is the least common multiple of the numbers in $\{1, 2, 3, \ldots, n\}$ is guaranteed to work, but there are typically much smaller values.

Using Trains to Model Recurrence Relations

Benjamin Sinwell

Northwood High School

Summary

This project gives students a hands-on approach to experiment with recurrence relations. By building trains using cars of different lengths students are given a concrete model with which to represent and better understand the concept of recursion.

Notes for the instructor

A brief introduction to building trains of different lengths using cars of different lengths is all that is necessary for students to complete this assignment. For example, a train of length four can be made from two cars of length one and one car of length two in three different ways.

Cuisenaire rods or colored paper (cut to match the different train sizes) could be used, but are not necessary, for students to complete this project. Students can easily record the trains on graph paper. That said, the use of manipulatives forces students to build the trains rather than simply represent the trains symbolically. This project could be given as a homework assignment or be used as a group activity during class. The first six problems should take approximately an hour for students to work through. The problems could be worked during class time or outside of class. Providing a forum (i.e., class discussion or an online discussion) for students to share their thinking on how they went about solving the problems would provide an additional and valuable learning opportunity.

Bibliography

[1] Benjamin, A.T. and J.J. Quinn, *Proofs That Really Count: The Art of Combinatorial Proof*, Dolciani Mathematical Expositions, Volume 27, Mathematical Association of America, 2003.

Worksheet on Using Trains to Model Recurrence Relations

Each of the following questions will involve building trains of different lengths using cars of different lengths. The following shows how a train of length 4 can be made from two cars of length 1 and one car of length 2 in three different ways.

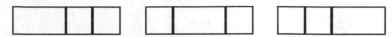

When asked how many trains of a given length you can make, simply count the number of *different* trains you can make.

1. Using only length 1 cars and length 2 cars, build and record all the trains that you can make for each of the following lengths: 1, 2, 3, 4, 5, and 6. Predict how many length 10 trains you can make. Explain how you determined your answer for length 10 trains. How do you know that you have counted all the trains?

2. Using only length 1 cars and now two different colors of length 2 cars, build and record all the trains that you can make for each of the following lengths: 1, 2, 3, 4, and 5. Predict how many length 6 trains you can make. Explain how you obtained your answer for the number of length 6 trains. The following are different trains.

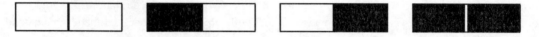

3. Using only length 1 cars and two different colors of length 3 trains, build and record all the trains that you can make for each of the following lengths: 1, 2, 3, 4, and 5. Predict how many length 6 trains you can build. Explain your prediction using only the drawings of the different length 3 trains and length 5 trains.

4. For each of #1, #2, and #3, write a rule that would tell you how many ways there are to make a length n train if you already knew the number of ways to make trains of length less than n.

5. Predict how many trains of length 1, 2, 3, 4, and 5 you can make using only length 2 cars and two different colors of length 1 cars. Verify your predictions. How can you determine the number of length 6 trains?

6. What types of cars (give the length and/or number of colors) would you need to use in order to create a pattern where the number of length n trains is equal to three times the number of length $n - 1$ trains plus two times the number of length $n - 2$ trains?

Extensions:

7. Using only cars of lengths 1, 2, and 3, how many trains can you make of each of the following lengths: 1, 2, 3, 4, 5, and 6? Explain how many ways are there to make a length 7 train using what you know about the number of ways to make trains of length less than 7.

8. Using cars of any length except 1, how many trains can you make of each of the following lengths: 1, 2, 3, 4, 5, and 6? How many ways are there to make a length 7 train? Have you seen these numbers before? Compare your results to #1. What is happening?

9. Using cars of any length except 2, how many trains can you make of the following lengths: 1, 2, 3, 4, 5, and 6? Explain how to find how many ways there are to make a length 7 train.

10. What types of cars would you need to produce the following table?

Train length	1	2	3	4	5	6	7
number of trains	0	2	0	4	0	8	0

Solutions

1. Using cars of lengths 1 and 2, the number of ways to make each of the first six train lengths is 1, 2, 3, 5, 8, and 13, respectively. There are 89 trains of length 10. This is the sum of the number of all length 9 trains and the number of all length 8 trains. To make a train of length n, add a length 1 car to the end of each length $n-1$ train and add a length 2 car to the end of each length $n-2$ train. If the number of trains one can make of length 1 and 2 are correct, it follows that one has counted all of the trains.

   Let t_n be the number of length n trains. We have the following recurrence relation:

   $$t_n = t_{n-1} + t_{n-2} \text{ with initial conditions } t_1 = 1, t_2 = 2.$$

   This is the same recurrence relation as the Fibonacci numbers, but with different starting values.

   The diagram below shows the trains 1, 2, 3, 4, and 5 that can be made from length 1 and 2 cars. Notice the order of the trains in columns for $n = 3, 4, 5$: first are the trains whose rightmost car is length 1, added to the trains of length $n - 1$, and below those are the trains whose rightmost car is length 2, added to the trains of length $n - 2$.

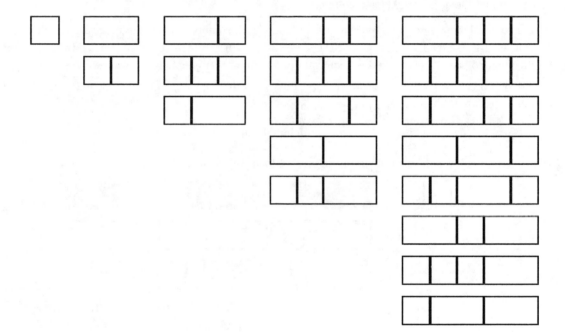

2. Using cars of length 1 and two types of cars of length 2, the number of ways to make each of the first five train lengths is 1, 3, 5, 11, and 21, respectively. There are 43 trains of length 6. To build them, add a length 1 car to each of the length 5 trains and add one of each color of length 2 cars to the length 4 trains.

   These satisfy the following recurrence relation:

   $$t_n = t_{n-1} + 2t_{n-2} \text{ with initial conditions } t_1 = 1, t_2 = 3.$$

   The diagram below shows all the trains that can be made of length 1, 2, 3, and 4.

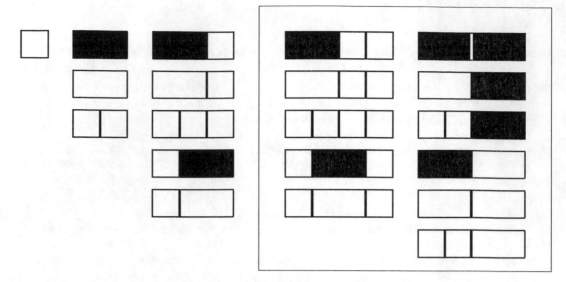

The trains of length 4 are boxed in the last two columns. They can be built by adding a car of length 1 to each train of length 3 (left hand column), a black car of length 2 to each train of length 2, and a white car of length 2 to each train of length 2 (right hand column).

3. Using cars of length 1 and two types of cars of length 2, the number of ways to make each of the first five train lengths is 1, 1, 3, 5, and 7, respectively. There are 13 trains of length 6. To build them, add a length 1 car to each of the length 5 trains and add one of each color of length 3 cars to the length 3 trains.

   The diagram below illustrates how this can be done using the drawings of trains of length 5 and length 3.

   These satisfy the following recurrence relation:

   $$t_n = t_{n-1} + 2t_{n-3} \text{ with initial conditions } t_1 = 1, t_2 = 1, t_3 = 3.$$

   The diagram below shows all the trains that can be made of length 1, 2, 3, 4, and 5.

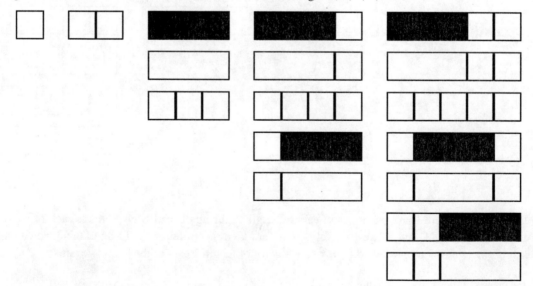

4. (Recurrence relations and initial conditions are given above.)

5. Using cars of length 2 and two types of cars of length 1, the number of ways to make each of the first five train lengths is 2, 5, 12, 29, and 70, respectively. There are 13 trains of length 6. To build them, add a length 1 car to each of the length 5 trains and add one of each color of length 3 cars to the length 3 trains.

The number of ways to make each of the first five train lengths is: 2, 5, 12, 29, and 70. To make all length 6 trains, double the number of length 5 trains (because one can add two different cars of length 1 to each of them) and add that to the number of length 4 trains (because one can add a car of length 2 to each of those, giving $2 \cdot 70 + 29 = 169$ length 6 trains.

These satisfy the following recurrence relation:

$$t_n = 2t_{n-1} + t_{n-2} \text{ with initial conditions } t_1 = 2, t_2 = 5.$$

6. Using three different colors of length 1 cars and two different colors of length 2 cars will produce such a pattern.

7. Using cars of lengths 1, 2, and 3, the number of ways to make each of the first six train lengths is 1, 2, 4, 7, 13, and 24, respectively. There are 44 length 7 trains: to make them all, add a length 1 car to all length 6 trains, a length 2 car to all length 5 trains, and a length 3 car to all length 4 trains.

These satisfy the following recurrence relation:

$$t_n = t_{n-1} + t_{n-2} + t_{n-3} \text{ with initial conditions } t_1 = 1, t_2 = 2, t_3 = 4.$$

8. Using cars of any length except 1, the number of ways to make each of the first six train lengths is 0, 1, 1, 2, 3, and 5, respectively. There are 8 length 7 trains: to make them all, extend the last car in each length 6 train by one unit and add a length 2 car to each length 5 train. These are the same numbers as problem #1, just shifted over two terms. They follow the same recurrence relation as those in problem #1, $t_n = t_{n-1} + t_{n-2}$, but have different initial values, $t_1 = 0$ and $t_2 = 1$.

9. Using cars of any length except 2, the number of ways to make each of the first six train lengths is 1, 1, 2, 4, 7, and 12, respectively. There are 21 length 7 trains: to make them all, add a length 1 car to each length 6 train, extend the last car in each length 5 train by two units, and add a length 4 car to each length 3 train. (Since none of the length 5 train can end in a length 2 car, none of their extended versions end in a length 4 car.)

These satisfy the following recurrence relation:

$$t_n = t_{n-1} + t_{n-2} + t_{n-4} \text{ with initial conditions } t_1 = 1, t_2 = 1, t_3 = 2, t_4 = 4.$$

10. Using just two different colors of length 2 cars produces the pattern. An explicit formula for the number of length n trains in this setting is

$$\frac{1}{2}\left[(\sqrt{2})^n + (\sqrt{-2})^n\right].$$

(Determining the number of certain types of length n trains could be used to motivate finding explicit formulas from recurrence relations.)

Codon Classes[1]

Brian Hopkins
Saint Peter's College

Summary

This project explores an application of equivalence relations to bioinformatics. RNA can be considered as a string over a four-letter alphabet. Cells use triples of these bases to regulate the production of protein by signaling the sequence of amino acids. But there are far fewer than 64 amino acids; is there a system behind the redundancy?

Students examine the equivalence classes arising from six difference equivalence relations on the set of codons, from simple to somewhat intricate. Then they research the actual genetic code; it is "closest" to one of the simpler relations, although it is far from an exact match.

Notes for the instructor

This project was motivated by student interest in bioinformatics and my difficulty finding engaging applications of equivalence relations and classes. The set of 64 codons is small enough to be tractable, but large enough to encourage labor-saving devices. While some of the relations have no biological motivation (e.g., number of A's), they do encourage the use of standard counting techniques. Students are instructed to "describe the equivalence classes" for each relation; this can prompt discussions of a good order for listing set elements and the proper use of ellipses.

I typically put the students in random groups of three (or two) for a project and give them a class period to get started, with the expectation that they will need to work outside class to complete the project and write it up. The examples usually suffice to clarify the verbal description of the relation; sometimes the distinction between #5 (the same three bases, a fixed number of each) and #6 (the same bases in whatever nonzero quantity) requires further examples. A more formal statement for #4 is "The two codons have the same number of bases from the set {C,U}." Students will need a biology or genetics text, or the Internet to answer #7. Also, the notion of "close" in #7 is not defined — two interpretations are mentioned in the Solutions; this could lead to a discussion of possible precise definitions of "close" in the context of relations and equivalence classes.

It is interesting that the actual correspondence, the "standard genetic code," is a rough approximation of one of the simpler relations. One could probably construct a not-too-complicated relation that gave the 21 classes with the proper number of codons in each, but it seems nature tends to parsimony and adjusts a very simple relation. This can lead to a discussion of the interplay between mathematical elegance and modeling.

---

[1] Thanks to Frances Raleigh and Katherine Wydner of the Saint Peter's College Department of Biology for assisting with the scientific content of this project.

Worksheet on Codon Classes

Ribonucleic acid (RNA) is used by cells to produce proteins, among other tasks. RNA can be thought of as a string of nucleotide bases drawn from the set adenine, cytosine, guanine, and uracil. A codon is an ordered triple of bases that signals for a certain amino acid to be included in the sequence; for example, the codon AUG signals for methionine.

There are $4^3 = 64$ possible codons, and all occur in nature. However, there are not that many amino acids. In this project, you will explore various equivalence classes on the set of all codons and then compare the correspondence that occurs in nature.

For each of the following relations on codons, determine the number of equivalence classes, the size of each equivalence class, and describe the equivalence classes (you don't necessarily need to fully write out each equivalence class).

1. The two codons have the same middle base (e.g., ACU $\sim$ GCU).

2. The two codons have the same first two bases (e.g., ACC $\sim$ ACU).

3. The two codons have the same number of A's (e.g., AUA $\sim$ GAA).

4. The two codons have the same number of C's or U's (e.g., ACU $\sim$ GUU).

5. The two codons are made up of the same three bases (e.g., AUA $\sim$ UAA).

6. The two codons use the same bases (e.g., AUA $\sim$ UUA; this is different from #5).

In fact, the codons fall into 21 classes, indicating twenty basic amino acids and a signal to stop the genetic process called translation. This is not quite an easily described pattern; it is "close to" one of the previous equivalence relations.

7. Find and record the 21 actual subsets of codons determined by the amino acid they signal (and "stop"). Which relation above is this close to?

Solutions

1. This relation partitions the 64 codons into 4 classes, each with 16 elements. Verifying the count, $4 \cdot 16 = 64$. One such class is {AAA, AAC, AAG, AAU, CAA, CAC, CAG, CAU, GAA, GAC, GAG, GAU, UAA, UAC, UAG, UAU}.

2. This relation partitions the codons into 16 classes, each with 4 elements. Verifying, $16 \cdot 4 = 64$. The classes are {AAA, AAC, AAG, AAU}, {ACA, ACC, ACG, ACU}, ..., {UGA, UGC, UGG, UGU}, {UUA, UUC, UUG, UUU}.

3. This relation partitions the codons into 4 classes, with 27, 27, 9, and 1 elements, respectively. Verifying, $27 + 27 + 9 + 1 = 64$. (Semicolons in the following sets suggest counting arguments that could justify the set sizes without writing them out.)

   no A's: {CCC, CCG, CCU; CGC, CGG, CGU; ...; UGC, UGG, UGU; UUC, UUG, UUU}

   1 A: {ACC, ACG, ..., AUG, AUU; ...; CCA, CGA, ..., UGA, UUA}

   2 A's: {AAC, AAG, AAU; ACA, AGA, AUA; CAA, GAA, UAA}

   3 A's: {AAA}

4. This relation partitions the codons into 4 classes, with 8, 24, 24, and 8 elements, respectively. Verifying, $8 + 24 + 24 + 8 = 64$.

   no C's or U's: {AAA, AAG, AGA, AGG, GAA, GAG, GGA, GGG}

   1 C or U: {AAC, AAU; AGC, AGU; ...; CAG, UAG; CGG, UGG}

   2 C's or U's: {ACC, ACU, AUC, AUU; CAC, CAU, UAC, UAU; ...; CCG, CUG, UCG, UUG}

   3 C's or U's: {CCC, CCU, CUC, CUU, UCC, UCU, UUC, UUU}

5. This relation partitions the codons into 20 classes: 4 with 1 element each, 12 with 3 elements each, and 4 with 6 elements each. Verifying, $4 \cdot 1 + 12 \cdot 3 + 4 \cdot 6 = 4 + 36 + 24 = 64$.

   1 base: {AAA}, {CCC}, {GGG}, {UUU}

   2 bases: {AAC, ACA, CAA}, {ACC, CAC, CCA}, ..., {GUU, UGU, UUG}

   3 bases: {ACG, AGC, CAG, CGA, GAC, GCA}, ..., {CGU, CUG, GCU, GUC, UCG, UGC}

6. This relation partitions the codons into 14 classes: 4 with 1 element each, 6 with 6 elements each, and 4 with 6 elements each. Verifying, $4 \cdot 1 + 6 \cdot 6 + 4 \cdot 6 = 4 + 36 + 24 = 64$. The classes are as in #5 except for the case of 2 bases, where those 12 classes combine into 6 as follows.

   2 bases: {AAC, ACA, ACC, CAA, CAC, CCA}, ..., {GGU, GUG, GUU, UGG, UGU, UUG}

7. The "standard genetic code" partitions the codons into 21 classes: 2 with 1 element each, 9 with 2 elements each, 2 with 3 elements each, 5 with 4 elements each, and 3 with 6 elements each. Verifying, $2 \cdot 1 + 9 \cdot 2 + 2 \cdot 3 + 5 \cdot 4 + 3 \cdot 6 = 2 + 18 + 6 + 20 + 18 = 64$.

   alanine: {GCA, GCC, GCG, GCU}

   arginine: {AGA, AGG, CGA, CGC, CGG, CGU}

   asparagine: {AAC, AAU}

   aspartic acid: {GAC, GAU}

   cysteine: {UGC, UGU}

   glutamic acid: {GAA, GAG}

   glutamine: {CAA, CAG}

glycine: {GGA, GGC, GGG, GGU}

histidine: {CAC, CAU}

isoleucine: {AUA, AUC, AUU}

leucine: {CUA, CUC, CUG, CUU, UUA, UUG}

lysine: {AAA, AAG}

methionine: {AUG}

phenylalanine: {UUC, UUU}

proline: {CCA, CCC, CCG, CCU}

serine: {AGC, AGU, UCA, UCC, UCG, UCU}

threonine: {ACA, ACC, ACG, ACU}

tryptophan: {UGG}

tyrosine: {UAC, UAU}

valine: {GUA, GUC, GUG, GUU}

stop: {UAA, UAG, UGA}

Although the relation of #5 is closest in the number of equivalence classes, the relation of #2 is closest structurally. Whenever an amino acid corresponds to four or fewer codons, the first two bases of the codons are the same. For the three amino acids with six codons, each contains an entire set with some two initial bases and half of another such quartet. There are chemical reasons that the code allows for variation in the last base; this is the "wobble effect" hypothesized by Francis Crick in 1966. For related reasons, amino acids with two codons have codons that end in A and G, or C and U. For more information, see a biology or genetics text, such as *Genetics: Analysis of Genes and Genomes* by Daniel L. Hartl and Elizabeth W. Jones (6th ed., Jones and Bartlett Publishers, 2005).

How to change coins, M&M's, or chicken nuggets:
The linear Diophantine problem of Frobenius

Matthias Beck
San Francisco State University

Summary

Let's imagine that we introduce a new coin system. Instead of using pennies, nickels, dimes, and quarters, let's say we agree on using 4-cent, 7-cent, 9-cent, and 34-cent coins. The reader might point out the following flaw of this new system: certain amounts cannot be exchanged, for example, 1, 2, or 5 cents. On the other hand, this deficiency makes our new coin system more interesting than the old one, because we can ask the question: "which amounts can be changed?" In the next section, we will prove that there are only finitely many integer amounts that *cannot* be exchanged using our new coin system. A natural question, first tackled by Ferdinand Georg Frobenius and James Joseph Sylvester in the 19th century, is: "what is the *largest* amount that cannot be exchanged?" As mathematicians, we like to keep questions as general as possible, and so we ask: given coins of denominations $a_1, a_2, \ldots, a_d$, which are positive integers without any common factor, can you give a formula for the largest amount that cannot be exchanged using the coins $a_1, a_2, \ldots, a_d$? This problem is known as the *Frobenius coin-exchange problem*. One of the appeals of this famous problem is that it can be stated in every-day language and in many disguises, as the title of these notes suggests. To be precise, suppose we're given a set of positive integers

$$A = \{a_1, a_2, \ldots, a_d\}$$

with gcd $(a_1, a_2, \ldots, a_d) = 1$ and we call an integer k *representable* (in terms of A) if there exist nonnegative integers $m_1, m_2, \ldots, m_d$ such that

$$k = m_1 a_1 + \cdots + m_d a_d .$$

In the language of coins, this means that we can exchange the amount k using the coins $a_1, a_2, \ldots, a_d$. The Frobenius problem (often called the *linear Diophantine problem of Frobenius*) asks us to find the largest integer that is not representable. We call this largest integer the *Frobenius number* and denote it by $g(a_1, \ldots, a_d)$. In the worksheet questions we will outline a proof for the folklore result for $d = 2$:

$$g(a, b) = ab - a - b .$$

This simple-looking formula for $g(a, b)$ inspired a great deal of research into formulas for the Frobenius number $g(a_1, a_2, \ldots, a_d)$, with limited success: While it is safe to assume that the $d = 2$ solution has been known for more than a century, no analogous formula exists for $d \geq 3$. The case $d = 3$ is solved algorithmically, i.e., there are efficient algorithms to compute $g(a, b, c)$ [7, 9, 10], and in form of a semi-explicit formula [8, 14]. The Frobenius problem for fixed $d \geq 4$ has been proved to be computationally feasible [1, 11], but not even an efficient practical algorithm for $d = 4$ is known.

A second classic theorem for the case $d = 2$, which Sylvester posted as a math problem in the *Educational Times* [18], concerns the *number* of non-representable integers. Sylvester proved that exactly half of the integers between 1 and $(a - 1)(b - 1)$ are representable (in terms of a and b). In other words, there are exactly $\frac{1}{2}(a - 1)(b - 1)$ non-representable integers. We will also outline a proof of Sylvester's Theorem.

Notes to the instructor

The first nine worksheet questions are suitable for any course in which the students discuss gcd's and the Euclidean algorithm. The next few questions assume some basic number theory, in particular, knowledge about the greatest-integer function and inverses in $\mathbb{Z}_n$. The different projects naturally vary in depth. Most problems in the Euclidean algorithm section are elementary; the slightly more complicated ones have a hint attached to them. The problems in the counting function section are a bit more advanced but should be doable in, e.g., an elementary number theory class. Further extensions and student research projects are discussed below.

The idea of the proofs hidden in the projects of the Euclidean algorithm section first appeared in [12], to the best of my knowledge. The questions in the counting function section are from [4].

Extensions: Beyond $d = 2$

This section includes an outline of what is known for the general Frobenius problem and some open problems, many of which are suitable for computational exploration and undergraduate research projects. One such extension was already mentioned in Question 11. For more, we refer to the research monograph [15]; it includes more than 400 references to articles written about the Frobenius problem.

To give the state of the art for the case $d = 3$ and beyond, we define the *generating function* of all representable integers, given some fixed parameters $a_1, a_2, \ldots, a_d$ with no common factor, as $F(x) := \sum_{k \text{ representable}} x^k$. One can prove that this generating function can always be written as a rational function of the form

$$F(x) = \sum_{k \text{ representable}} x^k = \frac{p(x)}{(1 - x^{a_1})(1 - x^{a_2}) \cdots (1 - x^{a_d})} .$$

Furthermore, in the case $d = 2$ one can show that $F(x) = 1 - x^{a_1 a_2}$. Denham [8] recently discovered the remarkable fact that for $d = 3$, the polynomial p in the numerator has either 4 or 6 terms. He gave semi-explicit formulas for p, from which one can deduce a semi-explicit formula for the Frobenius number $g(a, b, c)$. This formula was independently found by Ramírez-Alfonsín [14]. As we already remarked in the introduction, there is no "easy" formula for $d = 3$ that would parallel Theorem 2. However, Denham's theorem implies that the Frobenius number in the case $d = 3$ is quickly computable, a result that is originally due, in various guises, to Herzog [10], Greenberg [9], and Davison [7].

As much as there seems to be a well-defined border between the cases $d = 2$ and $d = 3$, there also seems to be such a border between the cases $d = 3$ and $d = 4$: Bresinsky [6] proved that for $d \geq 4$, there is no absolute bound for the number of terms in p, in sharp contrast to Denham's theorem.

On the other hand, Barvinok and Woods [1] proved recently that for fixed d, the rational generating function F can be written as a "short" sum of rational functions; in particular, F can be efficiently computed when d is fixed. A corollary of this fact is that the Frobenius number can be efficiently computed when d is fixed; this theorem is due to Kannan [11]. On the other hand, Ramírez-Alfonsín [13] proved that trying to efficiently compute the Frobenius number is hopeless if d is left as a variable. While these results settle the theoretical complexity of the computation of the Frobenius number, practical algorithms are a completely different matter. Both Kannan's and Barvinok-Woods' ideas seem complex enough that nobody has yet tried to implement them. The fastest known algorithm is due to Beihoffer, Nijenhuis, Hendry and Wagon [5]; it is currently being improved by Einstein, Lichtblau, and Wagon.

We conclude with a few projects. These differ distinctively from the questions of the worksheet in that they constitute open research problems. I list them in what I find decreasing order of difficulty (an estimate that is naturally subjective); the later projects are most suitable for undergraduate research and computational experiments that should bring new insights.

Project 1. Come up with a new approach or a new algorithm for the Frobenius problem in the $d \geq 3$ cases.

Project 2. There is a very good lower [7] and several upper bounds [15, Chapter 3] for the Frobenius number. Come up with improved upper bounds.

Project 3. Study vector generalizations of the Frobenius problem [16, 17], which seem for the most part unexplored.

Project 4. There are several special cases of $A = \{a_1, a_2, \ldots, a_d\}$ for which the Frobenius problem is solved, for example, arithmetic sequences [15, Chapter 3]. Extend these special cases and come up with new ones.

Project 5. Study the generalized Frobenius number g_j (defined in Question 11): Derive formulas for special cases, e.g., arithmetic sequences.

Bibliography

[1] Alexander Barvinok and Kevin Woods, *Short rational generating functions for lattice point problems*, J. Amer. Math. Soc. **16** (2003), no. 4, 957–979 (electronic), arXiv:math.CO/0211146.

[2] Matthias Beck, Ricardo Diaz, and Sinai Robins, *The Frobenius problem, rational polytopes, and Fourier-Dedekind sums*, J. Number Theory **96** (2002), no. 1, 1–21, arXiv:math.NT/0204035.

[3] Matthias Beck and Sinai Robins, *Computing the continuous discretely: Integer-point enumeration in polyhedra*, Undergraduate Texts in Mathematics, Springer-Verlag, New York, 2007.

[4] Matthias Beck and Sinai Robins, *A formula related to the Frobenius problem in two dimensions*, Number theory (New York, 2003), Springer, New York, 2004, pp. 17–23, arXiv:math.NT/0204037.

[5] Dale Beihoffer, Jemimah Hendry, Albert Nijenhuis, and Stan Wagon, *Faster algorithms for Frobenius numbers*, Electron. J. Combin. **12** (2005), no. 1, Research Paper 27, 38 pp. (electronic).

[6] Henrik Bresinsky, *Symmetric semigroups of integers generated by* 4 *elements*, Manuscripta Math. **17** (1975), no. 3, 205–219.

[7] J. Leslie Davison, *On the linear Diophantine problem of Frobenius*, J. Number Theory **48** (1994), no. 3, 353–363.

[8] Graham Denham, *Short generating functions for some semigroup algebras*, Electron. J. Combin. **10** (2003), Research Paper 36, 7 pp. (electronic).

[9] Harold Greenberg, *An algorithm for a linear Diophantine equation and a problem of Frobenius*, Numer. Math. **34** (1980), no. 4, 349–352.

[10] Jürgen Herzog, *Generators and relations of abelian semigroups and semigroup rings.*, Manuscripta Math. **3** (1970), 175–193.

[11] Ravi Kannan, *Lattice translates of a polytope and the Frobenius problem*, Combinatorica **12** (1992), no. 2, 161–177.

[12] Albert Nijenhuis and Herbert S. Wilf, *Representations of integers by linear forms in nonnegative integers.*, J. Number Theory **4** (1972), 98–106.

[13] Jorge L. Ramírez-Alfonsín, *Complexity of the Frobenius problem*, Combinatorica **16** (1996), no. 1, 143–147.

[14] Jorge L. Ramírez-Alfonsín, *The Frobenius number via Hilbert series*, preprint, 2002.

[15] Jorge L. Ramírez-Alfonsín, *The Diophantine Frobenius problem*, Oxford University Press, 2006.

[16] Les Reid and Leslie G. Roberts, *Monomial subrings in arbitrary dimension*, J. Algebra **236** (2001), no. 2, 703–730.

[17] R. Jamie Simpson and Robert Tijdeman, *Multi-dimensional versions of a theorem of Fine and Wilf and a formula of Sylvester*, Proc. Amer. Math. Soc. **131** (2003), no. 6, 1661–1671 (electronic).

[18] James J. Sylvester, *Mathematical questions with their solutions*, Educational Times **41** (1884), 171–178.

Worksheet on how to change coins, M&M's, or chicken nuggets: The linear Diophantine problem of Frobenius

I   The Euclidean algorithm and its consequences

We approach the Frobenius problem through the following important consequence of the Euclidean algorithm.

Theorem 1. *Suppose a and b are relatively prime positive integers. Then there exist $m, n \in \mathbb{Z}$ such that $1 = ma + nb$.*

What we really need is the fact that one can find such an *integral linear combination* of a and b for any integer:

Corollary 2. *Suppose a and b are relatively prime positive integers. Given an integer k, there exist $m, n \in \mathbb{Z}$ such that $k = ma + nb$.*

Students who just learned about the Euclidean algorithm might find the Frobenius problem amusing, since this last corollary *almost* solves the Frobenius problem: in the latter, we're "only" asking that $m, n \in \mathbb{Z}$ are *nonnegative*. It is this tiny additional condition that makes the Frobenius problem so hard (and interesting!). Let's put the Euclidean algorithm to good use.

Question 1. Suppose a and b are relatively prime positive integers. Show that a given integer k can be *uniquely* written as

$$k = ma + nb \, ,$$

where $m, n \in \mathbb{Z}$ and $0 \leq m < b - 1$.

This gives a simple but useful criterion for k to be representable — recall that this means that k can be written as a nonnegative integral linear combination of a and b.

Question 2. Suppose a and b are relatively prime positive integers, and write $k \in \mathbb{Z}$ as $k = ma + nb$ where $m, n \in \mathbb{Z}$ with $0 \leq m \leq b - 1$. Show that k is representable (in terms of a and b) if and only if $n \geq 0$.

This observation allows us to conclude, among other things, that the Frobenius problem is well defined:

Question 3. Suppose a and b are relatively prime positive integers. Show that every sufficiently large integer is representable (in terms of a and b).

Question 4. Prove that the general Frobenius problem is well defined. That is, show that, given relatively prime $a_1, a_2, \ldots, a_d$, every sufficiently large integer is representable (in terms of $a_1, a_2, \ldots, a_d$).

Question 2 can be taken a step further to solve the Frobenius problem for $d = 2$:

Question 5. Prove that $g(a, b) = ab - a - b$.

Hint: Try to maximize possible non-representable integers, using Question 2.

Question 2 can also be used to prove Sylvester's Theorem. We start with the following:

Question 6. Suppose a and b are relatively prime positive integers and $0 < k < ab$ is not divisible by a or b. Prove that k is representable (in terms of a and b) if and only if $ab - k$ is not representable.

Hint: Use Question 2 for a representable integer k. Think about how you can strengthen the conditions of Question 2 using the divisibility properties.

Question 6 allows us to prove Sylvester's Theorem:

Question 7. Prove that there are $\frac{1}{2}(a - 1)(b - 1)$ non-representable integers.

2   A counting function

Now we study the counting sequence

$$r_k = \# \left\{ (m, n) \in \mathbb{Z}^2 : m, n \geq 0, \, ma + nb = k \right\}$$

where a and b are fixed relatively prime positive numbers. In words, r_k counts the representations of $k \in \mathbb{Z}_{\geq 0}$ as nonnegative linear combinations of a and b. Question 3 states that this sequence has only finitely many r_k's that are 0, and the Frobenius problem asks for the largest among the r_k's that is 0.

Question 2 gives us the following almost-periodicity identity for r_k.

Question 8. Suppose a and b are relatively prime positive integers, and let r_k be given as above. Then

$$r_{k+ab} = r_k + 1 \, .$$

Remark: There is no analogous formula in the general case of d parameters $a_1, a_2, \ldots, a_d$. This is one reason why the Frobenius problem seems to be intractable for $d \geq 3$.

Let's take a moment to look at a geometric interpretation of r_k. As usual, fix two relatively prime positive integers a and b. Consider the line segment $L_k = \{(x, y) \in \mathbb{R}^2 : x, y \geq 0, \, ax + by = k\}$. The parameter k acts like a dilation factor of the line segment L_1 given by

$$L_1 = \left\{ (x, y) \in \mathbb{R}^2 : x, y \geq 0, \, ax + by = 1 \right\} .$$

Our counting sequence r_k enumerates integer points in $\mathbb{Z}^2$ that lie on the line segment L_k. As k increases, the line segment gets dilated. It is not too far fetched[1] to expect that the likelihood for an integer point to lie on the line segment L_k increases with k. In fact, one might even guess that this "probability" increases linearly with k, as the line segments are one-dimensional objects. Below we will give a formula (Theorem 3) which shows that this is indeed the case. Figure 1 shows the geometry behind the counting function r_k for the first few values of k in the case $a = 4, b = 7$. Note that

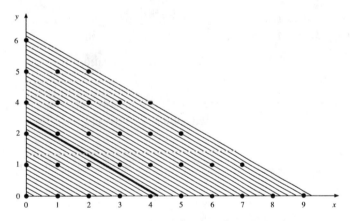

Figure 1. $4x + 7y = k$, $\qquad k = 1, 2, \ldots$

the thick line segment for the Frobenius number $k = 17 = 4 \cdot 7 - 4 - 7$ is the last one that does not contain any integer point.

Similar geometric pictures can be associated to the general Frobenius counting functions

$$\# \left\{ (m_1, m_2, \ldots, m_d) \in \mathbb{Z}^d : \text{ all } m_j \geq 0, \, m_1 a_1 + \cdots + m_d a_d = k \right\} .$$

Now the line segments get replaced by triangles ($d = 3$), tetrahedra ($d = 4$), and higher-dimensional simplices, but the general picture, namely that these counting functions enumerate integer points in $\mathbb{Z}^d$ in dilates of nice geometric

---

[1] However, one should be careful with such a statement—we invite the reader to prove that if a and b are not relatively prime, there are infinitely many line segments L_k that do not contain any integer point.

objects, stays the same. This geometric interpretation gives a glimpse into a subfield of discrete geometry called *Ehrhart theory*. It concerns the study of integer-point enumeration in *polytopes*, of which line segments, triangles, tetrahedra, etc., are special cases. The reader interested in these topics may consult [3]. There one can find a proof of the following beautiful formula for r_k due to Tiberiu Popoviciu, which we will use to re-derive some results on the Frobenius problem. First we need to define the *greatest-integer function* $\lfloor x \rfloor$, which denotes the greatest integer less than or equal to x. A close sibling to this function is the *fractional-part function* $\{x\} = x - \lfloor x \rfloor$.

Theorem 3 (Popoviciu). *If a and b are relatively prime, the counting function*

$$r_k = \#\left\{(m,n) \in \mathbb{Z}^2 : m,n \geq 0, \, ma + nb = k\right\}$$

is explicitly given by

$$r_k = \frac{k}{ab} - \left\{\frac{b^{-1}k}{a}\right\} - \left\{\frac{a^{-1}k}{b}\right\} + 1 \,,$$

where $b^{-1}b \equiv 1 \bmod a$ and $a^{-1}a \equiv 1 \bmod b$.

Remark: There are analogous formulas for the general Frobenius counting functions

$$\#\left\{(m_1, m_2, \ldots, m_d) \in \mathbb{Z}^d : \text{ all } m_j \geq 0, \, m_1 a_1 + \cdots + m_d a_d = k\right\}$$

but they are not as simple as in Popoviciu's Theorem, even if $d = 3$. These "higher-dimensional" counting functions, nevertheless, give rise to generalized *Dedekind sums*, finite arithmetic sums that appear in various other mathematical contexts [2].

Question 9. Using Popoviciu's Theorem 3, give an alternative proof of formula $g(a,b) = ab - a - b$ for the Frobenius number by proving that $r_{ab-a-b} = 0$ and that $r_k > 0$ for every $k > ab - a - b$.

Hint: Use the periodicity of $\{x\}$ and the inequality $\left\{\frac{m}{a}\right\} \leq 1 - \frac{1}{a}$ for integers m, a.

Question 10. Using Popoviciu's Theorem 3, give an alternative proof that $r_k + r_{ab-k} = 1$ for any integer $1 \leq k \leq ab - 1$ that is not divisible by a or b (cf. Question 6), and use this to give another proof of Sylvester's Theorem.

Recall that Question 6 allowed us to prove Sylvester's Theorem, so Question 10 gives an alternate proof of Sylvester's Theorem.

Question 11. Given two relatively prime positive integers a and b, we say the integer k is *j-representable* if there are exactly j solutions $(m,n) \in \mathbb{Z}_{\geq 0}^2$ to $ma + nb = k$. We define g_j as the largest *j*-representable integer. (So g_0 is the Frobenius number.) Prove:

(a) g_j is well defined.

(b) $g_j = (j+1)ab - a - b$.

(c) Given $j \geq 2$, the smallest *j*-representable integer is $ab(j-1)$.

(d) There are exactly $ab - 1$ integers that are uniquely representable.

(e) Given $j \geq 2$, there are exactly ab *j*-representable integers.

Solutions

1. We mentioned already in Corollary 2 that any integer k can be written as

$$k = ma + nb$$

for some $m, n \in \mathbb{Z}$. From this representation we get others, for example,

$$k = (m + b)a + (n - a)b \qquad \text{or} \qquad k = (m - b)a + (n + a)b .$$

In fact, because a and b are relatively prime, *all* possible representations of k as integral linear combinations of a and b are given precisely by the expressions

$$k = (m + jb)a + (n - jb)b , \qquad j \in \mathbb{Z} .$$

By choosing j accordingly, we can force the coefficient of a to be in the interval $[0, b - 1]$.

2. If $n \geq 0$, then k is representable by definition, since both coefficients m and n in $k = ma + nb$ are nonnegative. Conversely, suppose k is representable, say $k = ja + lb$ for some nonnegative integers j and l. If $0 \leq j \leq b-1$, we are done; otherwise, we subtract enough multiples of b from j such that $0 \leq m = j - qb \leq b - 1$. Then the coefficient l has to be adjusted to $n = l + qa$, which is positive.

3. Question 2 implies that every integer $k \geq ab$ is representable, since when writing $k = ma + nb$ with $0 \leq m \leq b - 1$, n has to be positive.

4. Given an integer k, the Euclidean algorithm asserts the existence of integers $m_1, m_2, \ldots, m_d$ such that k can be represented as $k = m_1 a_1 + m_2 a_2 + \cdots + m_d a_d$. With the same argument as in the solution to Question 1, we can demand that in this representation $0 \leq m_2, m_3, \ldots, m_d < a_1$, and by extension of Question 2, k is representable if and only if $m_1 \geq 0$. Hence certainly all integers beyond $a_1 (a_2 + a_3 + \cdots + a_d)$ are representable in terms of $a_1, a_2, \ldots, a_d$.

5. By Question 2, we have to maximize the integral coefficients m and n in

$$k = ma + nb ,$$

subject to $0 \leq m \leq b - 1$ and $n < 0$ (so that k is not representable). The maximal choice is apparently

$$k = (b - 1)a + (-1)b = ab - a - b .$$

6. Suppose k is representable, so by Question 2 we can write

$$k = ma + nb$$

for some nonnegative integers m and n with $0 \leq m \leq b - 1$. Since k is not divisible by a or b, we have $m \neq 0$ and n is not divisible by a; in particular, n is positive. But then

$$ab - k = ab - ma - nb = (b - m)a - nb ,$$

and we note that $0 < b - m < b$ and $n > 0$. This means that $ab - k$ can be written in the form $ab - k = ja + lb$ with $0 \leq j \leq b - 1$ and $l < 0$, and by Question 2, $ab - k$ is not representable.

7. Question 6 implies that, for k between 1 and $ab - 1$ and not divisible by a or b, exactly one of k and $ab - k$ is representable. There are

$$ab - a - b + 1 = (a - 1)(b - 1)$$

integers between 1 and $ab - 1$ that are not divisible by a or b. Finally, if k is divisible by a or b then it is representable, simply by writing k as a multiple of a or b. Hence the number of nonrepresentable integers is $\frac{1}{2}(a - 1)(b - 1)$.

8. Question 2 implies that if k is representable then it can be written as

$$k = ma + nb$$

for some nonnegative integers m and n with $0 \le m \le b - 1$. If $n \ge a$ then we get another representation, namely,

$$k = (m + b)a + (n - a)b \ .$$

We can continue the process of adding b to the coefficient of a and subtracting a from the coefficient of b, until the latter becomes negative, and those will be precisely the different representations of k. Suppose j is the largest integer such that $n - ja \ge 0$. That is, k has the $j + 1$ representations

$$k = ma + nb = (m + b)a + (n - a)b = (m + 2b)a + (n - 2a)b = \cdots = (m + jb)a + (n - ja)b \ .$$

Then $k + ab$ has the $j + 2$ representations

$$k + ab = ma + (n + a)b = (m + b)a + nb = (m + 2b)a + (n - a)b = \cdots = (m + (j + 1)b)a + (n - ja)b \ ,$$

precisely one representation more than k has.

9. We have to show that $r_{ab-a-b} = 0$ and that $r_{ab-a-b+n} > 0$ for any positive n. To prove the first assertion, we compute with Popoviciu's Theorem 3,

$$r_{ab-a-b} = \frac{ab - a - b}{ab} - \left\{ \frac{b^{-1}(ab - a - b)}{a} \right\} - \left\{ \frac{a^{-1}(ab - a - b)}{b} \right\} + 1$$

$$= 2 - \frac{1}{a} - \frac{1}{b} - \left\{ \frac{-b^{-1}b}{a} \right\} - \left\{ \frac{-a^{-1}a}{b} \right\} \ .$$

Since $b^{-1}b = 1 + ja$ for some integer j, $\left\{ \frac{-b^{-1}b}{a} \right\} = \left\{ \frac{-1}{a} \right\} = 1 - \frac{1}{a}$. With essentially the same argument, we conclude that $\left\{ \frac{-a^{-1}a}{b} \right\} = 1 - \frac{1}{b}$, which implies that $r_{ab-a-b} = 0$.

To prove that $r_{ab-a-b+n} > 0$ for $n > 0$, we note that for any integer m, $\left\{ \frac{m}{a} \right\} \le 1 - \frac{1}{a}$. Hence Popoviciu's Theorem 3 gives for any positive integer n,

$$r_{ab-a-b+n} \ge \frac{ab - a - b + n}{ab} - \left(1 - \frac{1}{a} \right) - \left(1 - \frac{1}{b} \right) + 1 = \frac{n}{ab} > 0 \ .$$

10. By Popoviciu's Theorem 3,

$$r_{ab-k} = \frac{ab - k}{ab} - \left\{ \frac{b^{-1}(ab - k)}{a} \right\} - \left\{ \frac{a^{-1}(ab - k)}{b} \right\} + 1$$

$$= 2 - \frac{k}{ab} - \left\{ \frac{-b^{-1}k}{a} \right\} - \left\{ \frac{-a^{-1}k}{b} \right\}$$

$$\overset{(\star)}{=} -\frac{k}{ab} + \left\{ \frac{b^{-1}k}{a} \right\} + \left\{ \frac{a^{-1}k}{b} \right\}$$

$$= 1 - r_k \ .$$

Here, $(\star)$ follows from the fact that $\{-x\} = 1 - \{x\}$ if $x \notin \mathbb{Z}$.

11. (a) Since every integer beyond $ab - a - b$ has at least one representation, every integer beyond $(j + 1)ab - a - b$ has at least $j + 1$ representations, by Question 8.

    (b) As we just showed, every integer beyond $(j + 1)ab - a - b$ has at least $j + 1$ representations. Furthermore, by the formula for $g(a, b)$ and Question 8, $(j + 1)ab - a - b$ has exactly j representations, and so $g_j = (j + 1)ab - a - b$.

(c) Let n be a nonnegative integer. Then

$$r_{ab(j-1)-n} = \frac{ab(j-1)-n}{ab} - \left\{\frac{b^{-1}(ab(j-1)-n)}{a}\right\} - \left\{\frac{a^{-1}(ab(j-1)-n)}{b}\right\} + 1$$

$$= j - \frac{n}{ab} - \left\{\frac{-b^{-1}n}{a}\right\} - \left\{\frac{-a^{-1}n}{b}\right\}.$$

If $n = 0$, this equals j. If n is positive, we use the fact that $\{x\} \geq 0$ to see that

$$r_{ab(k-1)-n} \leq j - \frac{n}{ab} < j.$$

(d) In the interval $[1, ab]$, there are, by Sylvester's Theorem and the fact that ab is the smallest 2-representable integer,

$$ab - \frac{(a-1)(b-1)}{2} - 1$$

1-representable integers. With Question 8 and again Sylvester's Theorem, we see that there are

$$\frac{(a-1)(b-1)}{2}$$

1-representable integers above ab. Hence there is a total of $ab - 1$ uniquely representable integers.

(e) It suffices to prove this result for $j = 2$; then the general statement follows by induction with Question 8. By the previous proof and Question 8, there are $ab - \frac{(a-1)(b-1)}{2} - 1$ integers with two representations in the interval $[ab+1, 2ab]$, and $\frac{(a-1)(b-1)}{2}$ such integers beyond $2ab$. Hence, together with the 2-representable integer ab, there are precisely ab integers with two representations.

Calculator Activities for a Discrete Mathematics Course

Jean M. Horn and Toni T. Robertson
Northern Virginia Community College

Summary

Using the symbolic capabilities of the TI-89 calculator can enhance discrete mathematics. Our mathematics and computer science majors take discrete mathematics as a post calculus course. Typical students come to the first class with their TI-89 in hand, both wanting and expecting to be able to use it. In response to their expectations we have developed a series of calculator activities which can be used to supplement and enhance some of the topics of discrete mathematics. They are designed to improve the basic understanding of course content, and to provide guidance for students as they transition from the concrete to the abstract.

Many of our students want the calculator to give them all of the answers. Their first experiences with the calculator reinforce this idea. As they advance through the curriculum, they begin to experience situations in which the calculator does not return the anticipated result. Our task is to take students from the level of dependency to the place where they can use the calculator as an analytical tool. As instructors, one way in which we can accomplish this purpose is by providing calculator activities which ask questions requiring interpretation of the output. The calculator cannot answer why, nor does it flag an erroneous answer.

Developing calculator skills is a direct way to improve analytical thinking. The calculator allows students to gain a sense of what is happening in a problem. It allows them to ask how their calculators can be used to help investigate problems and formulate solutions in a more concrete fashion. The guidance which an instructor can provide through activities like these will help students polish these skills while simultaneously reinforcing the concepts of discrete mathematics.

While these activities are intended for use by students outside of class, a demonstration in class tends to increase participation. The added benefit is the assurance that all students are using their calculators correctly. This encourages students to work independently and to delve into extensions of the activity outside of class.

Notes for the instructor

These activities were written for the TI-89 series calculators. Thus they also work on the TI-89 Titanium and Voyage 200 as well. Many will work on the TI-83+, TI-84+ series calculators. These adaptations will be noted with each activity.

- Activity 1: Modular Arithmetic

  This activity introduces students to the mod function. Familiar concepts, like sequence graphing, should inspire students to explore the mod function further. This activity is often done in class to insure that all students are comfortable with the TI-89 calculator.

  The TI-83+, TI-84+ series calculator does not have the mod n function and will not support this activity.

- Activity 2: Help with Proofs

  Discrete mathematics is often the first course in which students are introduced to the concept of proofs. The calculator will not prove or disprove a statement but can be used as an aid to determine the suspected truth of a given statement. It has also proved beneficial for finding counterexamples. While students should be doing this type of activity on their own, experience suggests they do not. Students often need guidance to start experimenting.

  Items 1–4 of this activity can be worked using any calculator. Item 5 requires a TI-89 series calculator to handle the symbolic manipulation.

- Activity 3: The Floor and Ceiling Functions

  This activity includes sums, quotients, and composites of the floor and ceiling functions. On the calculator many of the graphs appear similar but have hidden features which distinguish them. The student must determine when and where these hidden features exist.

  This activity can be adapted for the TI-83+ or TI -84+ by using $y = [\![x]\!]$ for $y = \lfloor x \rfloor$ and $y = [\![x + 1]\!]$ for $y = \lceil x \rceil$. The greatest integer function is found in the **MATH NUM** menu on these calculators as **y = int (x)**. Students need to indicate which endpoints are included on their graphs.

- Activity 4: Growth of Functions

  This activity helps students understand the concept of "big-oh" and "little-oh." While some students will have encountered these in prior computer science courses, it will be a first exposure for others. Students are asked to interpret data based on changing calculator window size, changing constants, and forming a graphical inspection of the limit. While we hope students will try graphing functions like those done in this activity, many do not think of doing so on their own.

  This activity is compatible with any graphing calculator.

Other similar calculator activities and work sheets for discrete mathematics are available by contacting the author.

Bibliography

[1] Anderson, J. A. *Discrete Mathematics with Combinatorics*, Prentice Hall, Inc., 2001

[2] Epp, Susanna S. *Discrete Mathematics with Applications*, third edition, Belmont, CA: Brooks/Cole Publishing Company, 2004.

[3] Johnsonbaugh, R. *Discrete Mathematics, Fifth Edition*, Prentice Hall, Inc., 2001

[4] Rosen, Kenneth H. *Discrete Mathematics and Its Applications*, fourth edition, McGraw-Hill, 1999.

Calculator worksheet on modular arithmetic

For any integer $n > 1$ we can define congruence modulo n as an equivalence relation on the set of integers. Distinct equivalence classes of the relation are the sets $[a] = \{y \in \mathbb{Z} \mid y \equiv a \bmod n\}$ for $a = 1, 2, \ldots, n - 1$. The graph of $y \equiv a \bmod n$ will show these equivalence classes on the TI-89 calculator. Be sure to put your calculator in sequence mode for this activity.

1. Graph $y \equiv a \bmod 4$ by entering $\begin{cases} u1 = iPart(mod(n, 4)) \\ ui1 = 1 \end{cases}$ and using the window $n\text{min} = 1$, $n\text{max} = 24$, $x\text{min} = -1$, $x\text{max} = 14$, $y\text{min} = -4$, $y\text{max} = 8$. Record your graph.

2. Now **TRACE** the function to help you get the sense of the behavior of the graph. As you move from one x-coordinate to another, estimate the n value and the corresponding y values. Using this display, substitute values for y and n in $y \equiv n \bmod 4$ and record the first five results.

3. There are two ways to think of this graph. In parts 1 and 2 you simply looked at consecutive values of n by moving from one value to the next. You should have observed the manner in which y values repeated. A second way to obtain information from the graph is to read across a horizontal line to find values of n for any given y. Reading horizontally on your graph, what values of n do you find when $y = 0$, $y = 1$, $y = 2$, $y = 3$? List the first five values for each y above. What do the values represent when listed this way? Record all answers.

4. Simultaneous congruences can also be solved graphically, such as $\begin{cases} y \equiv n \bmod 4 \\ y \equiv n \bmod 6 \end{cases}$. You already have $y \equiv n \bmod 4$ entered at $u1$. Leaving it there, enter $y \equiv n \bmod 6$ at $u2$. When you graph these two, you have two patterns showing on the display. Use **TRACE** to help you make sense of the graph and which pattern goes with which function. Look carefully for the first value where the graphs intersect. This occurs at $n = 1$ on $y \equiv n \bmod 4$. Use the up arrow to switch to tracing $y \equiv n \bmod 6$. You know you have found a common solution when the y value does not change. Now try switching between the two functions at $n = 7$. Does the cursor remain steady here? Record the value or values obtained. Is this a solution to both equations? Explore using the **TRACE** feature on the calculator to change between congruences. Solve the simultaneous congruences for $n = 3$ and record your results.

Calculator worksheet on help with proofs

Your calculator will not do proofs for you, but it can be very useful to you when you try to determine the truth of given statements. You can use it as an aid to get a sense of whether or not a given statement is true.

1. If asked whether every integer of the form $n^2 - n$ is divisible by 2, you may not see that this is true on a preliminary inspection. However, you can use your calculator to compute a few values of $n^2 - n$ and quickly conjecture that you are trying to prove a true statement. Try this by going to your **Y =** screen and entering $y1 = x^2 - x$. Then return to your home screen (press **HOME**) and enter $y1(\{-5, -2, -1, 0, 1, 5, 10, 101\})$ or some other integers of your choice. When you press **ENTER** the calculator will return a list to you. Record this list. You can then check that the list is divisible by 2 by clearing your entry line, arrowing up to highlight the list and pressing **ENTER**. After the list add $\div 2$ and press **ENTER**. Record this new list. Are all of the entries integers? What does your answer tell you about whether or not you are trying to prove a statement which is true? Note that this does not give you a *method* for proving or disproving the statement.

2. Consider the claim that $\frac{1}{5}n^5 + \frac{1}{3}n^3 + \frac{7}{15}n$ is an integer for every integer n. Examine this statement using a list of integers to conjecture the truth or falsity of the statement. Be sure to write down the list that you use, as well as the output when recording your answer and conclusion.

3. Consider the claim that $x^2 + x + 41$ always generates a prime number for all integers x. Examine this statement with the list $\{1, 5, 9, 39\}$. What can you conclude? Have you used enough values to draw a valid conclusion? Now try the list $\{1, 5, 9, 39, 41, 49, 115\}$. Is this list long enough? What do you conclude and why? Record all your answers.

4. Consider the claim that if x and y are odd, then $x^2 + y^2$ is even but not divisible by 4. Here, because of the two variables, you will need to switch the mode to **3D(MODE,graph,$\rightarrow$ to submenu,3D)**. You can then enter the three-dimensional function as $z1 = x^2 + y^2$. Then enter any odd number for x and a list for y to try this formula for a number of values. After entering the above expression and returning to the home screen, evaluate it at $(1, 3), (1, 7), (1, 25), (1, 101), (1, 57321)$ by entering **z1(1,{3,7,25,101,57321})**. Are all the entries in the returned list even? Are they divisible by 4? How can you tell if they are divisible by 4 using the calculator? Without using the calculator? Try this with some other values of your choosing. Based on this work, do you think the statement is true or false? Record *all* answers.

5. Your calculator can help you conjecture that $\binom{n}{r} = \binom{n}{n-r}$. This is entered as **nCr(n,r)** on your calculator. Can you prove this conjecture on your calculator? Show algebraically that $\binom{n}{r} = \binom{n}{n-r}$ is indeed true. Do not rely on your calculator here. Be sure to record your verification below and *explain the logic you used to reach your conclusion.*

Calculator worksheet on the floor and ceiling functions

The **Floor** of a real number x, denoted $\lfloor x \rfloor$, is the integer n such that $n \leq x < n + 1$ (it is also known as the greatest integer function). The floor function can be found in the **MATH Number (2nd 5 1: Number)** menu or in the catalog on the TI-89. The floor function can be graphed by entering **y = floor (x)** Be sure to put your calculator in dot mode (**F6,2** from the **Y=** screen) and graph the floor function in the standard window (**F2,6** from the **Y=** screen).

The **Ceiling** of a real number x, denoted $\lceil x \rceil$, is the integer n such that $n - 1 < x \leq n$. On the TI-89 the ceiling function can be found in the **MATH Number (2nd 5 1: Number)** menu or in the catalog. The ceiling function can be graphed by entering **y = ceiling (x)**. Be sure to put your calculator in dot mode and graph the ceiling function in the standard window.

Graph the following. Which ones are functions? Do they have any points of discontinuity? On what intervals are these functions continuous? Pay special attention to what happens at the integers. How do these graphs differ from those of $y = \lfloor x \rfloor$ or $y = \lceil x \rceil$? Describe the differences in your own words. How do they differ from the function you are taking the floor or ceiling of? Again, describe the difference in your own words. Use a standard graphing window of $-10 \leq x \leq 10$ for each graph.

1. $y = \lfloor x \rfloor$

2. $y = \lceil x \rceil$

3. $y = \lfloor x^2 \rfloor$

4. $y = \lceil x^2 \rceil$

5. $y = \lfloor x \rfloor + \lceil x \rceil$

6. $y = \lceil x \rceil - \lfloor x \rfloor$

7. $y = \lfloor \lceil x \rceil \rfloor$

8. $y = \lceil x \rceil / \lfloor x \rfloor$

9. $y = 1/\lfloor x \rfloor$

10. $y = 1/\lceil x \rceil$

Calculator worksheet on the growth of functions

The analysis of the complexity of algorithms in computer science makes use of concepts and techniques found in discrete mathematics courses. By a careful analysis of the number of steps an algorithm uses, we can say that one algorithm is faster than another. Mathematically, this requires a study of the growth of functions. We say that a function $f(x) = O(g(x))$ as x tends to infinity (read "$f(x)$ is big-oh of $g(x)$") where f and g are functions from the set of integers (or the set of real numbers) to the set of real numbers provided there are constants C and k such that $|f(x)| \leq C|g(x)|$ whenever $x > k$.

1. Big-oh notation can be used to talk about asymptotic growth of the function as x increases to infinity. What happens on any small interval is not important in determining whether $f(x)$ is $O(g(x))$. Set your calculator window to $0 \leq x \leq 1, 0 \leq y \leq 1$ and graph the functions $f(x) = x$ and $g(x) = x^2$. For this range which function is larger? What might this cause you to conjecture? Now enlarge your viewing window. What happens? What conclusion can you now make? Is $f(x) = O(g(x))$ or is $g(x) = O(f(x))$? Find C and k to support your conclusion.

2. By enlarging the viewing window enough we can get a good idea whether or not $f(x)$ is $O(g(x))$. However, we must remember that we only need to find one constant C such that $|f(x)| \leq C|g(x)|$. Let $f(x) = 3x^3 + 7x + 5$ and $g(x) = x^3$. Now graph both functions. Which is larger? What might you conclude from that? Change the constant in front of $g(x)$ and graph again. Try $C = 2, 3, 15$ regraphing each time. What conclusion can you draw from this?

3. To sort a list of size x, the merge sort takes about $f(x) = x(\log(x))$ steps, and the insertion sort takes about $g(x) = x^2$ steps. Is $f(x) = O(g(x))$? Is $g(x) = O(f(x))$?

4. One of the most useful characterizations of big-oh uses limits. We say that $f(x)$ is $O(g(x))$ if

$$\lim_{x \to \infty} \frac{f(x)}{g(x)} < \infty.$$

To actually evaluate the limits we will usually have to appeal to L'Hopital's Rule from calculus, but by graphing $f(x)/g(x)$ on a large enough window you can get some idea as to whether or not the ratio goes to a finite limit. Try $f(x) = 34x^{100}$ and $g(x) = 2^x$. Graph the ratio. Does it go to infinity? What does this tell you about the relationship between $f(x)$ and $g(x)$? Try $f(x) = x(\log(x))$ and $g(x) = x(\log(\log(x^{1000})))$. What conclusion can you draw here?

5. We say that $f(x)$ is $o(g(x))$ (read $f(x)$ is little-oh of $g(x)$) if for any constant c there exists some k such that $f(x) \leq cg(x)$ for $x > k$. This says that $f(x)$ grows more slowly than $g(x)$ and is a stronger statement than saying $f(x)$ is $O(g(x))$. Consider the functions $f(x) = 10x$ and $g(x) = x^{1.1}$. Use your calculator to find where $f(x)$ and $cg(x)$ intersect. Find the k associated with $c = 0.1, 1, 2, 5, 10$.

Solutions

- Activity 1: Modular Arithmetic

    2. $1 \equiv 1 \bmod 4, 2 \equiv 2 \bmod 4, 3 \equiv 3 \bmod 4, 4 \equiv 0 \bmod 4, 5 \equiv 1 \bmod 4$
    3. $y = 0$: $\{4, 8, 12, 16, 20\}$, $y = 1$: $\{1, 5, 9, 13, 17\}$, $y = 2$: $\{2, 6, 10, 14, 18\}$, $y = 3$: $\{3, 7, 11, 15, 19\}$. These are congruence classes.
    4. $n = 7$ is not a common solution. Solutions for $n = 3$ are $\{3, 15, 27, 39, 51, \ldots\}$.

- Activity 2: Help with Proofs

    1. Outputs for the given set are $\{30, 6, 2, 0, 0, 20, 90, 10100\}$ and $\{15, 3, 1, 0, 0, 10, 45, 5050\}$. The statement is true.
    2. Answers will vary. The statement is true.
    3. Outputs for the given sets are $\{43, 17, 131, 1601\}$ and $\{43, 71, 131, 1601, 1763, 2491, 13381\}$. Since 1763 is divisible by 41, the statement is false. (You may want to use the calculator's **Factor** command here.)
    4. The results will all be even and not divisible by 4. To test divisibility by 4, look for integer values after using $\div 4$ on the calculator, check if the last two digits divisible by 4 without the calculator. The statement is true.
    5. Answers will vary.

- Activity 3: The Floor and Ceiling Functions

    1. This is a function, discontinuous at the integers, continuous on $[n, n+1)$ for integers n. This floor function agrees with $y = x$ at the integers but has a flat step between integers rather than a diagonal line.
    2. Function, disc. at integers, cont. on $(n, n+1]$.
    3. Function, disc. at $\pm\sqrt{n}$, cont. on $(-\sqrt{2}, -1], (-1, 1), [1, \sqrt{2}), [\sqrt{2}, \sqrt{3})$, etc.
    4. Function, disc. at $\pm\sqrt{n}$, cont. on $[-\sqrt{2}, -1), [-1, 0), (0, 1], (1, \sqrt{2}], (\sqrt{2}, \sqrt{3}]$, etc.
    5. Function, disc. at integers, cont. on $(n, n+1)$.
    6. Function, disc. at integers, cont. on $(n, n+1)$.
    7. Function, disc. at integers, cont. on $(n, n+1]$.
    8. Function, disc. at integers, cont. on $(n, n+1)$ except $(0, 1)$.
    9. Function, disc. at integers, cont. on $[n, n+1)$ except $[0, 1)$.
    10 Function, disc. at integers, cont. on $(-n, n+1]$ except $(-1, 0]$.

- Activity 4: The Growth of Functions

    1. $x = O(x^2)$.
    2. $f(x) = O(g(x))$ and $g(x) = O(f(x))$; $C = 15$; $g(x)$ is strictly greater than $f(x)$ for $x \geq 1$.
    3. $x / \log(x) = O(x^2)$.
    4. $34x^{100} = O(2^x)$ even though x^{100} is greater than 2^x until x is quite large; $x(\log(\log(x))) = O(x(\log(x)))$.
    5. $k = (10/c)^{1/10}$.

Bulgarian Solitaire

Suzanne Dorée
Augsburg College

Summary

A player begins with coins arranged in piles. At each turn she rearranges the coins according to the following rule: remove the top coin from each pile, possibly eliminating piles, and form a new collected pile of coins. She repeats the process until she revisits a previously encountered arrangement, having reached a terminal cycle. Where are fixed points, if any? Are there any 2-cycles? Which states are cyclic? How can we visualize the process?

This process was dubbed "Bulgarian Solitaire" in a 1983 *Scientific American* column by Martin Gardner [11], though I first saw it in Doug West and John D'Angelo's text *Mathematical Thinking* [22]. In my sophomore level discrete mathematics class I use an activity on Bulgarian Solitaire called *The Coins Go 'Round 'n 'Round* to introduce students to partitions of integers, directed graphs, state graphs, and dynamical systems and to draw connections to the ubiquitous triangular numbers. In class I avoid using the name "Bulgarian Solitaire" since that would make it too easy for students to find answers on the Internet.

The activity takes two twenty-minute time blocks, or it can be done all as one entire class period. Each pair of students needs a dozen coins. It works best when I introduce the process prior to Part I and the graphs prior to Part II. I provide this introductory material below.

I expect students have seen the triangular numbers

$$1, 3, 6, 10, 15, 21, \ldots, t_k = \sum_{i=1}^{k} i = \frac{k(k+1)}{2}$$

earlier in the course, preferably including their representation as triangles of dots. It is helpful, but not necessary, for students to have seen the partitions of an integer. I do not expect that students have seen any graph theory prior to the activity. My students haven't and they are able to pick up on the terminology, notation, and ideas when needed.

My experience is that there are parts of the activity that challenge my strongest students, most of which I save as extra credit, but most parts are easily completed by all of my students. They really get into "playing the game," and are curious about what happens.

Notes for the instructor

I begin the activity by demonstrating the process by moving coins on an overhead projector while on the blackboard I record each state visited and connect it to the subsequent state with an arrow. Rather than piling the coins vertically, which would be difficult to see on the overhead screen, I extend them up the screen as in Figure 1. I literally remove the top coin from each pile and form the collected pile each time. For example, starting with $(5,1,1,1)$ as on the worksheet I show students how to get $(4,4)$, $(3,3,2)$, $(3,2,2,1)$, $(4,2,1,1)$, $(4,3,1)$, and then $(3,3,2)$ again. Along the way I point out how the singleton piles were eliminated and that for convenience we keep the piles in increasing order by size. By the end of this introduction on the blackboard I have drawn the progression

$$(5, 1, 1, 1) \rightarrow (4, 4) \rightarrow (3, 3, 2) \rightarrow (3, 2, 2, 1) \rightarrow (4, 2, 1, 1) \rightarrow (4, 3, 1) \rightarrow \text{back to } (3, 3, 2).$$

I explain that we have repeated a state and so we're going to stop. Without my recording these states on the board, the class probably would not recognize the repeat. Students usually need to stop and think about why the process would just repeat from there and so I have them turn to their neighbor to discuss why. I also ask them to decide if we'll ever return to (5,1,1,1) again, which of course we won't. I casually introduce the terminology "4-cycle" and refer to the repeated elements as "cyclic". These terms are defined later on the worksheet, but I find students get the idea from this one example.

At this point I have students work on Exercises 1-4.

Thus far in the activity students have only drawn the progression from one state. I begin the second part by drawing the first few state graphs on the board, G_4 on four coins and G_5 on five coins as appear in Figure 2. I point out that each vertex has out-degree one since there is a unique state to which it goes, but a vertex can have in-degree more than one (or zero).

If students have not seen the partitions of an integer, I show them how to systematically list the partitions of five to make sure we have all of the vertices of G_5. Before I set them off to work on the remaining exercises, I also list out the partitions of six. A word of warning: once students have the list, some are tempted to draw the graph by first calculating where each partition on the list is sent and then trying to connect the whole kit and caboodle. In addition to taking far more time than necessary, it also results in a tangled picture of the graph. Instead, I encourage them to pick a particular partition, such as $(1, 1, 1, 1, 1, 1)$, draw the progression from there, then pick a missing partition, and draw its progression, etc., each time connecting the new progression to the graph as soon as possible. I encourage them to work on scratch paper and later redraw the graph in an organized fashion without any crossings.

Now I have students work on Exercises 5-8.

Bibliography

[1] Akin, Ethan and Morton Davis. "Bulgarian solitaire," *The American Mathematical Monthly* 92 (1985) 237–250.

[2] Bancroft, Tim. "Bulgarian Exchange: Where Does It End?" Senior Honors Project, Augsburg College, Minneapolis, MN, 2004.

[3] Bending, Thomas. "Bulgarian Solitaire," *Eureka* 50 (1990) 12–19.

[4] Bentz, Hans-J. "Proof of the Bulgarian solitaire conjectures," *Ars Combinatoria* 23 (1987) 151–170.

[5] Brandt, Jørgen. "Cycles of partitions," *Proceedings of the American Mathematical Society* 85 (1982) 483–486.

[6] Broline, Duane M. and Daniel E. Loeb. "The combinatorics of Mancala-type games: Ayo, Tchoukaillon, and 1/pi," *UMAP Journal* 16 (1995) 21–36.

[7] Brualdi, Richard A. *Introductory Combinatorics*, Elsevier North Holland Inc., 1977.

[8] Campbell, Paul J. and Darrah P. Chavey. "Tchuka Runa solitaire," *UMAP Journal* 16 (1995) 343–365.

[9] Cannings, C. and J. Haigh. "Montreal solitaire," *Journal of Combinatorial Theory, Series A* 60 (1992) 50–66.

[10] Etienne, Gwihen. "Tableaux de Young et solitaire bulgare," *Journal of Combinatorial Theory, Series A* 58 (1991) 181–197.

[11] Gardner, Martin. "Mathematical Games (a.k.a. Bulgarian Solitaire and Other Seemingly Endless Tasks)," *Scientific American* (1983) 249.

[12] Griggs, Jerrold R. and Chih-Chang Ho. "The cycling of partitions and compositions under repeated shifts," *Advances in Applied Mathematics* 21 (1998) 205–227.

[13] Hobby, J. D. and D. Knuth. "Problem 1: Bulgarian Solitaire," A Programming and Problem-Solving Seminar, Department of Computer Science, Stanford University, Stanford, December 1983, 6–13.

[14] Hopkins, Brian and Michael A. Jones. "Shift-Induced Dynamical Systems on Partitions and Compositions," *Electronic Journal of Combinatorics* 13 (2006) R80, 19pp.

[15] Igusa, Kiyoshi. "Solution of the Bulgarian solitaire conjecture," *Mathematics Magazine* 58 (1985) 259–271.

[16] Loeb, Daniel E. "Combinatorial properties of Mancala," *Abstract of the American Mathematical Society* 96 (1994) 471.

[17] Nicholson, Al. "Bulgarian Solitaire," *Mathematics Teacher* 86 (1993) 84–86.

[18] Popov, Serguei. "Random Bulgarian Solitaire," *Random Structures and Algorithms* 27 (2005) 310–330.

[19] Rosen, Kenneth H. *Discrete Mathematics and Its Applications*, McGraw-Hill Inc., 2003.

[20] Servedio, Rocca and Yeong Nan Yeh. "A bijective proof on circular compositions," *Bulletin of the Institute of Mathematics, Academia Sinica* 23 (1995) 283–293.

[21] Sieve, Maria. "The Coin Game and Its State Graphs," Senior Honors Project, Augsburg College, Minneapolis, MN, 2002.

[22] West, Douglas B. and John P. D'Angelo. *Mathematical Thinking: Problem-Solving and Proofs*, Prentice-Hall, Inc., 1997.

[23] West, Douglas B. *Introduction to Graph Theory*, Prentice Hall, Inc., 1996.

[24] Yeh, Yeong-Nan. "A remarkable endofunction involving compositions," *Studies in Applied Mathematics* 95 (1995) 419–432.

Worksheet on The Coins Go 'Round 'n 'Round

Part I: A never-ending process

Take some coins and arrange them into piles. You may make any number of piles and have any number of coins in each pile. The piles may be different heights and a pile may have as few as one coin in it. Now, from that starting arrangement pick up the top coin from each pile and make one new pile out of those coins. Notice that any pile that had just a single coin was eliminated. You now have a new arrangement of the coins. Repeat this process to the new arrangement of coins, i.e. pick up the top coin from each new pile and make one new pile out of those collected coins. For convenience we list the piles in increasing order by size. Keep repeating the process. Although this process never ends, we'll call it quits if we get back to a place we've already been.

For example, suppose I take eight coins and place them into one pile of five coins and three piles of one coin each, which we denote $(5,1,1,1)$. At the first step I remove one coin from each pile, eliminating all but the tallest pile. This leaves a pile of four coins and the four collected coins, forming two piles of four coins each that we denote $(4,4)$. At the next step I remove one coin from each pile, leaving three coins in each and two collected coins that form a new pile. The resulting arrangement is $(3,3,2)$. These steps are illustrated in Figure 1.

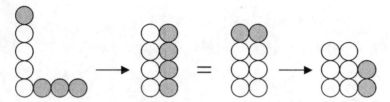

Figure 1. $(5, 1, 1, 1) \rightarrow (4, 4) \rightarrow (3, 3, 2)$.

When I repeat the process I get $(3,2,2,1)$, $(4,2,1,1)$, $(4,3,1)$, and then $(3,3,2)$ again, as you can check. Even though I did not get back to my original arrangement, since I have returned to an arrangement from earlier, I'll call it quits. If I were to continue, it would just repeat through those same four states over and over again. We represent this progression as

$$(5, 1, 1, 1) \rightarrow (4, 4) \rightarrow (3, 3, 2) \rightarrow (3, 2, 2, 1) \rightarrow (4, 2, 1, 1) \rightarrow (4, 3, 1) \rightarrow \text{back to } (3, 3, 2).$$

Exercises

1. Now it's your turn. Perform the process beginning with the given arrangement and record the progression until you repeat an arrangement.

    (a) $(2,1,1,1)$

    (b) (4)

    (c) $(3,3)$

    (d) $(4,2,1)$

    (e) $(2,2,2,2)$

2. Although we do not always return to our original arrangement, we will necessarily return to *some* state we have visited before. Explain why. Also explain why once we return to a previously-visited state the process repeats.

An arrangement of coins that is eventually repeated is called *cyclic* and the number of steps from a cyclic state to its next repeat is called the *length* of the cycle. For example, with eight coins the elements $(3,3,2)$, $(3,2,2,1)$, and $(4,2,1,1)$, and $(4,3,1)$ are cyclic states that form a cycle of length four, called a 4-cycle for short. An arrangement in a 1-cycle is called a *fixed point*, i.e. a fixed point is unchanged by the process.

3. Fixed points

    (a) Give an example of a fixed point from your earlier work. Give another example of a fixed point.

    (b) Make a conjecture about which states are fixed points under this process.

    (c) Make a conjecture about the number of coins in a fixed point state.

    (d) Prove your conjectures. *Hint: Be sure to both show that these states are fixed points and that any fixed point is one of these states.*

4. 2-cycles

    (a) Give an example of a 2-cycle from your earlier work. Give another example of a 2-cycle.

    (b) Make a conjecture about which states are in 2-cycles under this process.

    (c) ⋆ Make a conjecture about the number of coins in a 2-cycle state.

Note: The star (⋆) indicates a more difficult problem.

Part II: The state graphs

Thus far we have drawn the progression from one beginning state. For a fixed number of coins, n, we can visually represent the entire process using a *state graph* denoted G_n. The vertices (dots) are all possible arrangements of the coins, which we continue to list as a tuple ordered by size, regardless of the physical order of the piles. These labels are known as the *partitions* of the integer n. We draw an arrow from each vertex to the vertex we get by performing the process once. There is a unique arrow out of each vertex, but there may be one, several, or no arrows into a vertex. For example, the state graphs on four and five coins are shown in Figure 2.

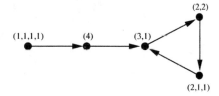

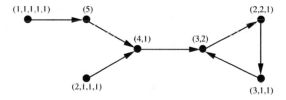

Figure 2. The state graphs G_4 and G_5.

Exercises

5. State Graphs

    (a) Draw G_6, the state graph on six coins. *Hint: Be sure to include all 11 possible arrangement of six coins.*

    (b) Draw G_7, the state graph on seven coins. *Hint: There are 15 vertices.*

6. Garden of Eden States

A state having no predecessors, i.e. no arrows into the vertex, is called a *Garden of Eden* state.

    (a) Show that the state (4,4,3,3) on fourteen coins is *not* Garden of Eden.

    (b) The state (4,3,3,2,1,1) on fourteen coins *is* Garden of Eden. Explain why.

    (c) ⋆ State and prove a conjecture about which states on n coins are Garden of Eden.

7. Ten Coins

   (a) Start with an arrangement of ten coins, perform the process, and draw the progression.

   (b) Start with an arrangement of ten coins that you haven't encountered yet, perform the process, and connect this progression to your graph.

   (c) Start with another arrangement of ten coins that you haven't encountered yet, perform the process, and connect this progression to your graph.

   (d) Make a conjecture about how the process with ten coins will always end.

   (e) Generalize your conjecture.

8. Eleven Coins

   (a) List the cyclic states on eleven coins. *Hint: There is a unique end cycle.*

   (b) ⋆ Suppose that n is an integer such that there is a fixed point with n coins. Consider the process on $n + 1$ coins. Make a conjecture about which states are cyclic.

Solutions and Notes

1. The Process

    (a) $(2, 1, 1, 1) \to (4, 1) \to (3, 2) \to (2, 2, 1) \to (3, 1, 1) \to$ back to $(3, 2)$

    (b) $(4) \to (3, 1) \to (2, 2) \to (2, 1, 1) \to$ back to $(3, 1)$

    (c) $(3, 3) \to (2, 2, 2) \to (3, 1, 1, 1) \to (4, 2) \to (3, 2, 1) \to$ back to itself

    (d) $(4, 2, 1) \to (3, 3, 1) \to (3, 2, 2) \to (3, 2, 1, 1) \to$ back to $(4, 2, 1)$

    (e) $(2, 2, 2, 2) \to (4, 1, 1, 1, 1) \to (5, 3) \to (4, 2, 2) \to (3, 3, 1, 1) \to$ back to $(4, 2, 2)$

2. We will necessarily return to some state we have visited before because there are only finitely many states. The process will repeat from there because the process and state are the same as when we were there before.

3. Fixed Points

    (a) A fixed point from earlier work is (3,2,1). Other examples include (1) on one coin, (2,1) on three coins, (4,3,2,1) on ten coins, etc.

    (b) Conjecture: A state is fixed if and only if it is of the form $(k, k-1, \ldots, 3, 2, 1)$. *I find it necessary to remind students that a conjecture should be a complete sentence.*

    (c) Conjecture: There is a fixed point state for n coins if and only if n is a triangular number.

    (d) Proof: First we check that these states are fixed by performing the process and observing that each pile is reduced by one coin and the smallest pile is eliminated, but there are k collected coins, which reinstates the largest pile. Conversely, suppose we had a fixed state with largest pile k. Since the process reduces the size of each pile, the only way we can get a pile of k coins is as the collected pile and so there must be exactly k piles. The pile of k coins is reduced to $k-1$ coins by the process and so there must have been a pile of $k-1$ coins. Similarly that pile of $k-1$ coins is reduced to $k-2$ coins by the process and so there must have been a pile of $k-2$ coins. The argument continues until we note that the pile of two coins is reduced to one coin by the process and so there must have been a pile of one coin. This gives us k piles in total, and so there are no additional piles. We therefore have the triangle state $(k, k-1, k-2, \ldots, 3, 2, 1)$, with a triangular number of coins. *For formal proof see [22].*

4. 2-Cycles

    This exercise can easily be left as homework or omitted altogether if short on time.

    (a) A 2-cycle from earlier work is $(4,2,2) \leftrightarrow (3,3,1,1)$ on eight coins. Other examples include $(2) \leftrightarrow (1,1)$ on two coins and $(6,4,4,2,2) \leftrightarrow (5,5,3,3,1,1)$ on eighteen coins.

    (b) Conjecture: A state is in a 2-cycle if and only if it is of the form

    $$(2t, 2t - 2, 2t - 2, \ldots, 4, 4, 2, 2) \text{ or } (2t - 1, 2t - 1, 2t - 3, 2t - 3, \ldots, 3, 3, 1, 1)$$

    for some integer t. *My students usually guess the pattern but have difficulty stating it. This result follows from the characterization of cycles in the work of Akin and Davis [1]. While it is easy to check that states of this form are in a 2-cycle, it takes a bit more work to prove that these are the only 2-cycles and so I do not ask my students for proof.*

    (c) ⋆ *I make this part **extra credit** since it relies on a clearly-stated conjecture in the previous part.* Adding the number of coins from the previous conjecture in the odd case we get

    $$\sum_{i=1}^{t} 2(2t - 1) = 2\,[t(t + 1) - t] = 2t^2.$$

    Conjecture: A two-cycle occurs exactly when the number of coins is twice a square. *As far as I can determine the first direct proof of this fact was given by one of my students, Augsburg College undergraduate Maria Sieve, [21].*

5. State Graphs

   (a) The state graph G_6 is shown in Figure 3.

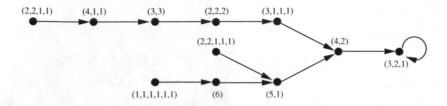

Figure 3. The state graph G_6.

   (b) The state graph G_7 is shown in Figure 4.

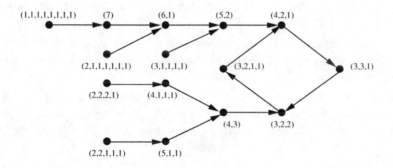

Figure 4. The state graph G_7.

The first seven state graphs are connected, but G_8 has two connected components. There are graphs with even more components [1].

6. Garden of Eden States

   (a) $(4,4,3,3)$ is *not* a Garden of Eden state because $(5, 4, 4, 1) \rightarrow (4, 4, 3, 3)$.

   (b) Proof: To see why $(4,3,3,2,1,1)$ *is* a Garden of Eden state, suppose this state had a predecessor. There are six piles of coins in $(4,3,3,2,1,1)$, only one of which could be the collected coins from a predecessor. Therefore a predecessor would have to include at least five piles of coins. (It might have additional singleton piles.) But then there would be at least five collected coins and so our state would include a pile with at least five coins in it. Since $(4,3,3,2,1,1)$ does not have any pile with more than four coins, the existence of a predecessor is impossible. Thus $(4,3,3,2,1,1)$ is Garden of Eden, as claimed.

   (c) ⋆ *I make this part **extra credit** since it requires a clear understanding of the previous part.* Conjecture: A state is Garden of Eden if and only if each pile is at least two coins smaller than the number of piles. Proof: That these states are Garden of Eden follows as in the proof for $(4,3,3,2,1,1)$. For the converse, given a state that does *not* meet the criteria, the largest pile must be no smaller than one less than the number of piles. So, we can construct a predecessor by picking up that largest pile, distributing one coin to each of the remaining piles (which we definitely have enough coins to do), and placing any remaining coins from the largest pile as new singleton piles.

7. Ten Coins

   (a-c) Any progression on ten coins leads to the fixed point (4,3,2,1). *I have sometimes done this part of the activity as an entire class, where each pair of students tries an arrangement of ten coins at once and then all report what happens.*

   (d) Conjecture: Any arrangement of ten coins progresses to the fixed point (4,3,2,1).

   (e) Conjecture: For a triangular number of coins, any arrangement progresses to the fixed point. *This result is proved in [5] and [1]. Students may try to generalize further but they should be reminded of the example of eight coins that has both a terminal 2-cycle and 4-cycle, as evidence that the outcome might depend on the initial state.*

8. Eleven coins

   (a) The cyclic states on eleven coins are (5,3,2,1), (4,4,2,1), (4,3,3,1), (4,3,2,2), (4,3,2,1,1). The easiest way to find these states is to begin with *any* partition of eleven and apply the process until you reach the 5-cycle. *Without the hint students would not know that these are the* only *cyclic elements. The pattern is that any cyclic state on eleven coins is formed from (4,3,2,1) by adding a coin to some pile or to the right of the last pile [1].*

   (b) ⋆ *I make this part* **extra credit** *since it is so difficult that few of my students get it. The question is stated a bit awkwardly to avoid revealing the answer to Exercise 3. You may want to restate it as "n + 1 coins when n is triangular".* Conjecture: For a triangular number of coins, n, a state with $n + 1$ coins is cyclic if and only if it is formed from the triangular base of n coins with the addition of a single coin. This result is proved in [1].

   For *any* number of coins, the cyclic states are formed by a triangular base with at most one coin atop *each* pile and possibly one coin to the right of the last pile. If the base triangle has k piles, and thus $t_k = \frac{k(k+1)}{2}$ coins, then there are $k + 1$ slots for possible additional coins. The cycle length will be a divisor of $k + 1$ and there are cycles of each divisor's length [1].

   A somewhat different question that intrigued Gardner was the maximal number of steps needed to get from a triangular number of coins to the fixed point [11]. The answer he posed became known as the "Bulgarian Solitaire Conjecture" though it has now been settled, for example in [10].

Can you make the geodesic dome?

Andrew Felt and Linda Lesniak

University of Wisconsin - Steven's Point *Drew University*

Summary

This project uses the geodesic dome made famous by Buckminster Fuller to extend students' understanding of Eulerian graphs. Students are directed to build a dome and, after determining that the graphical representation of the dome contains neither an Euler cycle nor trail, to find the minimal number of repeated edges that are necessary to visit each edge.

Notes for the instructor

I have done this construction six times with classes of up to 30 students, and each time the entire activity took less than two 50-minute class periods. There are many jobs to be done: corners to be held, rope to be strung through pipe, pipes to be labeled, desks to be moved, and supervisory work. All of these jobs can be assigned to students.

The 0.884 ratio cited for edge lengths comes from [1] which cites spherical trigonometry for the derivation. The decimal approximation is precise enough for this construction.

Use rope that is at least as strong as clothesline, and pull it taut as you proceed. Start from the bottom (the outer ring) and work up; it also helps to duct tape the bottom to the floor. The construction will be flimsy until the final pipe is tied into place. Preparing the materials will take some effort, but students are very engaged by the activity.

Bibliography

[1] Lloyd Kahn, *Domebook One*, Pacific Domes, 1970.

[2] Hugh Kenner, *Geodesic Math and How to Use It*, University of California Press, 2003.

[3] Thomas Zung (editor), *Buckminster Fuller: Anthology for the New Millennium*, St. Martin's Press, 2001.

Worksheet for Can you make the geodesic dome?

A geodesic dome is a three dimensional structure that looks roughly spherical, but is made of smaller triangles.

Geodesic domes are very strong and light. There are many different designs. Suppose you want to use plastic pipe to make a geodesic dome like the one in Figure 1. There are two different edge lengths. The ratio of the lengths of the dotted edges to solid edges is about 0.884. The drawing is not to scale.

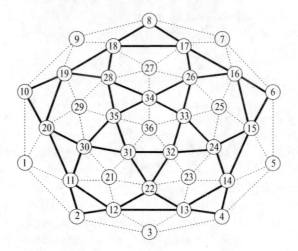

Figure 1. The design we will work with.

The circles in Figure 1 represent the joints at which the pipes must be fastened together. To join the pipes, you want to string a rope through each pipe, as shown in Figure 2. Note that the second and third times through the joint, the rope is looped around the pre-existing rope.

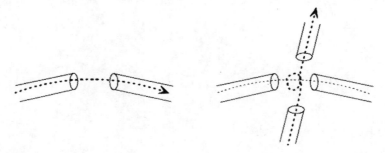

Figure 2. How the rope forms the joints: The second time through a joint the new rope loops around the existing rope.

The rope does not necessarily need to travel directly across a joint. It can take any other pipe out of the joint. See Figure 3.

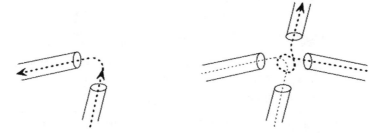

Figure 3. How the rope forms the joints: Loop for the second time through a joint when rope does not go across.

So all you need to do is to string the rope at least once through each pipe. You'd like to use the shortest rope possible. Your job will be to find out how much rope you need and what route the rope should take through the pipes.

Below is a small example of a different, two dimensional shape.

We could use the following route to string the rope exactly once through each pipe, starting at node b:

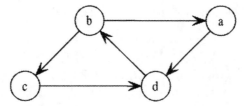

If each of the pipes above is 2 feet long, then we would need a rope about 10 feet long. (We ignore the length required to do the looping at the joints.)

The following questions relate to the design in Figure 1.

1. Find the nodes of odd degree and list them.

2. Is there a way to string a single rope such that the rope goes through each pipe exactly once, with the ends of the rope at the *same* node? Explain your answer.

3. Is there a way to string a single rope such that the rope goes through each pipe exactly once, with the ends of the rope at *different* nodes? Explain your answer.

4. If the ends of the rope are at different nodes, what is the fewest number of pipes through which the rope must travel twice? Which nodes should have the rope ends?

5. Find a route through the pipes which minimizes the amount of rope needed. Draw it on Figure 1. (Bonus: There are many correct answers. Try to make yours easy to construct. Is it best to wander around or to focus on a single area first?)

Here is a picture of a geodesic dome made by a small class following these instructions.

Solutions

1. There are six nodes of odd degree: 21, 23, 25, 27, 29, and 36.

2. It is not possible to have the rope pass through each pipe exactly once with ends of the rope at the same node. This question is equivalent to asking if an Eulerian cycle (or circuit) exists on the graph in Figure 1. Since there are nodes of odd degree, Euler's theorem tells us that no Eulerian cycle exists.

3. It is not possible to have the rope pass through each pipe exactly once with ends of the rope at different nodes. This question is equivalent to asking if an Eulerian trail (tour, path) exists on the graph in Figure 1. Since there are not exactly two nodes of odd degree (but rather six), Euler's theorem tells us that no such trail exists.

4. If we use two of the odd degree nodes for the rope ends (that is, the start and end of the trail), then the other four odd degree nodes must be made even degree by having the rope go through some pipes twice. Suppose we use nodes 21 and 36 for the rope ends. Then we have rope go through pipes 29-28, 28-27, 25-24 and 24-23 twice each, so we must travel through four of the dotted (shorter) pipes twice each. There are many other ways to do this, but four repeated edges is the minimal number possible.

5. One optimal path is shown below. Here is the order in which the rope visits the nodes: 36, 32, 31, 36, 35, 34, 36, 33, 34, 28, 35, 31, 22, 32, 33, 26, 34, 27, 28, 29, 35, 30, 31, 21, 22, 23, 32, 24, 33, 25, 24, 23, 24, 25, 26, 27, 28, 29, 30, 21, 12, 22, 13, 23, 14, 24, 15, 25, 16, 26, 17, 27, 18, 28, 19, 29, 20, 30, 11, 12, 13, 14, 15, 16, 17, 18, 19, 20, 11, 2, 12, 3, 13, 4, 14, 5, 15, 6, 16, 7, 17, 8, 18, 9, 19, 10, 20, 1, 2, 3, 4, 5, 6, 7, 8, 9, 10, 1, 11, 21. Note that the upper part of the geodesic dome is constructed first in this answer, with each successively lower layer added in order.

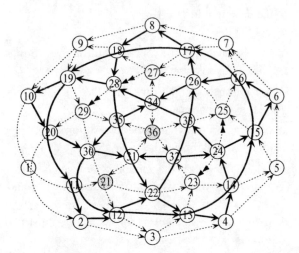

Figure 4. The solution detailed above. End vertices are gray. When the path does not use an adjacent edge, as in Figure 2, often the path is drawn over a vertex. Instances of rope passing twice through a pipe are shown with double headed arrows.

Exploring Polyhedra and Discovering Euler's Formula[1]

Leah Wrenn Berman and Gordon Williams
Ursinus College

Summary

The activities and exercises collected here provide an introduction to Euler's formula, an introduction to interesting related topics, and sources for further exploration. While Euler's formula applies to any planar graph, a natural and accessible context for the study of Euler's formula is the study of polyhedra.

The article includes an introduction to Euler's formula, four student activities, and two appendices containing useful information for the instructor, such as an inductive proof of Euler's theorem and several other interesting results that may be proved using Euler's theorem. Each activity includes a discussion of connections to discrete mathematics and notes to the instructor. Three of the activities have worksheets at the end of this article, followed by solutions and a template for a toroidal polyhedron.

A brief introduction to Euler's formula

Theorem (Euler's formula, polyhedral version). *Let P be any polyhedron topologically equivalent to a sphere. Let V be the number of vertices, E the number of edges and F the number of faces of P. Then $V - E + F = 2$.*

A graph is said to be a *simple graph* if it is an undirected graph containing neither loops nor multiple edges. A graph is a *plane graph* if it is embedded in the plane without crossing edges. A graph is said to be *planar* if it admits such an embedding.

Theorem (Euler's formula, graph version). *Let G be any simple plane graph. Let V be the number of vertices, E the number of edges and F the number of faces (alternately, the number of connected sets in the complement) of G. Then $V - E + F = 2$.*

Note that the number of faces F includes what is sometimes called the outside face, the connected component containing the unbounded region in the complement of the graph. The supposition that G is a simple graph is not necessary for the proof (the result holds for graphs with loops and multiple edges), but only simple graphs arise in the polyhedral context, and adding this assumption simplifies the inductive proof. Note that the adjective "simple" means different things for graphs and polyhedra; simple polyhedra will be defined later.

Throughout this treatment we are defining polyhedra as embedded compact two-dimensional polygonal surfaces with planar, nonself-intersecting faces. While some of these criteria are dropped in other contexts (the Kepler-Poinsot polyhedra are either immersed surfaces or have self-intersecting faces), omitting them introduces complexities we do not feel are appropriate to the activities collected here.

---

[1] Thanks to Doris Schattschneider and Gary Gordon who provided ideas and templates for early versions of some of these activities.

1   Activity: Polyhedra Exploration

The *Polyhedra Exploration* worksheet is the centerpiece of an activity designed to lead to the discovery of, and build confidence in, Euler's formula. This activity is a discovery-based activity in which students are asked to build models of various polyhedra, given a description of some feature of the polyhedron, and then asked to collect data on the number of vertices, edges, and faces of their polyhedron. After all students have contributed data, the class is asked to make conjectures about relationships between the numbers of vertices, edges, and faces. The obvious conjecture is Euler's formula. The worksheet is included at the end of this article.

Connection to discrete mathematics

One of the most important aspects of writing a proof is determining the appropriate proposition to prove. Thus, practice making conjectures is a useful part of a discrete mathematics course. The activity also leads into the presentation of a proof of Euler's formula; one of our favorite proofs of this formula is by induction on the number of edges in a graph. This is an especially nice proof to use in a discrete mathematics course, because it is an example of a nontrivial proof using induction in which induction is done on something other than an integer.

Notes for the instructor

We have had the most success with this activity, and it takes the least amount of time, if students have access to large numbers of *Polydrons* or *GeoFix* shapes. When they are available, this activity can take as little as 20 minutes. Alternatively, tape and polygons made from card stock may be used, though assembly in this case will usually take more time. If you are using homemade polygons it is important that the length of every side is the same for all polygons. At a minimum you will need 76 triangles, 29 squares, 14 pentagons, 20 hexagons and 6 octagons. We build a tetrahedron (polyhedron 0 in the worksheet list) as a demonstration.

At the beginning of the activity students are assigned (either individually or in groups) one of the items on the list to investigate. Students who finish early may either be assigned a second item from the list or additional assembly problems, given below. We usually handle problem assignments in both cases by printing polyhedron descriptions on half sheets of paper or index cards passed out randomly or by randomly assigning numbers to groups. The completed table is provided in the Solutions section, including the common names for the various polyhedra.

The additional assembly problems below are particularly important for the activities described in the next two Activities. If you plan on using the additional assembly problems, you will need a large selection of the above shapes, including at least 30 more triangles, 18 more squares and 34 more pentagons.

Additional assembly problems

The reader may be surprised at how many different examples students will generate if they are challenged to come up with unique examples subject to minimal constraints. Here are some supplementary exercises for the *Polyhedra Exploration* worksheet.

- Using only triangles, make two distinct polyhedra.
- Using only squares, make two distinct polyhedra.
- Using only pentagons, make two distinct polyhedra.
- Using triangles and squares, make a polyhedron where four triangles and a square come together at each corner. Can you make two different polyhedra?[2]
- Pick any two polygon shapes and make a polyhedron.
- Make a polyhedron using as many different shapes as you can.

Note that these additional assembly problems are, with the exception of the snub cube, open-ended. If students make spherical polyhedra, then as with the other worksheet exercises, they will find that $V + F$ is two more than E. However, be on the lookout for the clever students who decide to make toroidal polyhedra!

---

[2] The snub cube, which comes in two mirror image forms.

2 Activity: "Some of these things are not like the others"

This activity asks students to determine features that various polyhedra have in common, and it encourages them to carefully articulate those features.

Depending on the number of terms selected and the temperament of the students, this activity can take as little as 20 minutes, or as long as a full class. Depending on time availability, it may either be used directly as a follow on to the *Polyhedra Exploration*, or the models generated during the *Polyhedra Exploration* may be saved for a later class period for use in this activity.

Connection to discrete mathematics

Precision in vocabulary is an important feature in learning how to communicate mathematics effectively; not only are students exposed to vocabulary about polyhedra, but they should begin to recognize why having the vocabulary is useful. Students also make and test conjectures in the process of articulating features that various groups of polyhedra have.

Notes for the instructor

Once a large collection of polyhedral models has been assembled, it is possible to explore a number of questions about geometric and combinatorial properties of polyhedra. Our preference is to do this before students have been exposed to very much vocabulary about polyhedra.

We gather the students around a table where all of the interesting looking distinct models have been collected. Then we sort the models into two different piles. All of the polyhedra in one pile will have a chosen property, and all of the polyhedra in the remaining pile will not have this property. Part of the goal here is to provide students a context in which they can be challenged with several important mathematical problems: pattern recognition, geometric reasoning, hypothesis testing and development of appropriate nomenclature. Students often struggle with each of these, and only after students are able to describe in words correctly what distinguishes the chosen collection of objects are they told the formal term they have just defined. A selection of such terms and related definitions is included below.

It is important when doing this activity to wait a long time for students to make suggestions, to steer as little as possible, and to avoid giving any hints, although you should feel free to make more precise definitions that students have created. The level of depth and precision for these definitions should be fitted to the other instructional needs of the course but may also provide nice opportunities to emphasize the importance of making sure that a definition says what its author thinks it does (a common problem in geometry). Be careful not to reject examples of polyhedra or define them in ways that would artificially limit your ability later to talk about any of the definitions given. The goal is to have the students thinking as creatively and broadly as possible.

Regular Polyhedron/Platonic Solid This is a term that can be defined several ways. For convex polyhedra, it suffices to define a regular polyhedron as one in which all of the faces are regular polygons of exactly one type and each vertex has the same number of polygons around it. (It would be more precise to say that they are the polyhedra that are isogonal, isotoxal and isohedral; however there exists a combinatorial (nonconvex) polyhedron that satisfies all three of these conditions without meeting the modern definition of regularity.[3]) In most modern contexts, to be regular means that the polyhedron is "flag transitive." A flag is a vertex, edge and face of the polyhedron, all mutually incident, and to be flag transitive is to require that any two such flags may be mapped to each other by a symmetry of the polyhedron.

Convex Polyhedron Loosely, this means that the polyhedron lacks dents. Precisely, a polyhedron is convex if every pair of points on or in the polyhedron determines a segment contained entirely on or inside the polyhedron.

Symmetric Polyhedron Any polyhedron with a nontrivial symmetry mapping the polyhedron to itself.

---

[3]This is the Petrie dual (or Petrial, or Petrie polyhedron) of the cuboctahedron. For more information about Petrie duals we recommend Cromwell's discussion of Petrie polyhedra in [3]. More technical treatments are available in Coxeter's [2], Grünbaum's [4], and McMullen and Schulte's [6].

Isogonal Polyhedron Any vertex of the polyhedron may be mapped to any other vertex under a symmetry of the polyhedron. The triangular prism is isogonal, while the square pyramid is not isogonal, since the vertex at the apex of the pyramid can't be mapped to any vertex at the base.

Isohedral Polyhedron Any face of the polyhedron may be mapped to any other face under a symmetry of the polyhedron. Nonregular examples include the triangular bipyramid (or any of the other Catalan solids); Archimedean solids, such as the cuboctahedron, are not isohedral.

Isotoxal Polyhedron Any edge of the polyhedron may be mapped to any other edge under a symmetry of the polyhedron. The cuboctahedron is a good example of this phenomenon, while the triangular bipyramid does not have this property.

Uniform or Semiregular Polyhedron An isogonal polyhedron in which all of the faces are regular polygons, not all congruent.

Vertex Star The collection of polygons incident to a vertex. The combinatorial type of a vertex star is a listing between parentheses, separated by periods, in cyclic order, of the number of sides of the polygons incident to the vertex. A single such sequence is sufficient to characterize a uniform polyhedron. For example, a uniform polyhedron with vertex stars of type (5.6.6) must necessarily be the truncated icosahedron (soccer ball) in which every vertex is surrounded by a pentagon and two hexagons.

Archimedean Polyhedron A polyhedron in which all of the vertex stars are congruent, and the faces are all regular polygons.[4]

Prism A polyhedron with two parallel congruent faces connected by parallelograms. If the congruent faces may be obtained one from the other by translating in a direction perpendicular to the plane in which each lies, and so the remaining faces are rectangles, it is a *right* prism. If the rectangles are in fact squares, and the parallel polygons are regular, then the prism is also uniform.

Antiprism A convex uniform polyhedron with two parallel congruent regular n-gons and $2n$ equilateral triangles. The $2n$ triangles form a strip connecting the two parallel congruent n-gons. The vertex stars are all of combinatorial type $(3.3.3.n)$.

Spherical polyhedron A polyhedron that is topologically equivalent to a sphere.

Toroidal polyhedron A polyhedron that is topologically equivalent to a torus.

Simple polyhedron A polyhedron in which the number of edges (*degree*) at each vertex is three. Like other occurrences of the word "simple" in mathematics, some authors may mean something different when they use this term to describe a polyhedron, such as what we refer to as spherical.

Simplicial polyhedron A polyhedron made up entirely of triangles. Note that the dual of a simplicial polyhedron is simple and vice versa.

3   Activity: Symmetric Polyhedra and Angle Deficiency

A central theme in discrete mathematics is counting things, especially when counting things known to be equal by two different methods yields a useful equality. The activity on *Symmetric Polyhedra and Angle Deficiency* has two parts. The first part, which is discovery-based, asks students to make a conjecture about the total angle deficiency of a convex polyhedron, and the second part asks them to prove their conjecture using Euler's formula. The worksheet is included at the end of this article.

---

[4] Many sources use Archimedean and uniform as synonyms. However, there exists an example of an Archimedean polyhedron that is *not* uniform, the pseudorhomicuboctahedron (Johnson solid J_{37}).

Connection to discrete mathematics

An important class of proofs is "proofs by counting" — i.e., combinatorial proofs. This activity leads students through a nontrivial proof that proceeds by counting things two ways, as well as providing an application of Euler's formula.

Notes for the instructor

Exercise 5 on the *Symmetric Polyhedra and Angle Deficiency* worksheet (included at the end of the article) provides a good example of proving results by counting things two ways.

Here are two additional exercises you may want to use with this activity. Exercise 0 is a natural question to ask as a precursor for the worksheet, and Exercise 6 is a natural extension of the questions presented in the worksheet. Note that Exercise 6 may be answered positively, with prisms and antiprisms providing the remaining possible examples.

Exercise 0. Prove, using induction, that the sum of the angles in a convex n-gon is $180(n-2)$. Using this fact, determine the measure of a single angle in a regular n-gon.

Exercise 6. Using your answer from Exercise 0, determine whether there are any polyhedra other than those listed in the table of Exercise 3 of the *Symmetric Polyhedra and Angle Deficiency* worksheet that: (i) have faces that are regular polygons where every corner "looks" the same and (ii) has at least two kinds of faces.

Solutions for these exercises and the questions on the worksheet are provided in the solutions section that follows the worksheets.

4 Activity: Poincare's formula and higher genus polyhedra

Euler's formula only applies to polyhedra topologically equivalent to a sphere. This activity encourages students to construct nonspherical polyhedra, notice that Euler's formula does not hold, and make conjectures about what an appropriate generalization might be. It then asks students to think about a complicated polyhedron and, after making a guess, determine the genus using Poincaré's generalization of Euler's formula.

Connection to discrete mathematics

Paying attention to hypotheses is important; Euler's formula does not hold for nonspherical polyhedra! What is an appropriate generalization?

Notes for the instructor

Instruct students to construct "donuts" using the template provided on the last page of this article. If they then count up the numbers of vertices, edges and faces, they will find that the results do not obey the basic formulation of Euler's formula. Instead, the formula needs to take into account the genus of the object. Define $\chi(g) = 2 - 2g$ where g is the genus of the surface in question; Euler's formula generalizes to Poincaré's formula $V - E + F = \chi(g)$. The worksheet for this activity consists of two exercises that formalize this process, and solutions are provided in the solutions section.

Appendix A: Inductive proof of Euler's theorem using graphs

Every spherical polyhedron corresponds to a *simple* (no loops or multiple edges) *plane* (no edges cross) *connected* (you can get to any vertex from any vertex) graph. One way to see this is to imagine that the polyhedron is drawn on the surface of an extremely flexible balloon, with the opening of the balloon (where you blow) in the interior of a face. Pull the opening of the balloon wide and flatten the balloon to a disk in the plane; then the vertices, edges and faces of the polyhedron will form a graph in the plane. As long as we regard the outside of the graph as a face, there is a one-to-one correspondence between the vertices, edges and faces of the original polyhedron (on the blown-up balloon) and the vertices, edges, and faces (or regions) of the flattened-out balloon. Alternately, we may view the polyhedron as a transparent model, position an eye close enough to a face and trace the edges and vertices we see in that face (forming a *Schlegel diagram*). We can then transcribe the resulting figure into the plane and now the outside face corresponds to the face we traced the other faces into during our tracing process.

Therefore, if we can prove the stronger result that Euler's formula holds for *any* connected simple plane graph, we will have shown that Euler's formula must hold for the subclass of simple plane graphs which correspond to polyhedra.

Note that for students who have had exposure to proofs by induction, the following method of proving Euler's theorem, if approached casually, is a nice discovery-based learning topic, which can be used to introduce graph theory and also serves as an interesting, non-trivial example of a proof by induction, especially in classes where most of the proofs by induction deal only with properties of sequences of integers.

Theorem 1. *Suppose G is a connected simple plane graph with V vertices, E edges, and F faces (or regions). Then*

$$V - E + F = 2.$$

Proof. The proof is by (strong) induction on the number of edges, E. We assume that the "outside" of the graph is a face.[5] Let G be any connected simple plane graph with V vertices, E edges, and F faces (or regions).

Base Case $E = 0$. If $E = 0$ and G is a graph, then G is the unique connected graph with a single vertex and a single face, so $V = 1$, $E = 0$, $F = 1$ and $V - E + F = 2$.

Induction Hypothesis Suppose $E \geq 0$, and suppose for any k with $0 \leq k \leq E$ that if Q is any connected graph with V_Q vertices, $E_Q = k$ edges and F_Q faces, then $V_Q - E_Q + F_Q = 2$.

Let G be an arbitrary connected simple plane graph with $n + 1$ edges, and let e be an arbitrary edge of G. Removing e yields a graph H—not necessarily connected—with n edges. There are two cases to consider (see Figure 1).

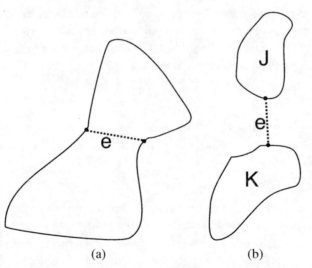

(a)   (b)

Figure 1. A graph G with a chosen edge e indicated as a dashed line segment; removal of e forms a graph H. (a) Case 1: H is connected; (b) Case 2: H consists of two connected components.

Case 1: H is connected, so e connects two vertices in H (Figure 1(a)) Note that since H is connected and has n edges, by the induction hypothesis, $V_H - E_H + F_H = 2$. In this case, because G can't have edges that cross, e must connect two vertices that share a face in H. When e is added back in to H to form G, the face in H becomes two faces in G. Therefore, $V = V_H$, $E = E_H + 1$, and $F = F_H + 1$, so

$$V - E + F = V_H - (E_H + 1) + (F_H + 1)$$
$$= (V_H - E_H + F_H) - 1 + 1$$
$$= 2.$$

---

[5]The Jordan Curve Theorem allows us to use the graph embedding to divide the plane into an "inside" and an "outside."

Case 2: H consists of two connected components (Figure 1(b)) Suppose that removing e from G disconnects the graph, and H consists of two connected components, J and K, with E_J and E_K edges respectively. (Note that one of the connected components could consist of a single vertex, so that one of E_J or E_K could equal zero!) Since $E_J + E_K = n$, it follows that $0 \le E_J \le n$ and $0 \le E_K \le n$, so we can apply the induction hypothesis to each component individually to conclude that $V_J - E_J + F_J = 2$ and $V_K - E_K + F_K = 2$. Note that $V = V_J + V_K$ and $E = E_J + E_K + 1$. However, F_J counts the outside face, and F_K also counts the outside face, so $F = F_J + F_K - 1$. Therefore,

$$
\begin{aligned}
V - E + F &= (V_J + V_K) - (E_J + E_K + 1) + (F_J + F_K - 1) \\
&= (V_J - E_J + F_J) + (V_K - E_K + F_K) - 2 \\
&= 2 + 2 - 2 \\
&= 2.
\end{aligned}
$$

Therefore, by induction on the number of edges, every connected simple plane graph with V vertices, E edges and F faces satisfies $V - E + F = 2$. □

Appendix B: Other interesting applications of Euler's formula

Euler's formula provides a number of nice opportunities to discuss a variety of methods suitable to a discrete mathematics course. Its proof provides a wonderful nontrivial example of the use of induction as a proof method. In application, Euler's formula is a rich source of examples of the classic combinatorial argument involving counting things two different ways. We have collected here some of our favorite examples; they make excellent demonstrations or homework exercises depending on the level and preparation of the students. Throughout this section we assume all polyhedra are spherical (so that Euler's theorem applies).

Our first example is a classical result known since antiquity. Note that every convex polyhedron is spherical, so that Euler's theorem applies to proofs involving convex polyhedra.

Theorem 2. *There are no more than five regular convex polyhedra.*

Proof. We start by observing that there must be at least three polygons at a vertex and that each polygon must have at least three vertices and three edges. Let P be a regular convex polyhedron, let d be the number of edges emanating from each vertex (the degree) and let k be the number of sides to a face. (These must all be the same since P is regular.) Let F be the number of faces, let E be the number of edges, and let V be the number of vertices. If we count all the edges around each face, we will have counted each edge twice (once for each edge it appears on) so $Fk = 2E$ or $F = 2E/k$. Likewise, if we count the number of edges coming out of each vertex, we will have again counted each edge twice (once for each vertex at its ends) so $Vd = 2E$ or $V = 2E/d$. Since P is a polyhedron, Euler's formula applies, so $V - E + F = 2$. Substituting for V and F in Euler's formula we see that $\frac{2E}{d} - E + \frac{2E}{k} = 2$, so if we divide both sides of the equation by $2E$ we obtain $\frac{1}{d} - \frac{1}{2} + \frac{1}{k} = \frac{1}{E}$. Note that $\frac{1}{E} > 0$ since $E > 1$, so $\frac{1}{d} + \frac{1}{k} > \frac{1}{2}$. We therefore need only check to see for which values of d and k the inequality holds, as in Table 1, using the fact that both numbers have to be at least three. Observe that only the first 5 rows are admissible cases (each corresponding to a known regular polyhedron), while the last three exhaust all other possibilities, completing the proof. □

It helps to have models of these polyhedra around when discussing this argument, particularly to demonstrate that five Platonic solids do exist. The tetrahedron, cube (hexahedron), octahedron, dodecahedron and icosahedron form the complete set of regular convex polyhedra.

Euler's theorem also has applications to the study of polyhedra that lack the high degree of symmetry seen in the Platonic solids. One interesting class consists of polyhedra whose faces all have five or more sides. Surprisingly, Euler's theorem constrains the number of pentagons in such polyhedra.

Theorem 3. *Any polyhedron made up entirely of polygons with five or more sides must have at least 12 pentagons.*

d	k	$\frac{1}{d} + \frac{1}{k}$	Polyhedron
3	3	2/3	Tetrahedron
3	4	7/12	Cube
4	3	7/12	Octahedron
3	5	8/15	Dodecahedron
5	3	8/15	Icosahedron
3	6	1/2	Hexagonal tessellation
6	2	1/2	Tessellation by equilateral triangles
4	4	1/2	Square tessellation
$d > 4$	$k > 4$	$< 1/2$	

Table 1. Table of admissible and inadmissible values for d and k.

Proof. Let f_n be the number of n-gons in our polyhedron P. Note that f_3 and f_4 are both zero by assumption. The number of faces F in the polyhedron is then $F = \sum_{n=5}^{\infty} f_n$. If we count all the edges around each face and add them up we will have counted each edge twice, so $\sum_{n=5}^{\infty} n\, f_n = 2E$, where E is the number of edges in P. The degree at each vertex of the polyhedron must be greater than or equal to three (as in Theorem 2), so if we count up the number of edges around each vertex, again we would have counted each edge twice, so $3V \leq 2E$ or $V \leq 2E/3$. The numbers of vertices, edges and faces of P must obey Euler's formula, so $V - E + F = 2$. Substituting for the number of vertices and faces from our calculations above we see that

$$2 \leq \frac{2}{3}E - E + F$$

$$2 \leq -\frac{1}{3}E + F$$

$$2 \leq -\frac{1}{3} \cdot \left(\frac{1}{2} \sum_{n=5}^{\infty} n\, f_n \right) + \sum_{n=5}^{\infty} f_n$$

$$12 \leq -\sum_{n=5}^{\infty} n\, f_n + \sum_{n=5}^{\infty} 6 f_n$$

$$12 \leq -5f_5 - 6f_6 - \sum_{n=7}^{\infty} n\, f_n + 6f_5 + 6f_6 + \sum_{n=7}^{\infty} 6 f_n$$

$$12 \leq f_5 - \sum_{n=7}^{\infty} (n - 6) f_n.$$

Since $(n - 6)f_n \geq 0$ for all $n \geq 7$, the only way the inequality can hold is if $f_5 \geq 12$. $\square$

This problem can be made simpler for students by limiting the types of polygons to pentagons and hexagons.

Theorem 4. *Any polyhedron made up entirely of regular pentagons and hexagons must have exactly 12 pentagons.*

Proof. The proof follows from use of the notion of deficiency at a vertex, namely that at each vertex, the sum of the angles around that vertex must be less than 360°. Therefore, the degree at each vertex cannot be more than three since the angle contribution of each polygon at a vertex is either 108° or 120°. Let P be a polyhedron made up entirely of regular pentagons and hexagons; then the number of faces F is the sum of the number of pentagons p and the number of hexagons h. Note that since P is a polyhedron, the degree at each vertex d must be at least 3 and due to the deficiency constraint it can be no more than 3, so $d = 3$. If we count up all the edges around all the vertices we will have counted each edge twice, so $3V = 2E$ or $V = 2E/3$. As in the proof of Theorem 3, $5p + 6h = 2E$. Since P is a polyhedron, the number of vertices V, edges E and faces F of the polyhedron must satisfy $V - E + F = 2$.

Substituting for V and F we obtain

$$2 = \frac{2E}{3} - E + F$$

$$2 = -\frac{1}{3}E + F = -\frac{1}{3} \cdot \frac{1}{2}(5p + 6h) + (p + h)$$

$$12 = -5p - 6h + 6p + 6h = p$$

completing the proof. □

Theorem 4 also holds if we allow nonregular pentagons and hexagons, but require the polyhedron to be simple. On the other hand, there exist nonsimple polyhedra with nonregular pentagonal and hexagonal faces that have more than 12 pentagonal faces.

A simple polyhedron made up entirely of (not necessarily regular) pentagons and hexagons is called a *fullerene*; certain fullerenes have important applications in chemistry. A famous example is buckminsterfullerene, C_{60}, also known as buckyballs, which was first synthesized in 1985 by Robert F. Curl [1]. Every fullerene contains exactly 12 pentagons, but Euler's formula places no constraints on the number of hexagons. Long before the first buckyballs were studied in the lab, H. S. M. Coxeter raised the question of what values were possible for f_6 in a simple polyhedron made up of only pentagons and hexagons. This question was answered in 1963 by Branko Grünbaum and Theodore Motzkin, who were able to demonstrate that while it was impossible to build a fullerene with exactly one hexagon, it is possible to build one with any other number of hexagons [5]. Readers interested in related topics should consider either [1] or [5].

An argument very similar to that used in Theorems 3 and 4 may be used to demonstrate that every simple polyhedron satisfies the equality

$$\sum_{n=3}^{\infty}(6 - n) \cdot f_n = 12.$$

This sum, in turn, is very useful for determining constraints on the number of faces of polyhedra subject to constraints about what kinds of faces are available. For example, any simple polyhedron must have at least some faces with five or fewer sides, any simple polyhedron made up of quadrangles must be combinatorially equivalent to a cube, and in any figure made up of pentagons, hexagons and heptagons, the number of pentagons and heptagons will differ by 12.

Bibliography

[1] Chung, Fan and Shlomo Sternberg. "Mathematics and the buckyball," *American Scientist* 81 (1993) no. 1, 56–71.

[2] Coxeter, H. S. M. *Regular polytopes*, Dover Publications, 1973.

[3] Cromwell, Peter. *Polyhedra*, Cambridge University Press, 1999.

[4] Grünbaum, Branko. "Regular polyhedra—old and new," *Aequationes Math.* 16 (1977) no. 1–2, 1–20.

[5] Grünbaum, Branko and Theodore S. Motzkin, "The number of hexagons and the simplicity of geodesics on certain polyhedra," *Canadian Journal of Mathematics* 15 (1963) 744–751.

[6] McMullen, Peter and Egon Schulte, *Abstract regular polytopes*, Cambridge University Press, 2002.

Worksheet: Polyhedra Exploration

Given the polygon pieces, follow the directions below assigned to you by the instructor for assembling one or more polyhedra. After you make the construction(s), follow the directions on the next page. We will then inventory the results for the whole class.

0. (Demonstration) Given: 4 equilateral triangles. Make a polyhedron.

1. Given: 14 equilateral triangles. Make two polyhedra, following the directions in the next two parts.

   (a) At every corner, make four triangles come together.
   (b) Make a polyhedron that has at least two different numbers of faces meeting at a corner using the remaining pieces.

2. Given: 8 regular hexagons and 4 equilateral triangles.

   (a) Make a polyhedron using four hexagons and four triangles; every corner is the same.
   (b) Can you make a polyhedron using only regular hexagons? Give reasons for your answer.

3. Given: 7 squares and 4 equilateral triangles. Make two polyhedra:

   (a) Use exactly six squares.
   (b) Use the remaining square and four triangles.

4. Given: 6 squares and 8 equilateral triangles. Make a polyhedron with every corner the same: triangle, square, triangle, square, in that order.

5. Given: 12 regular pentagons. Make a polyhedron with these.

6. Given: 2 regular pentagons, 2 equilateral triangles, 8 squares. Make two polyhedra:

   (a) Use the triangles and exactly three squares; make every corner the same.
   (b) Use the pentagons and the remaining squares; make every corner the same.

7. Given: 20 equilateral triangles. Make a polyhedron with every corner the same: five triangles come together.

8. Given: 2 squares, 8 equilateral triangles. Make a polyhedron with these. Make every corner the same: three triangles and one square come together.

9. Given: 8 equilateral triangles and 6 octagons. Make every corner the same.

10. Given: 6 squares and 8 hexagons. Make every corner the same.

After you have constructed the polyhedra, follow these directions:

For each polyhedron you have constructed, count the number of its corners or *vertices* (V), the number of its edges (E), and the number of its faces (F). Enter these numbers in the chart as well as the sum $V + F$. If you know the name of the polyhedron, fill it in.

Polyhedron	Name	V (# vertices)	F (# faces)	E (# edges)	$V + F$
0	tetrahedron	4	4	6	8
1 (a)					
1 (b)					
2 (a)					
2 (b)					
3 (a)					
3 (b)					
4					
5					
6 (a)					
6 (b)					
7					
8					
9					
10					

What do you notice?

Worksheet: Symmetric Polyhedra and Angle Deficiency

Working in teams of 2 or 3, carry out the steps indicated below, and record your observations.

Part I, Some simple figures

Exercise 1. Take the pieces necessary to make up one corner of a cube. Split them along one of the edges, and lay them flat. What is the angle measure of the gap formed along the separated edge?

What you have just measured is the *angle deficiency* of a vertex of the cube. In general, the angle deficiency of a vertex of a polyhedron is the sum of the angles formed by the polygons that meet at that vertex, measured in radians, subtracted from 2π.

When describing the vertices of a symmetric polyhedron made up of regular polygons, a convenient way of cataloging them is to list the numbers of edges that belong to each face at a vertex. For example, in the cube, three squares meet at each vertex. We would denote such a figure by the symbol (4.4.4), because each square has four edges, and call the faces that make up that corner a *vertex star*. In general, a vertex star is a collection of faces that meet at a vertex.

The *total angle deficiency* of a polyhedron is the sum of the angle deficiencies of the vertices of the polyhedron. In other words, if a figure has 8 vertices (like the cube), you would add up the deficiencies at each of the vertices. In the case of the cube this would be 4π.

For the following exercises it may be helpful to recall that the interior angle of a regular n-gon is $\pi(n-2)/n$ radians.
Exercise 2. Build vertex stars for at least two of the polyhedra described below, collect the appropriate data and complete the table.

name	vertex symbol	# vertices	angle deficiency at a vertex	total angle deficiency
tetrahedron	(3.3.3)	4		
octahedron	(3.3.3.3)	6		
icosahedron	(3.3.3.3.3)	12		
cube	(4.4.4)	8		4π
dodecahedron	(5.5.5)	20		

Worksheet: Symmetric Polyhedra and Angle Deficiency

Part II, Some slightly more complicated figures

An important class of polyhedra have faces that are regular polygons, with vertices such that every vertex "looks the same," and at least two kinds of faces. These polyhedra have very high degrees of symmetry, and traditionally they are called *Archimedean solids*.

Exercise 3. Figure out the missing entries in the table below.

name	vertex symbol	# vertices	angle deficiency at a vertex	total angle deficiency
truncated tetrahedron	(3.6.6)	12		
truncated cube	(3.8.8)	24		
truncated octahedron	(4.6.6)	24		
truncated dodecahedron	(3.10.10)	60		
truncated icosahedron	(5.6.6)	60		
cuboctahedron	(3.4.3.4)	12		
icosidodecahedron	(3.5.3.5)	30		
snub dodecahedron	(3.3.3.3.5)	60		
rhombicuboctahedron	(3.4.4.4)	24		
great rhombicosidodecahedron	(4.6.10)	120		
rhombicosidodecahedron	(3.4.5.4)	60		
great rhombicuboctahedron	(6.4.8)	48		
snub cube	(3.3.3.3.4)	24		

Exercise 4. Based on the results from Exercise 3, what do you conjecture?

Exercise 5. Recall that the angle deficiency at a vertex of a polyhedron $\mathcal{P}$ is the sum of the angles at the corners of the faces that meet at that angle subtracted from 2π. Suppose the polyhedron $\mathcal{P}$ has V vertices, E edges and F faces, and suppose that

$$f_1, f_2, \ldots, f_F \text{ are the faces of } \mathcal{P},$$
$$v_1, v_2, \ldots, v_V \text{ are the vertices of } \mathcal{P}.$$

(a) Explain why $\displaystyle\sum_{i=1}^{V}(\text{sum of angles around } v_i) = \sum_{j=1}^{F}(\text{sum of angles in } f_i)$.

(b) Suppose that n_{f_j} is the number of sides of face f_j. What is the sum of the angles in face f_j, in radians?

(c) What is $\displaystyle\sum_{j=1}^{F} n_{f_j}$ in terms of the number of edges, E, of the polyhedron? Why?

(d) Note that by definition,

$$\text{total angle deficiency} = \sum_{i=1}^{V}(2\pi - (\text{sum of the angles around } v_i)).$$

Use the facts you showed from (a), (b) and (c) and Euler's theorem to determine the total angle deficiency of $\mathcal{P}$. Does the answer you get agree with your conjecture from Exercise 4?

Worksheet: Poincare's Formula and Higher Genus Polyhedra

Exercise 1. Using the template provided, construct a polyhedral "donut" (formally, a torus, plural tori).

(a) Count the number of vertices, edges and faces. What is $V - E + F$?

(b) Take two of the polyhedral tori, and glue them along an outer rectangle. You have now created a surface with two holes, called a two-holed torus. Count the number of vertices, edges and faces, ignoring the glued face. What is $V - E + F$?

(c) Take three of the polyhedral tori, and glue them along two of the outer rectangles. You have now created a surface with three holes, called a three-holed torus. Count the number of vertices, edges and faces, ignoring the glued faces. What is $V - E + F$?

(d) If you were to create a four-holed torus, what do you predict for the value of $V - E + F$?

(e) What about an n-holed torus?

Exercise 2. Consider a solid cube with holes punched out of it in the following way:

- Subdivide each face of a solid cube into nine equal squares whose sides are one third the length of the side of the cube.

- For each pair of opposing middle squares, cut out a square hole going all the way through the cube.

Without calculating anything, try to determine how many holes there are on the surface formed. Note that the three tubular holes all meet in a smaller cube ($1/27^{th}$ the size) at the center of the cube. Once you think you have a reasonable guess, use Poincaré's formula to check your intuition. Note that when you are counting faces of the punched-out cube, a face cannot itself have a hole in it!

Solutions to Worksheets and Additional Exercises

Worksheet Solutions: Polyhedra Exploration

Polyhedron	Name	V (# vertices)	F (# faces)	E (# edges)	V+F
0	tetrahedron	4	4	6	8
1 (a)	octahedron	6	8	12	14
1 (b)	triangular bipyramid	5	6	9	11
2 (a)	truncated tetrahedron	12	8	18	20
2 (b)	(impossible)				
3 (a)	cube	8	6	12	14
3 (b)	square pyramid	5	5	8	10
4	cuboctahedron	12	14	24	26
5	dodecahedron	20	12	30	32
6 (a)	triangular prism	6	5	9	11
6 (b)	pentagonal prism	10	7	15	17
7	icosahedron	12	20	30	32
8	square antiprism	8	10	16	18
9	truncated cube	24	14	36	38
10	truncated octahedron	24	14	36	38

We notice that in all these cases, $V + F = E + 2$.

Exercise and Worksheet Solutions: Symmetric Polyhedra and Angle Deficiency

Exercise 0, from Notes to the instructor. Prove that the sum of the angles in a convex n-gon is $180(n - 2)$. Determine the measure of a single angle in a regular n-gon.

Base case: For a single triangle ($n = 3$), the sum of the angles is $180°$.

Induction hypothesis: Suppose, for any convex k-gon with $k < n$, that the sum of the angles in the convex k-gon is $180(k - 2)$.

We need to show that given any $n > 3$, the sum of the angles in the convex n-gon is $180(n - 2)$. Label the vertices of the n-gon cyclically (say, counterclockwise) as $v_1, v_2, \ldots, v_n$, and construct the diagonal $v_1 v_3$; note that since the n-gon is convex, this diagonal must lie inside the polygon. The diagonal divides the n-gon into two smaller polygons, one triangle and one $(n - 1)$-gon. Applying the induction hypothesis to each piece, we conclude that the sum of the angles of the triangle is 180, and the sum of the angles of the $(n - 1)$-gon is $180(n - 3)$; adding the sums together, we see that the total angle sum of the n-gon is $180(n - 2)$, as desired.

With this, we see that a single angle of a regular n-gon is $180(n - 2)/n$ degrees.

Exercise 1, from Worksheet Part I. The angle measure of the gap is $90° = \frac{\pi}{2}$ radians.

Exercise 2, from Worksheet Part I.

name	vertex symbol	# vertices	angle deficiency at a vertex	total angle deficiency
tetrahedron	(3.3.3)	4	π	4π
octahedron	(3.3.3.3)	6	$\frac{2\pi}{3}$	4π
icosahedron	(3.3.3.3.3)	12	$\frac{\pi}{3}$	4π
cube	(4.4.4)	8	$\frac{\pi}{2}$	4π
dodecahedron	(5.5.5)	20	$\frac{\pi}{5}$	4π

Exercise 3, from Worksheet Part II.

name	vertex symbol	# vertices	angle deficiency at a vertex	total angle deficiency
truncated tetrahedron	(3.6.6)	12	$\frac{\pi}{3}$	4π
truncated cube	(3.8.8)	24	$\frac{\pi}{6}$	4π
truncated octahedron	(4.6.6)	24	$\frac{\pi}{6}$	4π
truncated dodecahedron	(3.10.10)	60	$\frac{\pi}{15}$	4π
truncated icosahedron	(5.6.6)	60	$\frac{\pi}{15}$	4π
cuboctahedron	(3.4.3.4)	12	$\frac{\pi}{3}$	4π
icosidodecahedron	(3.5.3.5)	30	$\frac{2\pi}{15}$	4π
snub dodecahedron	(3.3.3.3.5)	60	$\frac{\pi}{15}$	4π
(small) rhombicuboctahedron	(3.4.4.4)	24	$\frac{\pi}{6}$	4π
great rhombicosidodecahedron	(4.6.10)	120	$\frac{\pi}{30}$	4π
(small) rhombicosidodecahedron	(3.4.5.4)	60	$\frac{\pi}{15}$	4π
great rhombicuboctahedron	(6.4.8)	48	$\frac{\pi}{12}$	4π
snub cube	(3.3.3.3.4)	24	$\frac{\pi}{6}$	4π

Exercise 4, from Worksheet Part II. We conjecture that the total angle deficiency is 4π.

Exercise 5, from Worksheet Part II.

(a) Both sums $\sum_{i=1}^{V}$ (sum of angles around v_i) and $\sum_{j=1}^{F}$ (sum of angles in f_i) give the total measure of all the angles around all vertices. The left-hand sum does it by summing the angle measure around each vertex, while the right-hand sum does it by summing the angle measure around each face.

(b) Note that a if polygon face f_j has n_{f_j} edges, it also has n_{f_j} vertices. Therefore, the sum of the angles in face f_j is $\pi(n_{f_j} - 2)$.

(c) The sum $\sum_{j=1}^{F} n_{f_j}$ equals twice the number of edges. To see this, imagine each time a vertex is counted, a little bit of each edge that is incident with that vertex is highlighted. Since the sum counts all the vertices exactly once, and each edge is incident with two vertices, if we count the "highlighted" parts of the edges as we count the vertices, each edge will be counted twice. That is, $\sum_{j=1}^{F} n_{f_j} = 2E$.

(d) Note that

$$\text{total angle deficiency} = \sum_{i=1}^{V}(2\pi - (\text{sum of the angles around } v_i))$$

$$= 2\pi V - \sum_{i=1}^{V}(\text{sum of the angles around } v_i)$$

$$= 2\pi V - \sum_{i=1}^{F}(\text{sum of the angles in } f_i) \qquad (\text{ by (a) })$$

$$= 2\pi V - \left(\sum_{i=1}^{F}\pi(n_{f_j}-2)\right) \qquad (\text{ by (b) })$$

$$= 2\pi V - \left(\sum_{i=1}^{F}\pi n_{f_j} - \sum_{i=1}^{F}2\pi\right)$$

$$= 2\pi V - 2\pi E + 2\pi F \qquad (\text{by simplifying and (c)})$$

$$= 2(V - E + F)\pi$$

$$= 2(2)\pi \qquad (\text{by Euler's theorem}).$$

That is, the total angle deficiency equals 4π, which is what we had conjectured.

Exercise 6, from Notes for the instructor. Yes. The prisms and antiprisms each have two kinds of faces; an n-prism has each corner consisting of a regular n-gon and two squares, while an n-antiprism has each corner consisting of a regular n-gon and three equilateral triangles.

5   Worksheet: Poincare's Formula and Higher Genus Polyhedra

Exercise 1.

(a) There are 9 vertices, 9 faces, and 18 edges, so $V - E + F = 0$.

(b) Each torus has 9 vertices, 9 faces, and 18 edges. We need to remove entirely one face of each torus to glue them together, so the two-holed torus has $(9-1) + (9-1) = 16$ faces. Also, we need to remove four edges and four vertices from one of the tori, so the two-holed torus has $(9-4) + 9 = 14$ vertices and $(18-4) + 18 = 32$ edges. Therefore, the two-holed torus has $V - E + F = -2$.

(c) Reasoning as before, the three-holed torus has $(16-1) + (9-1) = 23$ faces, $(14-4) + 9 = 19$ vertices, and $(32-4) + 18 = 46$ edges, so $V - E + F = -4$.

(d) For a four-holed torus, we predict $V - E + F = -6$.

(e) For an n-holed torus, we predict $V - E + F = 2 - 2n$.

Exercise 2. The correct answer to this question is, somewhat surprisingly, five holes. One way to understand this is that the first square hole punches one hole out of the cube, but the remaining two punches each cut two more holes (since they pass through the space punched out by the first punch). If we break up the exterior so that each face that has been punched through now has eight squares on it, using Poincaré's formula we note that there are 64 vertices, 144 edges and 72 faces, so we have $64 - 144 + 72 = -8 = 2 - 2(5)$, or genus 5. Alternately, one may cut the exterior faces that had the holes punched through them along their diagonals into two nonconvex c-shaped pieces, in which case we have 40 vertices, 84 edges and 36 faces, yielding the same conclusion.

It is worth noting that this construction is the first iteration in the construction of the Menger Sponge.

Template for a paper torus

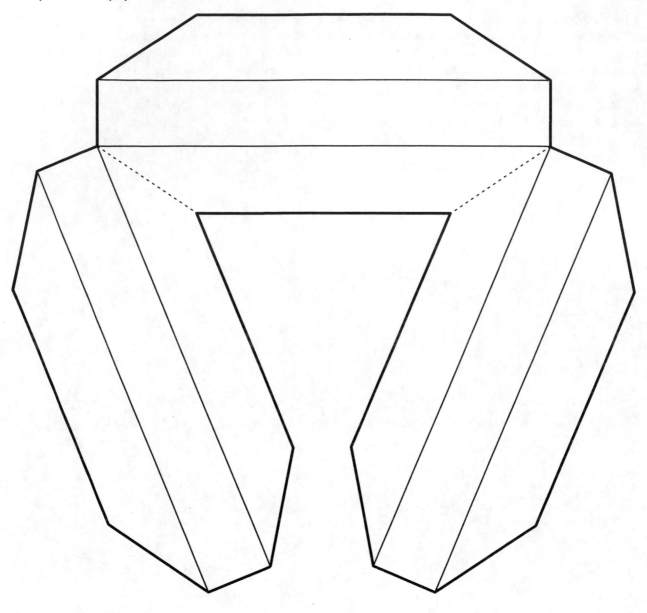

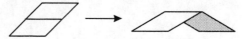 **cut**

—————— **mountain fold (fold away from you)**

- - - - - - - - **valley fold (fold towards you)**

Further Explorations with the Towers of Hanoi

Jon Stadler
Capital University

Summary

This project is a supplement to the usual introduction to the Towers of Hanoi problems that appear in most discrete mathematics texts. Many concepts from discrete mathematics are discussed, including Hamiltonian paths, representation of integers in other bases, graph theory, and number theory.

Notes for the instructor

Puzzles are a great way to pass time. However, they can also be an excellent source for mathematical applications. There are few puzzles more famous than the Towers of Hanoi. The point of this project is not to teach the students how to solve the puzzle, nor is it designed to teach a formula for the minimum number of moves required to solve the puzzle. In fact, it is highly probable that you have already discussed these things with your class. Instead, the purpose of this project is to tie together many of the topics that you have taught in your discrete mathematics course.

If you have not already done so, introduce the students to the Towers of Hanoi in your textbook. If your textbook does not discuss the puzzle, write to the author and demand that it be included in the next edition of the book and then find a reference on how to solve the puzzle on the Internet. Students will be ready when they know how to solve the puzzle and a formula for the minimum number of moves required to solve a puzzle with n disks. This project works well as a long-term project. It is possible to assign different parts of it throughout the term as the appropriate topics are introduced.

Required knowledge by project component

- *Hanoi's Hypercubes*

  - Hamiltonian paths
  - Definition of factor or divides

- *The Code of Hanoi*

  - Representation of integers in non-decimal bases

- *The Graph of Hanoi*

  - Hamiltonian paths
  - Recursion

- *The kth Position of the Towers of Hanoi*

  - Modular arithmetic
  - Representation of integers in non-decimal bases

Bibliography

[1] Berlekamp, E., J. Conway, and R. Guy. *Winning Ways for Your Mathematical Plays, Vol. 4 (2nd ed.)*, A. K. Peters, LTD, 2004.

[2] Crowe, D. W. "The *n*-dimensional cube and the tower of Hanoi," *Amer. Math. Monthly* 63 (1956) 29–30.

[3] Gardner, M. *Mathematical Puzzles & Diversions*, Simon and Schuster, 1959.

[4] Hinz, A. M. "Pascal's triangle and the Tower of Hanoi," *Amer. Math. Monthly* 99 (1992) no. 6, 538–544.

[5] Poole, D. G. "The towers and triangles of Professor Claus (or, Pascal knows Hanoi)," *Math. Mag.* 67 (1994) no. 5, 323–344.

Worksheet on Hanoi's Hypercubes

Let us suppose that we are trying to solve a Towers of Hanoi puzzle that has n disks. To keep track of our moves, label the top disk 0, the next disk 1 and so forth, so that the largest disk is labeled $n - 1$. For example, suppose $n = 3$. At the beginning of the puzzle, disk 0 will be on top of disk 1, which rests on top of disk 2, all of which are stacked on the left-most peg.

As you solve the puzzle, record the number of the disk that you move. With $n = 3$, first move disk 0, then disk 1, disk 0 (onto disk 1), disk 2, disk 0 (onto the empty peg), disk 1 and finally disk 0. In short, the disks moved are 0102010. Your first exercise involves the relationship between the puzzle's solution and paths on an n-dimensional cube.

1. On the cube below, note that there are three directions in which we can travel; left/right, front/back, up/down. Also notice that if we start at any vertex, we can travel in each of the three listed directions. Start at the vertex v. Now, read your solution to the 3-disk puzzle and create a path using the following rules. Each time you encounter a 0, move left or right. Each time you encounter a 1, move forwards or backwards. Each time you encounter a 2, (you guessed it) move up or down. What kind of path is the resulting path?

2. Move on to a puzzle with four disks. What is the sequence of moves that solves this puzzle?

3. The figure below is a 4-dimensional cube, also known has a *hypercube* or *tesseract*.

   You now have four directions in which you can travel. Add to the left/right, up/down and front/back motions the direction of in/out, represented by the line segments whose slopes are negative. Also add to the rules that if you encounter a "3" in your solution code, move in the in/out direction. Trace out the solution to the 4-disk Towers of Hanoi problem, moving through the hypercube according to the rules. What kind of path is the resulting path? How do you think this generalizes to the solution to the n-disk puzzle and paths on the n-dimensional cube?

4. Describe how to create a solution to the $n + 1$-disk puzzle from the n-disk puzzle using the numbering system described here.

5. Let $e(n)$ be the largest power of 2 that is a factor of n. For example, $e(40) = 3$ since 2^3 is a factor of 40, but 2^4 is not. Make a table of values for $e(n)$ for $1 \leq n \leq 15$. How does the sequence $e(1), e(2), \ldots, e(15)$ relate to the solution to the Towers of Hanoi?

Worksheet on the Code of Hanoi

When using the system developed in *Hanoi's Hypercubes*, one is only told *which* disk to move, but not *where* it should be relocated. When it is time to move any disk with number 1 or higher, there is only one choice (why is this?). However, the 0 disk can always be moved to either of the other two pegs. (Hint — never place an even-numbered disk onto an odd-numbered disk or vice versa!) We continue to use the disk numbering system from *Hanoi's Hypercubes*, but also introduce a numbering system for the pegs, labeling the left peg 0, the middle peg 1 and the right peg 2.

To encode the position of the disks, we create a *ternary*, or *base 3* code for the configuration, known as the *state* of the puzzle. Let p_i denote the number of the peg on which disk i rests, $0 \le p_i \le 2$ for $0 \le i \le n - 1$. The state of any configuration is defined as the ternary string

$$p_{n-1} p_{n-2} \cdots p_2 p_1 p_0.$$

For example, the state of the configuration below is 0112021 since disk 6 is on peg 0, disk 5 is on peg 1, disk 4 is on peg 1, disk 3 is on peg 2, disk 2 is on peg 0, disk 1 is on peg 2, and disk 0 is on peg 1.

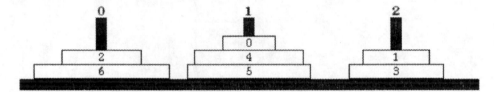

Conversely, given any state, it is possible to determine the disk configuration in the following manner. Determine which disks rest on each peg, stack disks from the same peg from largest to smallest, and then place them on the appropriate peg.

With this new method, it is possible to determine all states that can follow a given state. Note that the 0 disk can always be moved to either of the other two pegs even though it may lead us down the wrong path to the puzzle's solution. Thus, the last digit in the ternary string can always be changed to either of the other two numbers. With the exception of the positions in which all disks are on the same peg, there will be exactly one other disk that can be moved. This disk is the smallest disk (that is, with the smallest label) that is not on the same peg as disk 0. As an example, in state $v = 22011022$, disk 0 can move from peg 2 to pegs 0 or 1, so states 22011020 and 22011021 can follow v. Notice that in state v, disk 1 is on the same peg as disk 0 and thus, cannot be moved. However, disk 2 is on peg 0 so it can be moved. Since disks 0 and 1 are on peg 2, disk 2 must be moved to peg 1, so the other state that can follow v is 22011122.

1. Determine the state of each puzzle below.

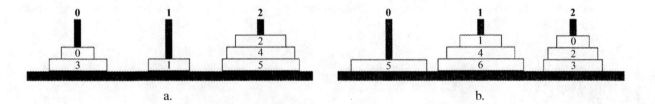

a.   b.

2. Draw the puzzle for each of the following states.

   (a) 0001112   (b) 212011

3. List all states that can follow each state below. If possible, use the method described in the last paragraph of this worksheet rather than draw the puzzle.

   (a) 0212011   (b) 22012000

Worksheet on the Graph of Hanoi

In many puzzles, it is possible to begin making mistakes that lead the solver away from the solution. The Towers of Hanoi is no exception. By creating the n-disk Hanoi graph, H_n, it will be possible to get yourself back on track. This section of the project describes how to create H_n.

The vertices of H_n will be arranged in a triangle and labeled with the 3^n possible ternary strings of length n. It is best to arrange them in a specific order and the arrangements for H_1, H_2 and H_3 are depicted below.

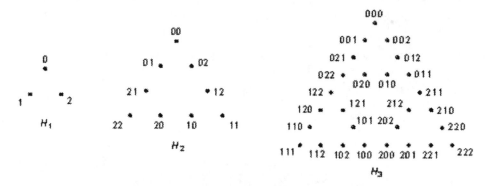

You will not be asked to do this, but to create the vertex arrangement for H_n from H_{n-1} in general, append a zero to the left of each vertex label of H_{n-1} to create the top portion of the triangle for H_n. On a copy of this top portion, add 1 modulo 3 to each number of every vertex label . For example, if the vertex label is 0201, we obtain 1012 when we add 1 modulo 3 to each number. If n is even, rotate this new arrangement 120° clockwise and place it to create the bottom *right* corner of H_n, otherwise, rotate it 120° counterclockwise to create the bottom *left* corner of H_n. With another copy of the top portion of the triangle, add 2 modulo 3 to each number of every vertex and rotate this arrangement 120° counterclockwise if n is even or clockwise if n is odd. Place this new arrangement next to the previous arrangement to complete the triangle.

Once the vertices are properly arranged, the edges are inserted according to the following rule. Vertices u and v are adjacent if and only if state u of the puzzle can be obtained from state v and vice versa.

1. Draw in all edges in H_1, H_2 and H_3 above.

2. Find a Hamiltonian path from 0 to 1 in H_1, from 00 to 11 in H_2 and from 000 to 111 in H_3.

3. How does a Hamiltonian path in H_n guarantee that a solution to the n-disk Towers of Hanoi puzzle exists from any legal configuration?

4. Someone left a partially-solved Towers of Hanoi puzzle in a store in state 011. Use your Hamiltonian path in H_3 to solve the puzzle.

5. Give a recursive method for finding a Hamiltonian path in H_n.

Worksheet on the kth Position of the Towers of Hanoi

In the fourth volume of their outstanding books *Winning Ways for Your Mathematical Plays*, Berlekamp, Conway and Guy describe a creative way of determining the state of the puzzle after k moves using the optimal solution. Their method is as follows.

To determine the state of the n-disk Towers of Hanoi puzzle after k moves:

- write k in binary, so that $k = b_{n-1}b_{n-2} \cdots b_2 b_1 b_0$, where $b_i \in \{0, 1\}$ for all i,

- carry out the ternary sum below, without carrying. In other words, add down the columns and record the result modulo 3.

If n is

even	odd
$1 \cdot b_0$	$2 \cdot b_0$
$21 \cdot b_1$	$12 \cdot b_1$
$122 \cdot b_2$	$211 \cdot b_2$
$2111 \cdot b_3$	$1222 \cdot b_3$
$12222 \cdot b_4$	$21111 \cdot b_4$
$\vdots$	$\vdots$

Example. Suppose we wish to determine the 141st move of the puzzle with eight disks. The binary representation of 141 is 10001101. Since 8 is even, we form the ternary sum

$$
\begin{array}{l}
1 \cdot 1 \\
21 \cdot 0 \\
122 \cdot 1 \\
2111 \cdot 1 \\
12222 \cdot 0 \\
211111 \cdot 0 \\
1222222 \cdot 0 \\
+21111111 \cdot 1 \\
\hline
\end{array}
\qquad \text{or simply} \qquad
\begin{array}{r}
1 \\
122 \\
2111 \\
+21111111 \\
\hline
21110012
\end{array} .
$$

When we add these four terms in base 3 without carrying, we obtain the string 21110012, which is the state of the puzzle after 141 moves with 8 disks. The rules are different when n is even or odd so that the puzzle will be solved when all disks are on peg 2.

1. Determine the state of the puzzle after k moves with n disks for the given k and n.

   (a) $k = 23, n = 6$

   (b) $k = 81, n = 7$

   (c) $k = 81, n = 8$

   (d) $k = 401, n = 10$

2. For the 10-disk puzzle, determine the states of the puzzle after 207 and after 208 moves. Which disk was moved? On which peg did the disk start and to which peg did it move?

Solutions

Hanoi's Hypercubes

1. The path is a Hamiltonian path.

2. 010201030102010

3. The path is a Hamiltonian path. The path on the n-dimensional cube corresponding to the solution of the n-disk puzzle will also be a Hamiltonian path.

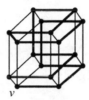

4. The solution to the $n + 1$-disk puzzle from the solution to the n-disk puzzle $\mathbf{z}$ is the sequence $\mathbf{z}n\mathbf{z}$.

5.

n	1	2	3	4	5	6	7	8	9	10	11	12	13	14	15
$e(n)$	0	1	0	2	0	1	0	3	0	1	0	2	0	1	0

The sequence is identical to the solution to the 4-disk puzzle.

The Code of Hanoi

1. (a) 220210 (b) 1012212

2. (a)

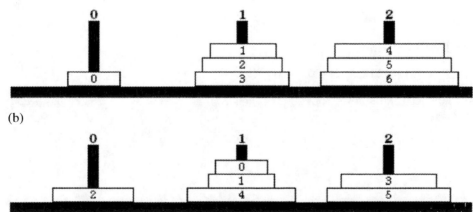

   (b)

3. (a) 0212010, 0212012, 0212211 (b) 22012001, 22012002, 22011000

The Graph of Hanoi

1. The three graphs are depicted below.

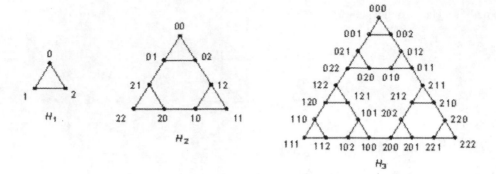

2. Hamiltonian paths on the three graphs are depicted below.

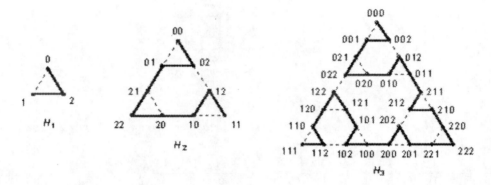

3. Since each vertex lies on the Hamiltonian path, there is a path from any vertex to the solution vertex $1 \cdots 1$. Following this path will list the moves in the solution.

4. 011, 211, 212, 210, 220, 222, 221, 201, 202, 200, 100, 102, 101, 121, 122, 120, 110, 112, 111 (this is not the optimal solution from state 011). *Instructor's note: Answers may vary.*

5. To create the Hamiltonian path in H_n from the Hamiltonian path in H_{n-1}, copy the path from H_{n-1} onto the top corner of H_n. Rotate another copy of the Hamiltonian path from H_{n-1} 120° clockwise and place it in the bottom left portion of H_n. Rotate a third copy 120° counterclockwise and place it in the bottom right portion of H_n. Finally, draw edges between vertices $011 \cdots 11$ and $211 \cdots 11$ as well as between vertices $100 \cdots 00$ and $200 \cdots 00$.

The kth Positions of the Towers of Hanoi.

1. (a) 12000 (b) 2102221 (c) 1201112 (d) 110012220

2. The states after the 207th and 208th moves are 22002222 and 22012222 respectively. Disk 4 has been moved from peg 0 to peg 1.

The Two Color Theorem

David Hunter
Westmont College

Summary

The purpose of this project is to help students prove that, under certain conditions, a map can be two-colored. There are at least four ways to approach this problem; the proofs vary in difficulty and use several techniques from discrete mathematics, including induction, structural induction, graphs, decision trees, and parity checks.

Notes for the instructor

There are two parts to this project, and each part guides the students through two different proofs of the result. Part I is fairly straightforward and could be assigned as a homework problem or as an in-class group activity. Part II is more challenging and is suitable for a longer-term (group) project, ideally after the students have completed Part I.

Solution I.1 is the easiest and most intuitive, and gives a good example of an inductive argument in a simple geometric setting. Solution I.2 avoids induction by using a parity check argument.

Solution II.1 involves some intuition about the topology of curves in the plane. To develop this intuition, students might want to experiment with loops of string (possibly even knotted). The trickiest part of the proof involves showing that a smoothing can always be chosen so that the curve remains connected. This property (and the basis of the inductive argument) can be discovered by constructing a binary decision tree, where each branch gives the result of smoothing a vertex in two different ways. This decision tree is related to the calculation of the Kauffman bracket polynomial, an important invariant in knot theory. Interested students could investigate related topics in knot theory in [3] or [1].

The only topological insight needed for solution II.2 is that each intersection yields a four-sided face, and that all faces of the graph are obtained in this way. Granting this, the graph theory argument is standard, though perhaps quite challenging for most students. This approach gives the most general characterization of a two-colorable map, and invites investigation of the converse, leading to connections with graph walking.

Bibliography

[1] Adams, Colin. *The Knot Book*, Freeman, 1994.

[2] Gardner, Martin. *Sixth book of mathematical games from Scientific American*, Freeman, 1971.

[3] Kauffman, Louis. *On Knots*, Princeton University Press, 1987.

Worksheet on the Two Color Theorem, Part I

The Four Color Theorem is a great example of a result that is simple to state, but very hard to prove. Yet under certain conditions, it is possible to show that two colors suffice. Consider any finite collection of (infinite) straight lines in the plane.

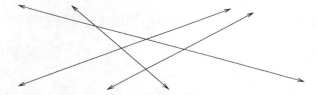

These lines divide the plane into regions, forming a *line map*. This assignment will help you write two proofs of the following result.

Theorem 1. *Any line map can be two-colored.*

Why do we need two proofs of this theorem? Well, we certainly don't "need" two proofs—one is enough to establish the validity of the theorem. But finding more than one proof for a result gives us practice with different proof techniques, and can sometimes show how different areas of mathematics are connected. It is also interesting to compare two different proofs of the same result: any similarities point to something fundamental in the theorem that makes it work.

Exercise 1. Prove Theorem 1 by induction on the number of lines.

Hint: The secret to constructing an inductive argument often lies in seeing the structures involved from an inductive point of view. In this case, observe that given a line map with n lines, removing any line will yield a line map with $n - 1$ lines. So, when writing your inductive argument, consider using the following inductive hypothesis.

Inductive Hypothesis: Any line map with $n - 1$ lines can be two-colored.

Suppose you are given a line map with n lines. If you remove some line l, you can apply the inductive hypothesis to the resulting graph. Then you need to explain how you can put l back and create a two-coloring for the original given line map. Remember that you also need a base case.

Exercise 2. Prove Theorem 1 without using induction.

Hint: One approach you might try is as follows. Pick a "good" side and a "bad" side for each line. It doesn't matter how you choose these, just keep track of each line's good side. For each region R in the map, count the number of lines that R is on the good side of. This assigns each region a number. Explain how these numbers can produce a valid two-coloring.

Exercise 3. What do your two proofs have in common?

Hint: What geometric properties of lines did you use in each proof?

Worksheet on the Two Color Theorem, Part II

A *closed curve* in the plane is a continuous loop, possibly with self-intersections. For example, a closed curve can be drawn by placing a pencil at point X, drawing any curve without lifting the pencil, and finishing the curve at point X.

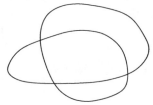

Any closed curve divides the plane into regions, forming a *closed curve map*. This project will help you construct two different proofs of the following.

Theorem 2. *Any closed curve map can be two-colored.*

For simplicity, assume that the curve is in general position (i.e., no multiple crossings, no self-tangents).

Exercise 4. Prove Theorem 2 using induction on the number of crossings.

Hint: The structure of this proof will be similar to the proof in Exercise 1 of Part I, where you used induction on the number of lines. In Exercise 1, you proved the inductive step by removing and adding back a line. In this exercise, you can prove the inductive step by removing and adding back a crossing.

A crossing can be removed by performing a *smoothing*:

Start by proving two facts about the smoothing process:

1. Smoothing (and reversing a smoothing) preserves a two-coloring of the map.

2. For any crossing, there is always a choice of smoothing that leaves the curve connected.

Exercise 5. Prove Theorem 2 using graph theory.

Hint: To convert a closed curve to a planar graph G, place a vertex in each region, and draw an edge between any two vertices whose regions share a common border (i.e., take the dual graph). Two-coloring the vertices of G is the same as two-coloring the regions of the closed curve map.

The key observation, which you need to justify, is that all the faces of G have an even number of edges. Using this fact, you should be able to finish the proof of the theorem using induction on the number of faces in G.

Solutions

I. Straight lines in the plane:

1. Use induction on the number of lines. If there are no lines, color the entire region white. Now let $n > 0$. Suppose as inductive hypothesis that any collection of $n-1$ lines can be two-colored. Given a collection of n lines, remove any line l and two-color the resulting map, by inductive hypothesis. Now put l back. Reverse the coloring on one side of the line l. Each side of l will be correctly two-colored, and it remains to show that any two regions whose border lies on l have opposite colors. But any two such regions must have been the same region when l was removed, so the reversal of colors guarantees that the new regions will have opposite colors.

2. (This solution is adapted from [2], which considers a map formed by a collection circles in the plane.) Choose an orientation (i.e., a preferred direction) on each line. For each region R, count the number $k(R)$ of lines l such that R is on the right side of l.

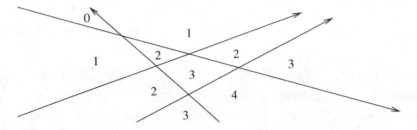

   If $k(R)$ is even, color R black; otherwise, color R white. By construction, this is a two-coloring, since regions on opposite sides of any boundary will have different colors.

3. Both proofs used the fact that a line divides the plane into two disjoint half-planes. This is sometimes called the *plane separation postulate* in modern treatments of geometry.

II. Closed curve:

4. At any given crossing, there are two ways to perform a smoothing. Therefore, you can represent the process of smoothing all the crossings as a binary decision tree. For example, the tree below illustrates the different ways to smooth the upper left crossing followed by the lower left crossing for the curve given in the problem statement.

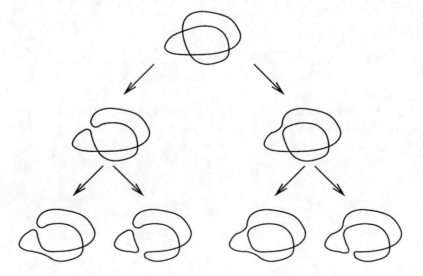

   Notice that a smoothing may change the number of components of the curve, but it is always possible to choose a smoothing that leaves the curve connected. To see this, consider any crossing in any given curve:

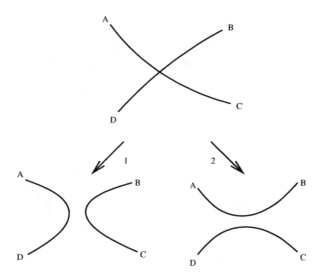

Suppose, without loss of generality, that when traversing the original curve starting at point A, you encounter points A, C, B, D, and then A, in order. This must happen (up to a choice of labels) because the original curve is closed and connected. Then, since C is connected to B directly (without passing through A), smoothing #2 must not disconnect the curve.

Now we can prove that any closed curve map can be two-colored by induction on the number of crossings. If there are no crossings, simply color the inside of the curve black and the outside white. (Note that we are using the Jordan Curve Theorem here.) Let $n > 0$. Suppose, as inductive hypothesis, that any closed curve graph with $n - 1$ crossings can be two-colored. Given a curve with n crossings, perform a smoothing on any crossing (and choose the smoothing so that the curve remains connected). By inductive hypothesis, the smoothed curve map can be two-colored. The following picture shows how to color the n-crossing curve map using the coloring of the $(n - 1)$-crossing curve map.

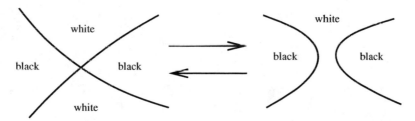

5. Given any closed curve map, construct a planar graph G as follows: place one vertex in each region, and connect any two vertices whose regions share a common border (i.e., take the dual graph of the graph whose vertices are the crossings of the curve, and whose edges follow the curve).

Claim. *All the faces of G have an even number of edges.*

Proof of claim. By duality, each face of G contains one crossing, and the number of edges of each face is the number of paths to the crossing. Since the curve is closed, each crossing has an even number of paths to it, because the curve cannot terminate at a crossing. (Note that this argument holds without assuming general position. If the given curve is in general position, then each face of G must have four edges.)

It now suffices to prove the following.

Lemma. *If every face of a planar graph has an even number of edges, then the vertices of the graph can be two-colored.*

Proof. We proceed by induction on the number of faces in the graph. If the graph has no faces, then it has no cycles, so it is bipartite, i.e., two-colorable. Let $n > 0$. Suppose as inductive hypothesis that every planar graph with $n - 1$ faces—all having an even number of edges—is vertex two-colorable. Let G be a planar graph with

n faces, all of which have an even number of edges. Since G is planar, it must have a face with an edge on the "border" of G, that is, an edge that is part of no other face. Let $X_1 X_2 \cdots X_{2k}$ be such a face, with edge $X_1 X_2$ on the border of G. Form a new graph G' by removing edge $X_1 X_2$ from G. By construction, G' has $n - 1$ faces, so it can be two-colored. Since the trail $X_2 X_3 \cdots X_{2k} X_1$ can be two-colored, the colors must alternate. And since this trail has an even number of vertices, the first vertex X_2 and the last vertex X_1 must have different colors. Hence the two-coloring of G' gives a valid two-coloring of G.

Counting Perfect Matchings and Benzenoids

Fred J. Rispoli
Dowling College

Summary

The connection between perfect matchings and benzene was discovered by the German chemist Kekulé in the mid 1800s. Subsequently, chemists have learned that the number of perfect matchings contained in a molecular model is an important parameter related to chemical stability. Hence, counting perfect matchings has been an important problem in chemistry for over 50 years. However, counting perfect matchings in general graphs is a computationally difficult problem. Consequently, chemists and graph theorists have developed efficient counting methods for certain classes of graphs that arise in modeling special hydrocarbons called benzenoids. Many of these methods involve counting principles usually discussed in discrete mathematics courses. In this article we discuss several of these methods and show how to implement a general determinant based formula.

Notes for the instructor

This project works well as an enrichment topic for an advanced discrete mathematics course focused on applications. Students should be familiar with counting techniques, graphs and determinants. I usually give the paper to students to read and then present a summary of the material at the end of the course. I spend roughly one class meeting on it. Exercises that reinforce and extend some key ideas are given in the last section, along with selected solutions.

Bibliography

[1] J. Aihara, "Why Aromatic Compounds Are Stable," *Scientific American* 266(3) (1992) 62–68.

[2] S. Cyvin and I. Gutman, *Kekulé Structures in Benzenoid Hydrocarbons*, Springer-Verlag, New York, 1988.

[3] T. Došlić, "Perfect Matchings, Catalan Numbers, and Pascal's Triangle." *Mathematics Magazine* 80 (2007) 219–225.

[4] R.A. Horn and C.R. Johnson, *Matrix Analysis*, Cambridge University Press, New York, 1985

[5] S. Iijima, "Helical Microtubules of Graphitic Carbon," *Nature* 354 (1991) 56–58.

[6] L. Lovász and M.D. Plummer, *Matching Theory*, North Holland, Amsterdam, 1986.

[7] P.W. Kasteleyn, "Graph theory and crystal physics," in F. Harary, editor, *Graph Theory and Theoretical Physics*, Academic Press, 1967.

[8] J. Propp, "Enumerations of Matchings: Problems and Progress," *New Perspectives in Geometric Combinatorics*, MSRI Publications 38 (1999) 255–291.

[9] F. Rispoli, "Counting Perfect Matchings in Hexagonal Systems Associated with Benzenoids," *Mathematics Magazine* 74 (2001) 194–200.

[10] H. Sachs, "Perfect Matchings in Hexagonal Systems," *Combinatorica* 4 (1984) 89–99.

[11] H. Sachs, P. Hansen and M. Zheng, "Kekulé count in tubular hydrocarbons," *MATCH Communications in Mathematical and in Computer Chemistry* 33 (1996) 169–241.

[12] N. Trinajstić, *Chemical Graph Theory*, Second Edition, CRC Press, 1992.

Perfect Matchings and Benzenoids

Given a graph G, a *perfect matching* is a subgraph M that contains every vertex of G such that all vertices in M have degree 1. Chemists, who use graphs to model hydrocarbon molecules, are interested in perfect matchings since they provide possible double-bond arrangements for carbon bonds. For example, Figure 1 contains two different graph models of benzene. The duplicate edges in the model on the left illustrate possible locations for double carbon bonds. Since every carbon atom must have exactly four bonds, there are actually two possible arrangements for the double carbon bonds. To study possible locations of double-bonds, chemists model benzene-like molecules called "benzenoids," by omitting the hydrogen bonds and use special hexagonal grid graphs such as the model on the right in Figure 1 (and also Figures 2-8). In these models perfect matchings are used to indicate the location of the double bonds. In the graph on the right side of Figure 1 the bold edges are used to illustrate the perfect matching corresponding to the double bonds in the graph on the left.

It is a well-established fact that, roughly speaking, for benzenoid molecules with the same number of hexagons, the chemical stability increases with the number of perfect matchings in its associated hexagonal graph. Surprisingly, when we consider special classes of benzenoids and count perfect matchings, many well-known numbers arise such as Fibonacci numbers and powers of 2. In this article we shall discuss techniques used to enumerate perfect matchings in hexagonal graphs.

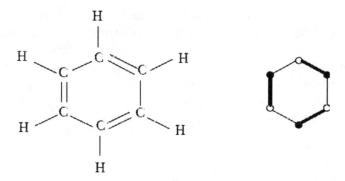

Figure 1. A graph model of benzene and its associated hexagonal graph.

A graph G is called 2-*connected* if it is connected and at least two vertices must be removed to make G disconnected. A *hexagonal system* is a 2-connected planar graph such that each interior face can be drawn as a regular hexagon. Notice that this condition forces all vertex degrees in a hexagonal system to be 3 or 2, and only 2 on the exterior boundary. Moreover, every pair of adjacent hexagons have exactly one edge in common. For convenience, all hexagonal systems illustrated in this article will be drawn with two vertical edges. We leave it as an exercise to show that every hexagonal system is bipartite which can be proved using induction on the number of hexagons in the hexagonal system. Hence, we may partition the vertices into subsets of black vertices $B = \{b_i\}$ and white vertices $W = \{w_j\}$ which is illustrated in Figure 2. Clearly G must have an even number of vertices in order for it to contain a perfect matching M. Moreover, if G has $2n$ vertices, then M must have exactly n edges.

A *hydrocarbon* is a substance consisting only of carbon and hydrogen atoms. A *benzenoid* is a special type of hydrocarbon that has a benzene like structure. Every benzenoid has a unique corresponding hexagonal system H obtained by removing the edges representing carbon-hydrogen bonds and letting the remaining edges of H represent either single or double carbon-carbon bonds. In Figure 2 the hexagonal system obtained from naphthalene is given. Observe that there are eight carbon-hydrogen bonds that have been suppressed, all resulting in degree 2 vertices. It is known that all hexagonal systems that arise from benzenoids contain a perfect matching.

For any graph G, the number of perfect matchings in G is denoted by $\phi(G)$. For example, in Figure 2 the bold edges illustrate a perfect matching, and the reader should confirm that $\phi(H) = 3$. For general graphs G, computing $\phi(G)$ is known to be an NP-hard problem (see [6]). That is, for an arbitrary graph with n vertices, there is no known algorithm to compute $\phi(G)$ involving $O(n^k)$ operations where k is a fixed constant. This remains true even when G is a bipartite graph. However, as we shall see, there are efficient methods that can be used to compute $\phi(G)$ for special classes of planar graphs such as hexagonal systems, and also explicit formulas for special types of hexagonal systems.

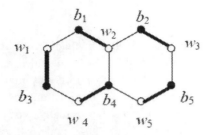

Figure 2. The hexagonal system associated with naphthalene.

Counting Perfect Matchings for Special Classes

A *linear chain of length h*, denoted by L, consists of h hexagons such that all adjacent pairs of hexagons share exactly one vertical edge and no nonvertical edges. A linear chain of length 2 is given in Figure 2. Observe that for any linear chain there is a one-to-one correspondence between perfect matchings in L and the vertical edges in L. Thus every linear chain of length h has exactly $h + 1$ perfect matchings. Linear chains are very important because they are used as building blocks of more complex hexagonal systems.

Given any hexagon in a hexagonal system, we refer to its six edges using the terms northeast, east, southeast, southwest, west and northwest. The *northern* edges consist of both the northwest and northeast edges; *southern* edges are defined similarly. In Figure 2 the bold edges are precisely the two northeast edges, the two southeast edges and the western edge in the western-most hexagon. A *rectangular hexagonal system*, denoted by $R[h, v]$, consists of v linear chains, $L_1, \ldots, L_v$, each of length h, together with $v - 1$ linear chains $\overline{L}_1, \ldots, \overline{L}_{v-1}$ each of length $h - 1$, such that for $i = 1, \ldots, v - 1$, all northern edges of $\overline{L}_i$ are southern edges of L_i, and all southern edges of $\overline{L}_i$ are northern edges of L_{i+1}. For example, a linear chain with h hexagons is $R[h, 1]$, and $R[5, 4]$ appears in Figure 3.

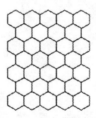

Figure 3. The rectangular hexagonal system $R[5, 4]$.

Theorem 1. *For a rectangular hexagonal system $R[h, v]$, we have $\phi(R[h, v]) = (h + 1)^v$.*

Proof. Every perfect matching of $R[h, v]$ contains only edges in the v linear chains $L_1, \ldots, L_v$ (i.e., the odd in-dexed/longer rows of hexagons), and none of the vertical edges in the $v - 1$ linear chains $\overline{L}_1, \ldots, \overline{L}_{v-1}$ (i.e., the even indexed/shorter rows). To see this suppose that we color every other vertex in the top linear chain L_1 white, and the remaining vertices black, where the vertical edges in $\overline{L}_1$ are incident to the black vertices. If one or more of the vertical edges from $\overline{L}_1$ is part of a perfect matching, then finding matching edges for the remainder of L_1 is impossible, since there will be more white vertices available than black. The fact that none of the vertical edges from $\overline{L}_1$ can be used in a perfect matching forces the same for $\overline{L}_2, \ldots, \overline{L}_{v-1}$. Hence, any perfect matching from L_i can be used with any perfect matching of any of the other linear chains L_j, $j \neq i$, and hence, $\phi(R[h, v])$ is the product of $\phi(L_1)$, $\phi(L_2)$, $\ldots, \phi(L_v)$. Since $\phi(L_i) = h + 1$, for all $i = 1, 2, \ldots, v$, we have that $\phi(R[h, v]) = (h + 1)^v$. $\qquad\square$

A *parallelogram hexagonal system*, denoted by $P[h, v]$, consists of v linear chains $L_1, \ldots, L_v$, each of length h, such that for $i = 1, \ldots, v - 1$, all southern edges of hexagons in L_i, except for the southeast edge of the eastern-most hexagon in L_i, are also northern edges of hexagons in L_{i+1}, except for the northwest edge of the western-most hexagon in L_{i+1}; Figure 4 shows $P[5, 3]$. To count perfect matchings in parallelogram hexagonal systems we need the well-known function for counting combinations, $C(n, r) = \frac{n!}{r!(n-r)!}$, where n and r nonnegative integers, with $0 \leq r \leq n$.

Figure 4. The parallelogram hexagonal system $P[5, 3]$.

Theorem 2. *For a parallelogram hexagonal system $P[h, v]$, we have $\phi(P[h, v]) = C(h + v, v)$.*

Proof. The proof is by induction on $k = h + v$. Clearly, the formula holds for $P[1, 1]$ (which corresponds to benzene and is given in Figure 1). Notice that $P[h, 1]$ and $P[1, h]$ are linear chains of length h. Hence, $\phi(P[h, 1]) = h + 1$ and the formula holds for both $P[h, 1]$ and $P[1, h]$. Assume that the result holds for all parallelogram hexagonal systems $P[h, v]$ with $h + v = k$.

Consider $P[h, v + 1]$ (the case $P[h + 1, v]$ is similar). Let L be the northern-most linear chain of $P[h, v + 1]$, and let M be a perfect matching of $P[h, v + 1]$. Suppose that M contains the eastern-most vertical edge of L (see the graph on the left in Figure 5). Then M must also contain the northwest edge of every hexagon in L. The remaining edges in M can be any perfect matching of the hexagonal system $P[h, v + 1]$ with L removed; that is, $P[h, v]$. By the inductive assumption, there are $C(h + v, v)$ such perfect matchings.

Suppose that M contains the eastern-most northeast edge of L (see the graph on the right in Figure 5.). Then M must also contain the southeast edge of every hexagon in the eastern-most hexagon of every linear chain in $P[h, v + 1]$. The remaining edges of M can now be any perfect matching of the hexagonal system $P[h, v + 1]$ with the eastern-most hexagon removed from every row; that is, $P[h - 1, v + 1]$ By the inductive assumption, there are $C(h + v, v + 1)$ such perfect matchings.

Every perfect matching in $P[h, v + 1]$ contains either the eastern-most vertical edge of L or the eastern-most northeast edge of L, but not both. Therefore, $\phi(P[h, v + 1]) = C(h + v, v) + C(h + v, v + 1)$. By Pascal's Identity, $\phi(P[h, v + 1]) = C(h + v + 1, v + 1)$. $\qquad\square$

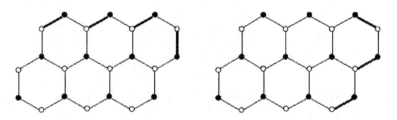

Figure 5. An illustration of the proof of Theorem 2.

Example 1.

(a) Use Theorem 1 to compute the number of perfect matchings for $R[5, 4]$.

(b) Use Theorem 2 to compute the number of perfect matchings for $P[5, 3]$.

(c) Determine the number of perfect matchings in Figure 6 which is obtained from $R[5, 4]$ by deleting the interior hexagons.

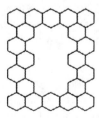

Figure 6. $R[5, 4]$ with the interior hexagons deleted.

Solution.

(a) $\phi(R[5, 4]) = 6^4 = 1296$.

(b) $\phi(P[5, 3]) = C(8, 3) = 56$.

(c) Consider the two hexagons adjacent to the linear chain on top, and label them H_1 and H_2, with H_1 on the left side. As in the proof of Theorem 1, none of the four vertical edges in H_1 and H_2 can be part of a perfect matching. Moreover, the southeast edge of H_1 must be in every perfect matching, and this forces a sequence of eight edges that must also be in every perfect matching until we get to the linear chain on the bottom. A similar sequence of eight edges starting with the southwest edge of H_2 must be in every perfect matching. But again, no vertical edges from the two hexagons adjacent to the linear chain on the bottom can be used in a perfect matching. Consequently, any perfect matching can be used for the linear chain on the bottom. Since there are six possible perfect matchings for the linear chain on top which can be used with any of the six perfect matchings for the linear chain on the bottom we get a total of $6^2 = 36$ perfect matchings.   □

A hexagonal system is called a *fibonaccene chain* if it consists of a chain of hexagons $H_1, \ldots, H_v$, with H_1 on top, and only the following shared edges: For i even and $1 < i < v$, H_i shares its northwest edge with H_{i-1} and its southwest edge with H_{i+1}; H_1 shares only its southeast edge and H_v shares only its northeast edge when v is odd, and only its northwest edge when v is even. The name is due to the fact that ϕ satisfies a Fibonacci-style recurrence relation, given in Theorem 3. The proof is left as an exercise.

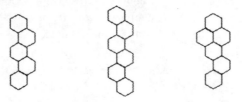

Figure 7. Fibonaccene chains with 5 and 6 hexagons, and a fibonaccene chain with an additional hexagon.

Theorem 3. *Let H be a fibonaccene chain with h hexagons and let $a_h = \phi(H)$. Then $a_1 = 2, a_2 = 3$, and $a_h = a_{h-1} + a_{h-2}$ for $h \geq 3$.*

Example 2.

(a) Use Theorem 3 to determine the number of perfect matchings for the two fibonaccene chains given on the left in Figure 7.

(b) Determine the number of perfect matchings for the fibonaccene chain with an additional hexagon given in Figure 7.

Solution.

(a) We need to find the terms a_5 and a_6 in the sequence 2, 3, 5, 8, 13, 21. Hence the chain on the left contains 13 perfect matchings and the chain in the middle of Figure 7 contains 21.

(b) Let H_1 be the additional hexagon. Every perfect matching must contain either the northwest edge of H_1 or its western vertical edge. If a perfect matching contains the northwest edge of H_1, then this forces eight edges that must be in the perfect matching, and leaves two possibilities for matching edges in the hexagon on the bottom. So there are two perfect matchings with this edge. If a perfect matching contains the western vertical edge, then any perfect matching of the fibonaccene chain with five hexagons can be used. Thus, there is a total of $2 + 13 = 15$ perfect matchings.   □

The final special class we consider arises from tubular benzenoids which were discovered in the early 1990s. A *tubulene* is a benzenoid whose carbon skeleton is a rectangular hexagonal system embedded in a cylinder with open ends (top and bottom). There are several different types of tubulene structures depending on how much they are "twisted," but here we consider only the untwisted variety (for more information on tubulenes and perfect matchings, see [5] and [11].) Let $T[h, k]$ denote the tubulene obtained from $R[h, v]$ embedded in a cylinder, where $k = 2v - 1$, i.e., k is the number of rows of hexagons. For example, if $T[5, 1]$ is drawn on paper it would look like a linear chain of length 5 with its left most vertical edge and its right most vertical edge "glued" together. Notice that as a result of the glued edges, the number of perfect matchings is reduced from 6 to 4. It is left for the reader to show that $\phi(T[5, 1]) = 4$. In Figure 8 an illustration of $T[5, 7]$ is given. The long vertical lines are used to indicate the edges that are glued together; or equivalently, the vertical line used to cut the cylinder.

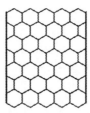

Figure 8. The tubulene hexagonal system $T[5, 7]$.

Theorem 4. *For a tubulene hexagonal system $T[h, k]$, we have $\phi(T[h, k]) = 2^{k+1}$.*

A proof of Theorem 4 is left as an exercise. By applying the theorem we see that $\phi(T[5, 7]) = 2^8 = 256$. It is interesting to note that $\phi(T[h, k])$ is independent of h, but dependent only on k. Thus $\phi(T[h, k])$ will increase if the tubulene is extended in the vertical direction, but does not change at all when it is extended horizontally. What does this imply in terms of chemical properties of tubulene?

To analyze stability levels and obtain a measure of energy, chemists have investigated a graph parameter based on ϕ defined as follows. Given a hexagonal system H, let $\eta(H)$ denote the number of hexagons in H. Then the *Kekulé index* is defined by $\kappa(H) = \frac{\log_2 \phi(H)}{\eta(H)}$. This index is considered by some chemists as an "average resonance energy per hexagon," and is known to satisfy $0 \leq \kappa(H) \leq 1$. For more details, see [11]. If we now compare the Kekulé index of $T[5, 7]$ and $R[5, 4]$, we see that $\phi(T[5, 7]) = 2^8$, which implies that $\kappa(T[5, 7]) = 8/35 = 0.2286$. From Theorem 1, we know that $\phi(R[5, 4]) = 6^4$, and consequently $\kappa(R[5, 4]) = 10.3399/32 = 0.3231$. As a general rule higher energy implies less stability. Since the benzenoid associated with $R[5, 4]$ has a higher energy level compared to $T[5, 7]$, we conclude that it is less stable than the benzenoid associated with $T[5, 7]$.

The Determinant Formula

Next we discuss a method of computing $\phi(H)$ for all hexagonal systems. Since every hexagonal system is bipartite, they can be represented with matrices defined as follows. Let H be a hexagonal system and let E denote the set of edges in H. Let $B \cup W$ be the set of vertices of H, where $B = \{b_1, \ldots, b_n\}$ and $W = \{w_1, \ldots, w_n\}$. We assume that B and W contain the same number of vertices since this is a necessary condition for the existence of a perfect matching. Now define the $n \times n$ *biadjacency matrix $A(H) = [a_{ij}]$*, by $a_{ij} = 1$ if $\{b_i, w_j\} \in E$ and $a_{ij} = 0$ if $\{b_i, w_j\} \notin E$. The biadjacency matrix for the hexagonal system in Figure 2 is given below where the columns represent $w_1, \ldots, w_5$, and the rows represent $b_1, \ldots, b_5$.

$$A(H) = \begin{bmatrix} 1 & 1 & 0 & 0 & 0 \\ 0 & 1 & 1 & 0 & 0 \\ 1 & 0 & 0 & 1 & 0 \\ 0 & 1 & 0 & 1 & 1 \\ 0 & 0 & 1 & 0 & 1 \end{bmatrix}$$

Recall that given an $n \times n$ matrix A, the *determinant* of A is defined by

$$det(A) = \sum_{\sigma} (sgn\ \sigma) a_{1\sigma(1)} a_{2\sigma(2)} \ldots a_{n\sigma(n)}$$

where the sum runs over all $n!$ permutations σ of $\{1, 2, \ldots, n\}$, and $sgn\ \sigma$ is $+1$ or -1, according to whether σ is an even permutation or an odd one (for a reference on determinants, see [4].) By storing the adjacency data of a hexagonal system H with $2n$ vertices in an $n \times n$ matrix, we can establish a one-to-one correspondence between perfect matchings in H and the nonzero terms in the expansion of $det(A(H))$. For example, the perfect matching M below corresponds to the permutation σ shown.

$$M = \{\{b_1, w_2\}, \{b_2, w_3\}, \{b_3, w_1\}, \{b_4, w_4\}, \{b_5, w_5\}\} \qquad \sigma = \begin{pmatrix} 1 & 2 & 3 & 4 & 5 \\ 2 & 3 & 1 & 4 & 5 \end{pmatrix}$$

Observe that when the term associated with σ is used in calculating the determinant of $A(H)$, the product of the bold numbers in the $A(H)$ given above is computed and gives a result of 1. This is true in general, the nonzero terms of $det(A(H))$ whose permutations correspond to a perfect matching will either be 1 or -1, and it can be proven that every nonzero term will have the same sign (for a proof, see [9].) The terms whose associated permutation do not correspond to a perfect matching will yield a 0 in the determinant expansion. Hence, $|det(A(H))|$ counts the number of permutations in H. This fact was first proven by Kasteleyn [7].

Theorem 5. *For a hexagonal system H, $\phi(H) = |det(A(H))|$.*

Theorem 5 may be implemented using many different technologies. Most of the time required is in labeling the vertices, finding the biadjacency matrix and entering data. For example, to compute $\phi(H)$ for the hexagonal system given in Figure 9, the computation requires finding the biadjacency matrix $A(H)$ which is a 32×32 matrix. Calculating the determinant gives $\phi(H) = 1,764$.

Example 3.

Use the determinant formula to compute the number of perfect matchings contained in the hexagonal system given in Figure 9.

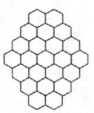

Figure 9. A hexagonal system with 1,764 perfect matchings.

We conclude this article by noting that computing $\phi(G)$ is an important problem in many areas of science and mathematics. For example, perfect matching enumeration is used in the famous dimer problem in physics and to help solve many tiling problems. Counting perfect matchings in the parallelogram hexagonal system $P[h, v]$ can also be used to count non-decreasing sequences of length v with elements from $\{0, 1, \ldots, h\}$, and lattice paths in a rectangular lattice (see [3]).

Exercises on Counting Perfect Matches and Benzenoids

1. (a) Draw all three perfect matchings contained in the linear chain of length 2 given in Figure 2.

    (b) Show that there is a one-to-one correspondence between the perfect matchings in the linear chain of length 2 and its vertical edges.

2. (a) Draw all perfect matchings in the rectangular hexagonal system $R[3, 2]$.

    (b) Draw all perfect matchings in the parallelogram hexagonal system $P[3, 2]$.

In exercises 3 to 7 determine the number of perfect matchings contained in each of the following hexagonal systems.

3.

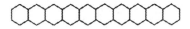

4.

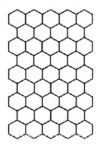

5.

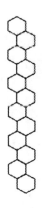

6.

7.

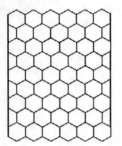

8. Determine if the given hexagonal systems contains a perfect matching. If so, determine exactly how many.

   (a)

   (b)

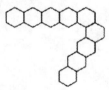

   (c)

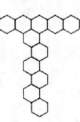

9. Use induction on the number of hexagons to show that every hexagonal system is bipartite.

In exercises 10, 11, 12 and 13 use the determinant formula to calculate the number of perfect matchings in the given hexagonal systems.

10.

11.

12.

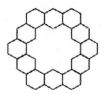

13.

14. (a) Draw all of the perfect matchings in a fibonaccene chain of length 4, 5 and 6.

    (b) Prove Theorem 3.

15. (a) Show that a linear chain with five hexagons contains six perfect matchings.

    (b) Show that the tubulene $T[5, 1]$ satisfies $\phi(T[5, 1]) = 4$.

    (c) Show that $\phi(T[h, 1]) = 4$, for all $h \geq 2$.

16. Prove Theorem 4.

17. Show that $\lim_{h \to \infty} \kappa(T[h, k]) = 0$ and $\lim_{k \to \infty} \kappa(T[h, k]) = \frac{1}{2h}$.

18. Find $\lim_{v \to \infty} \kappa(R[7, v])$.

Selected solutions

1. Notice that each vertical edge corresponds to a unique perfect matching.

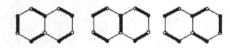

3. 11

4. 7,776

5. 377

6. 792

7. 1,024

8. (a) 0, (b) 29, (c) 0

9. Figure 1 shows that the result is true for a hexagonal system with one hexagon. For the inductive step remove a hexagon with edges on the boundary and consider cases defined by the number of boundary edges removed.

10. 105

11. 980

12. 130

14. (b) See [9].

15. (a) A linear chain with five hexagons contains six vertical edges, and hence, six perfect matchings. The tubulene $T[5, 1]$ contains the four perfect matchings given below (assume that the vertical edges on the ends are glued together.)

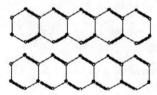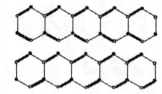

16. See [11].

17. The result follows from $\kappa(T[h, k]) = \frac{\log_2 2^{k+1}}{hk} = \frac{k+1}{hk}$.

18. $\frac{3}{13}$

Exploring Data Compression via Binary Trees[1]

Mark Daniel Ward
Purdue University

Summary

We investigate the Lempel-Ziv '77 data compression algorithm by considering an analogous algorithm for efficiently embedding strings in binary trees. This project includes a discussion of this comparison with two optional addenda on error correction and decompression, followed by exercises and solutions.

Notes for the instructor

Students in discrete mathematics often have a dual interest in computer science. This project succinctly combines these two areas. Data compression can be viewed as a discrete mathematics topic with many ramifications for computer scientists. Students who have completed one or two semesters of computer science (in particular, who are familiar with trees) may be eager to implement the algorithms discussed in C++, Java, or another object-oriented programming language.

The Lempel-Ziv '77 data compression algorithm was introduced in [1]. Analysis of the multiplicity matching parameter of suffix trees was presented in the present author's Ph.D. thesis; an abridged journal version with many more references to the literature can be found in [3]. An error correcting version of LZ'77 is outlined in [2].

Bibliography

[1] Lempel, A. and J. Ziv. "A universal algorithm for sequential data compression," *IEEE Transactions on Information Theory* 23 (1977) 337–343.

[2] Lonardi, S., W. Szpankowski, and M. D. Ward. "Error resilient LZ'77 data compression: algorithms, analysis, and experiments," *IEEE Transactions on Information Theory* 53 (2007) 1799–1813.

[3] Ward, M. D. and W. Szpankowski. "Analysis of the multiplicity matching parameter in suffix trees," *Discrete Mathematics and Theoretical Computer Science* AD (2005) 307–322, available online at http://www.dmtcs.org/proceedings/abstracts/dmAD0128.abs.html.

---

[1] The author thanks his Ph.D. advisor, Wojciech Szpankowski, for guidance and encouragement throughout graduate school at Purdue University (2001–2005).

Exploring Data Compression and Error Correction

We first discuss a standard method of embedding binary strings into a tree *retrieval* structure, often abbreviated as a "trie". A trie is a rather efficient tree structure. Every node has at most two children. When a node has no children, we refer to it as a leaf. A string is inserted at the minimal depth necessary; in other words, each string is inserted at the earliest depth that allows it to be distinguished from all other strings that currently reside in the trie. A "0" in a string corresponds to a left branch in the trie; a "1" corresponds to a right branch. Consider two strings:

$$S_1 = 0010111011$$
$$S_2 = 1001010101$$

When we embed S_1 and S_2 into a trie, we can immediately distinguish these two strings just by examining their first elements, which are 0 and 1, respectively.

If we embed another string, say $S_3 = 1110001010$, we note that the first bit of S_1 already distinguishes S_1 from S_3, but two bits of S_3 are necessary to make a distinction from S_2. After S_3 is embedded into the above trie, we have

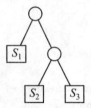

If S_4 has a long prefix in common with S_1, for instance, $S_4 = 1010111100$, then we have

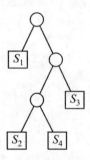

Currently four strings S_1, S_2, S_3, and S_4 have been inserted into the trie. Suppose that $S = S_i$ for some i. If S begins with 0, we know immediately that $S = S_1$. Otherwise, S begins with 1, so S is found to the right of the root of the trie. If S begins with 11, then $S = S_3$; otherwise, S begins with 10, and in this case, we proceed one level further into the trie to determine if $S = S_2$ (equivalently, S begins with 100) or $S = S_4$ (i.e., S begins with 101). This example illustrates an unambiguous way to efficiently embed binary strings into a binary tree. Many tree structures are found in the literature and have been extensively analyzed; the trie structure given above is one of the most popular and frequently analyzed.

For another example, consider the following eight strings:

$$S_1 = 0101010100$$
$$S_2 = 1001001001$$
$$S_3 = 1000000111$$
$$S_4 = 0111010000$$

$$S_5 = 1100101011$$
$$S_6 = 1010010110$$
$$S_7 = 0011001001$$
$$S_8 = 0100001000$$

The associated trie is

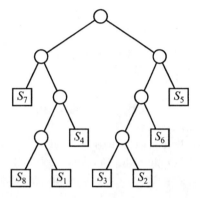

Next, we discuss a particular type of trie. In the examples above, the strings S_i were not dependent on each other; the strings were generated independently. We now consider strings that are generated as suffixes of a common word. We write X for a long binary string, for instance, $X = 01101001001110011010\ldots$. Then we write S_1 to denote X itself. We let S_2 denote X with only the first character removed (so S_2 is the second suffix of X). We let S_3 denote X with only the first and second characters removed (so S_3 is the third suffix of X). So we have

$$S_1 = 01101001001110011010\ldots$$
$$S_2 = 1101001001110011010\ldots$$
$$S_3 = 101001001110011010\ldots$$
$$S_4 = 01001001110011010\ldots$$
$$S_5 = 1001001110011010\ldots$$
$$S_6 = 001001110011010\ldots$$

Computer scientists often use the term "suffix tree" to denote the trie constructed from the suffixes of a word. So the suffix tree associated with $S_1, \ldots, S_6$ is constructed by the same method that the tries above were constructed. Therefore, the suffix tree built from the first six suffixes of X is

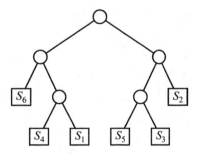

In the 1970s, Jacob Ziv and Abraham Lempel introduced a variety of sequential data compression algorithms. In particular, they presented LZ'77 and LZ'78, two schemes that still pervade every aspect of modern data compression theory and practice. Although thirty years have passed since these algorithms were introduced, these algorithms still remain popular and widely-used. Below, we describe a slightly simplified version of LZ'77.

Keeping in mind the suffix trees constructed above, we now discuss the motivating concept for LZ'77. Our goal is to compress a binary string, performing the compression one block at a time. To see the novel idea behind LZ'77, take a second look at the suffix tree associated with

$$X = 01101001001110011010\ldots$$

in the last example above. Notice that S_4 and S_1 both have several characters at the beginning in common. For this reason, S_1 and S_4 are placed close together in the suffix tree; both of these strings begin with the prefix 01. We can make the following three observations:

1. The string S_4 is very comparable to the string S_1.

2. Both of these strings have a common prefix of length 2.

3. The next character of S_4 after this common prefix is 0.

So we have described the first three characters of S_4 by noticing that there is a block of length 2 (namely, 01) in common with S_1, and then the third character of S_4 is 0. So $S_4 = 010\ldots$. This is the inspiration behind LZ'77. (As a side note, in step 3, we encode the next character explicitly so that the algorithm never gets "stuck." This reasoning for this extra character is crucial for the LZ'77 algorithm but admittedly will seem mysterious to those who are new to the algorithm. Nonetheless, we insist that step 3 above is crucial in LZ'77 for preventing the algorithm from getting stuck.)

Next we make a general statement of the algorithm; afterwards, we consider several illustrative examples.

We compress a binary string $X = X_1 X_2 X_3 \ldots$ in blocks, starting from the beginning. Consider the stage of the compression when the first n bits (namely, $X_1 X_2 \ldots X_n$) have already been compressed. Then $S_{n+1} = X_{n+1} X_{n+2} X_{n+3} \ldots$ remains uncompressed.

In order to compress the next block of S_{n+1}, we consider the various strings S_{i+1}, for $0 \leq i < n$. We aim to find the value of i such that S_{n+1} and S_{i+1} have the longest prefix in common. We find this desired value of i, and we write L to denote the length of the longest common prefix. Then we are able to compress the first $L + 1$ characters of S_{n+1} by noting that:

1. The string S_{n+1} is very comparable to S_{i+1}.

2. Both of these strings have a common prefix of length L.

3. The next character of S_{n+1} after this common prefix $X_{n+1} X_{n+2} \ldots X_{n+L}$ is exactly X_{n+L+1}.

Notice the similarity in these three steps to our earlier comparison of S_1 and S_4. Despite the notation, our comparison of $S_{n+1} = X_{n+1} X_{n+2} X_{n+3} \ldots$ to $S_{i+1} = X_{i+1} X_{i+2} X_{i+3}$ is just the same idea as found in our comparison of S_1 to S_4.

Finally, we note that the ingenuity behind Lempel and Ziv's LZ'77 scheme is that, rather than storing S_{n+1}, it is often more efficient to store the following three pieces of information:

1. a pointer to X_{i+1}, i.e., to the beginning of S_{i+1},

2. the length L of the common prefix,

3. the next character after this common prefix $X_{n+1} X_{n+2} \ldots X_{n+L}$, namely, X_{n+L+1}.

(In fact, this is a provably more efficient scheme when X is generated with a Markov dependency.) At any rate, it is most helpful to consider some examples.

Example 1. Suppose $X = 101101000110\ 00010100\ldots$ and $n = 12$ (the small space in X is just written for the clarity of the reader in this example; of course X does not contain any actual gaps or spaces). So we are considering the stage where the first twelve bits of X, namely $X_1 \ldots X_{12} = 101101000110$, have already been compressed. The rest of X, namely $S_{13} = X_{13} X_{14} \ldots = 00010100\ldots$, is still uncompressed.

In the notation given above, we have

$$S_{n+1} = X_{n+1}X_{n+2}\ldots = 00010100\ldots$$

We should compare this string to $S_{i+1} = X_{i+1}X_{i+2}\ldots$ for $i = 6$, because $S_7 = X_7X_8\ldots$ has a prefix of length $L = 4$ (namely, 0001) in common with S_{13}. Thus, the first five characters of S_{13} can be efficiently compressed by:

1. storing a pointer to X_7,

2. recording the length $L = 4$ of the common prefix between $S_7 = X_7X_8\ldots$ and $S_{13} = X_{13}X_{14}\ldots$,

3. noting that the next character after this common prefix $X_{13}X_{14}X_{15}X_{16}$ is $X_{17} = 0$.

By this method, $X_{13}X_{14}X_{15}X_{16}X_{17}$ efficiently gets compressed. The algorithm proceeds to the next step, with $X_1\ldots X_{17}$ already compressed; the remainder $S_{18} = X_{18}X_{19}\ldots$ still needs to be compressed.

Incidentally, when constructing the suffix tree built on $S_1, S_2, \ldots, S_{13}$, one notices that S_7 and S_{13} are siblings. This is not a coincidence; the general situation will be described at the end of the next example.

Example 2. Suppose $X = 0110110\ 01110100000100010\ldots$ and $n = 7$. So the first seven bits of X, namely $X_1\ldots X_7 = 0110110$, have already been compressed. The rest of X, namely $S_8 = X_8X_9\ldots = 01110100000100010\ldots$, has not yet been compressed.

We should compare S_8 to $S_{i+1} = X_{i+1}X_{i+2}\ldots$ for $i = 0$ or $i = 3$, because $S_1 = X_1X_2\ldots$ and $S_4 = X_4X_5\ldots$ each have a prefix of length $L = 3$ (namely, 011) in common with S_8. So, the first four characters of $S_{n+1} = X_{n+1}X_{n+2}\ldots$ can be efficiently compressed with the usual three steps:

1. storing a pointer to X_1 (or a pointer to X_4; either pointer is OK to use),

2. recording the length $L = 3$ of the common prefix between S_1 and S_8,

3. noting that the next character after this common prefix $X_8X_9X_{10}$ is $X_{11} = 1$.

By this method, $X_8X_9X_{10}X_{11}$ is efficiently compressed. The algorithm proceeds to the next step, with $X_1\ldots X_{11}$ already compressed, and $S_{12} = X_{12}X_{13}\ldots$ is still uncompressed.

Consider the suffix tree built on $S_1, S_2, \ldots, S_8$. We note that S_8's parent has S_1 and S_4 as descendants. This is true in general: the parent of the uncompressed string (in this case, the parent of S_8) will be an ancestor of the appropriate strings S_i used to perform the compression (in this case, S_1 and S_4). In fact, all descendants of the parent of the uncompressed string are appropriate candidates for performing the compression. Drawing the relevant suffix trees is extremely helpful for illustrating the situation.

Addendum about Error Correction

For those readers wishing to explore research in this vein: Whenever two or more values of i are valid (i.e., two or more strings S_i are available for use in the compression), then some error correction can be performed at this stage in the compression. In fact, this author dedicated his entire Ph.D. thesis to a precise study of how much error correction can be performed in such a situation. The number of S_i's available is called the multiplicity matching parameter; see the References section for more details.

At each stage of the compression, whenever there are multiple pointers to choose from, let M_n denote the number of valid values of i when compressing $X_{n+1} X_{n+2} \ldots$; in the example just given, $M_7 = 2$ because $i = 0$ or $i = 3$ could be used. Then we can embed $\lfloor \log_2 M_n \rfloor$ extra bits at no extra cost whatsoever. (Of course, for many n, we have $M_n = 1$, and no parity check can be performed.)

As an example, when $M_n = 2$, we can perform a parity check, choosing the first pointer if a "1" is required in our parity check, or selecting the second pointer if a "0" is required in our parity check. For example, the algorithm can check the parity of the bits compressed up until the present stage. If an error is detected, a warning message about the inconsistency could be given, or some error recovery could be performed. Such methods allow students a great degree of freedom and creativity, because there are many possible ways of implementing the algorithm in an object-oriented computer language.

Addendum about Decompression

Students who are quite eager to implement the LZ'77 algorithm with these extra features will certainly want to know how to write decompression algorithms too. Such files are uncompressed in an analogous way to the method by which they are compressed. The decompression is performed block-by-block, starting at the beginning of the file. In order to decompress the next block, we consider the stage at which n bits have already been decompressed, say $X_1 X_2 \ldots X_n$. In the compressed version of the file, we encounter a triple that describes a pointer, a length, and a next bit. The pointer to some X_{i+1} and the length L tell us to copy $X_{i+1} X_{i+2} \ldots X_{i+L}$ into the next L positions, namely $X_{n+1} X_{n+2} \ldots X_{n+L}$. The next bit is exactly X_{n+L+1}. By repeatedly decoding triples, we reproduce the original (uncompressed) file.

We note that students who are interested in both discrete mathematics and computer science will perhaps enjoy implementing the LZ'77 program in C++, Java, or another object-oriented programming language. Afterwards, they are able to measure the redundancy present in this aspect of LZ'77 by computing and tabulating M_n at each step during the execution of the program. They are also able to experiment with a variety of uses for the redundant bits that are available at most stages of the compression. Students can devise various error correction schemes or image embedding techniques. See the references for a published implementation of the LZ'77 algorithm with error correcting extensions.

Questions on Exploring Data Compression via Binary Trees

1. Construct the trie associated with the following eight strings:

$$S_1 = 0001000011$$
$$S_2 = 0011011111$$
$$S_3 = 0101010110$$
$$S_4 = 0010000101$$
$$S_5 = 1110000001$$
$$S_6 = 1001010001$$
$$S_7 = 0000010011$$
$$S_8 = 1100001001$$

2. Consider the string $X = 01011001111101001110\ldots$ Let $S_1, S_2, \ldots, S_8$ denote the first eight suffixes of X. Construct the suffix tree associated with these eight strings.

3. Consider the string $X = 110110001010\,10010011\ldots$ Suppose that the first twelve characters (namely, $X_1 \ldots X_{12} = 110110001010$) have been compressed and the remainder (i.e., $S_{13} = 10010011\ldots$) is uncompressed. What is the next block to be compressed in the LZ'77 algorithm?

4. Consider the string $X = 1111001101001101110110101\,001101101000011\ldots$ Suppose that the first twenty-five characters (namely, $X_1 \ldots X_{25} = 1111001101001101110110101$) have been compressed and the remainder (i.e., $S_{26} = 001101101000011\ldots$) is uncompressed. What is the next block to be compressed in the LZ'77 algorithm?

Solutions

1. The desired trie is

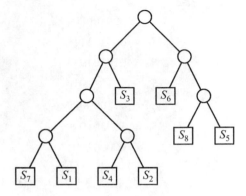

2. The eight strings are

$$S_1 = 010110011111101001110\ldots$$
$$S_2 = 10110011111101001110\ldots$$
$$S_3 = 0110011111101001110\ldots$$
$$S_4 = 110011111101001110\ldots$$
$$S_5 = 10011111101001110\ldots$$
$$S_6 = 0011111101001110\ldots$$
$$S_7 = 011111101001110\ldots$$
$$S_8 = 1111101001110\ldots$$

and the suffix tree associated with these eight strings is

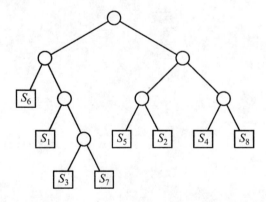

3. We see that $S_{13} = 10010011\ldots$ and $S_5 = 1000101010010011\ldots$ have a prefix of $L = 3$ characters in common (namely, $X_{13}X_{14}X_{15} = X_5X_6X_7 = 100$). So the next block has four characters, namely $X_{13}X_{14}X_{15} = 100$, followed by the next character, which is $X_{16} = 1$.

4. We see that $S_{26} = 001101101000011\ldots$ and $S_{11} = 0011011110110101001101101000011\ldots$ have a prefix of $L = 7$ characters in common (namely, $X_{26}\ldots X_{32} = X_{11}\ldots X_{17} = 0011011$). So the next block has eight characters: namely, $X_{26}\ldots X_{32} = 0011011$, followed by the next character, which is $X_{33} = 0$.

A Problem in Typography

Larry E. Thomas
Saint Peter's College

Summary

This project shows how a problem in computerized typesetting can be viewed as a problem in finding the shortest path from the beginning to the end of a weighted acyclic graph. Students are shown how to construct such a graph from the diagnostic information provided by Donald Knuth's typesetting program, TEX.

Notes for the instructor

This project uses a very special case of the general Knuth-Plass algorithm in order to make it manageable. The actual algorithm as implemented in [1] is general enough to handle unjustified text and mathematical expressions, but including all of the details would turn this into a course rather than a project. This treatment emphasizes the general idea of the algorithm, so the actual formulas for such things as badness and demerits are omitted. The details can be found in [2, 3, 4, 3, 6].

Bibliography

[1] Knuth, Donald E. *Computers & Typesetting, volume B. TEX: The Program*, Addison-Wesley, Reading, 1986.

[2] Knuth, Donald E. *The TEXbook*. Addison-Wesley, Reading, third edition, 1986.

[3] Knuth, Donald E. *Digital Typography*, CSLI Publications, Stanford, California, 1999.

[4] Knuth, Donald E. and Michael F. Plass. "Breaking paragraphs into lines," *Software—Practice and Experience* 11 (1981) 1119–1184.

[5] Salomon, David D. *The Advanced TEXbook*, Springer-Verlag, New York, 1995.

[6] Thomas, Larry E. "The Knuth-Plass Line-breaking Algorithm," to appear.

Worksheet on a Problem in Typography

1. The general problem

Mathematics is used behind the scenes in all sorts of places where you might not expect to find it. One of these places is in word processing programs. Such programs have to determine the best way to break text into lines for display on a printed page. In times past, when type was set by hand, a skilled typesetter made the decisions about where to break the lines and how much space to leave between words. These days almost all typesetting is done by computer programs, so there has to be some kind of algorithm for the program to use to make these decisions. In what follows we will look at one of the most celebrated of these algorithms, the *Knuth-Plass (KP) line-breaking algorithm*. The algorithm forms the basis of the method the typesetting program TEX uses to justify text and is therefore used by various other programs based on TEX. The algorithm is based on a very elegant use of graph theory.

To keep things manageable, we will consider a special case of the algorithm: the problem of setting a paragraph of fully justified text. This means that we want to arrange the lines of text on the page so that both the left and right margins are even. When we arrange the lines we have to answer two questions: Where should we break the text between two lines and how much space should we leave between the words.

As they stand these questions are not very well defined. In a typical text there are a lot of places where the line could be broken and there are a number of ways to put space between the words. What we are really looking for is the *best* way to arrange a paragraph of text on the page. But what do we mean by *best*? How can we tell if one way of formatting the paragraph is better than another?

In order to answer these questions, Knuth and Plass introduced *badness* and *demerits*. These two numerical quantities attempt to measure the aesthetic quality and readability of a paragraph. Generally speaking, *badness* measures whether the words on a line are too close together or too far apart, while *demerits* measure how well a given line fits in with the other lines in the paragraph. (For example, demerits are added if two adjacent lines of text end in discretionary hyphens.) Of course, there are precise formulas for badness and demerits, but an intuitive idea of these quantities should suffice for a general overview of the algorithm.

2. A specific problem

While working through the algorithm we will use a portion of the text that Knuth used in examples throughout [2]. The problem is to set the following text in a paragraph 2.5 inches wide:

```
Mr. Drofnats---or``R. J.,'' as he preferred to be called---was
happiest when he was at work typesetting beautiful documents.
```

The text is shown in a regular monospaced typewriter font. When we set the text in a proportional font using Knuth's TEX typesetting program this is the result:

> Mr. Drofnats—or "R. J.," as he pre-
> ferred to be called—was happiest when he
> was at work typesetting beautiful docu-
> ments.

Problem 1. Find out something about Donald E. Knuth. Use what you find to discern the hidden message in the name of his character Mr. Drofnats.

To achieve this result using the Knuth-Plass algorithm the program scans the text of the paragraph a character at a time and does all of the following:

- It chooses *feasible breakpoints* as places where it will consider breaking a line. Feasible breakpoints are chosen so that the badness of a line never exceeds a certain predefined *tolerance*[1]. So the feasible breakpoints are chosen so that the words in a line will not be too far apart or too close together.

---

[1] The tolerance can be changed by the user. The default value of the tolerance is 200 and its maximum value is 10,000, which is TEX's version of infinitely large. Lines with badness exceeding the tolerance will not be included in the final paragraph. When the tolerance is 10,000 the algorithm will tolerate any lines.

- For each feasible breakpoint it computes the demerits for the line ending at that breakpoint. The demerits measure how readable the line is as part of the entire paragraph. For example, the demerits of a line will increase if the line breaks at a hyphen.

- The program keeps a data structure in which the feasible breakpoints form an acyclic graph from the start of the paragraph to its end. The edges of the graph are weighted by the demerits.

- The program accumulates information about the graph so that it can find the path through the graph from the beginning to the end that minimizes the sum of the demerits along the path.

So we see that the algorithm structures the problem of formatting a paragraph as a problem of finding the shortest path from start to finish in a (weighted) graph. In practice the steps in the algorithm are carried out in a dynamic fashion. That is, as the paragraph is read the list of possible feasible breakpoints is continually adjusted, and data about the demerits is continually updated, so that the algorithm builds the shortest path as it goes along. The underlying method and data structures used are similar to those used in Dijkstra's algorithm. Unlike the algorithms used in many standard word processors, the Knuth-Plass algorithm considers the entire paragraph before it makes its final decision.

Problem 2. If there are n possible places to place line breaks (not necessarily feasible) in a paragraph, how many possible ways are there to divide the paragraph into lines? The short paragraph about Mr. Drofnats has about 30 possible places to place line breaks (including spaces between words and places to break words at hyphens). How many ways are there to divide the paragraph into lines?

Problem 3. If the *tolerance* for badness is increased will the number of feasible breakpoints in a paragraph increase or decrease?

It is time to return to Mr. Drofnats to see what the graph for his paragraph looks like. First, we can look at the feasible breakpoints as marked in the following version of the text. A superscript n indicates a feasible breakpoint that could be used to end line number n in the typeset version of the text.

```
Mr. Drofnats---or''R. J.,'' as he pre¹ferred to be called---was
happiest when² he² was at work typesetting beautiful doc³u³ments.⁴
```

This means that the only feasible breakpoint to end the first line is after `pre`. The second line could be broken after `when` or `he` without having too much or too little space between words in the second line. And the feasible choices for breaking the third line are after `doc` and `docu`. Which of these two possibilities is actually chosen depends on the choice of breakpoint in the previous line. This same information is better summarized in a graph as shown in Fig. 1. (The superscripts in the paragraph above correspond to the levels of the corresponding vertices in the graph. That is, all of the possible break-points for the first line are on the first level, those for the second line are on the second level, and so on.)

The vertices in the graph show the possible breakpoints; they are numbered from 0 to 6 for reference. The numbers on the edges are the demerits associated with breaking a line at a given point. For example, if the second line breaks after `he` and the third after `doc-`, line 3 has 103101 demerits. Notice that there is no edge from vertex 2 to vertex 5. This is because the badness of line 3 exceeds the tolerance for badness if the second line breaks after `when` and the third breaks after `u-`. In this case it means that the words in line 3 would be too close together if the breaks were at vertices 2 and 5.

The graph shows three possible ways to break the lines in the paragraph. The one we saw at the beginning of this section corresponds to following the path 0-1-3-5-6. This path, with total demerits 24,682, is the shortest path. Another path is 0-1-2-4-6, with total demerits 50,543, which would lead to the following setting of the text:

> Mr. Drofnats—or "R. J." as he pre-
> ferred to be called—was happiest when
> he was at work typesetting beautiful doc-
> uments.

The final possibility is 0-1-3-4-6, with total demerits 112,367, is the worst choice. It gives:

> Mr. Drofnats—or "R. J." as he pre-
> ferred to be called—was happiest when he
> was at work typesetting beautiful doc-
> uments.

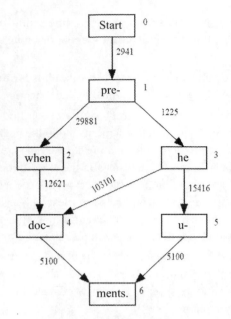

Figure 1. A Knuth-Plass graph.

If you compare these last two examples with the first one you will see that in one case the second line is a little too spread out and in the other the second line is a little too squeezed together.

3. Drawing the graph

As part of its diagnostic apparatus, TEX provides the information necessary to draw the graph for a paragraph. In our case it provided this rather cryptic information:

```
[]/cmr10/Mr. Drofnats---or``R. J.,'' as he pre-
@\discretionary via @@0 b=11 p=50 d=2941
@@1: line 1.2- t=2941 -> @@0
ferred to be called---was hap-pi-est when
@ via @@1 b=131 p=0 d=29881
@@2: line 2.0 t=32822 -> @@1
he
@ via @@1 b=25 p=0 d=1225
@@3: line 2.3 t=4166 -> @@1
was at work type-set-ting beau-ti-ful doc-
@\discretionary via @@2 b=1 p=50 d=12621
@\discretionary via @@3 b=291 p=50 d=103101
@@4: line 3.2- t=45443 -> @@2
u-
@\discretionary via @@3 b=44 p=50 d=15416
@@5: line 3.1- t=19582 -> @@3
ments.
@\par via @@4 b=0 p=-10000 d=5100
@\par via @@5 b=0 p=-10000 d=5100
@@6: line 4.2- t=24682 -> @@5
```

If you compare this listing with the graph in Figure 1 you can see that the listing shows a clever way to represent a graph without drawing any vertices or edges. The lines in the listing that begin with @@ show the feasible breakpoints, that is the vertices in the graph. Lines that begin with a single @ symbol mark the ways the following feasible breakpoints can be reached. For example, we can decode the following lines

```
ferred to be called---was hap-pi-est when
@ via @@1 b=131 p=0 d=29881
@@2: line 2.0 t=32822 -> @@1
he
```

to mean that the feasible breakpoint in vertex 2 is after when, that vertex 2 can be reached from vertex 1, and that the resulting line of text will have badness 131 and 29,881 demerits. This means that there is an edge in the graph from vertex 1 to vertex 2 and that edge will be labeled with 29,881 demerits. You are urged to go through the rest of the listing to see how 1 can be created.

By the way, the notation p=0 on the second line means that no *penalties* were assigned to the second line of text. Penalties measure the readability of the text and are used along with badness in the formula for demerits. Penalties are assigned if a break occurs at a hyphen, for example. Penalties are also assigned if a line with a lot of space between words is next to one with very little space between words. The full Knuth-Plass algorithm uses penalties to enhance the aesthetic appeal of the typeset text; they were omitted from this simplified version of the algorithm.

Problem 4. Setting the text

  Mrs. Johnson, the last living relative of Albert Johnson, was surprised when she inherited a small---but in the end adequate--- collection of etchings.

in lines 3 inches wide produces this listing

```
[]\tenrm Mrs. Johnson, the last living relative of Albert
@ via @@0 b=86 p=0 d=9216
@@1: line 1.3 t=9216 -> @@0
@secondpass
[]\tenrm Mrs. John-son, the last liv-ing rel-a-tive of Al-
@\discretionary via @@0 b=24 p=50 d=3656
@@1: line 1.1- t=3656 -> @@0
bert
@ via @@0 b=86 p=0 d=9216
@@2: line 1.3 t=9216 -> @@0
John-son, was sur-prised when she in-her-
@\discretionary via @@1 b=991 p=50 d=1014501
@@3: line 2.0- t=1018157 -> @@1
ited
@ via @@1 b=66 p=0 d=5776
@@4: line 2.1 t=9432 -> @@1
a
@ via @@1 b=0 p=0 d=100
@ via @@2 b=958 p=0 d=947024
@@5: line 2.2 t=3756 -> @@1
small---but in the end adequate---collection
@ via @@3 b=0 p=0 d=10100
@@6: line 3.2 t=1028257 -> @@3
of
@ via @@4 b=48 p=0 d=3364
@ via @@5 b=733 p=0 d=562049
@@7: line 3.1 t=12796 -> @@4
etch-ings.
```

```
@\par via @@6 b=0 p=-10000 d=100
@\par via @@7 b=0 p=-10000 d=100
@@8: line 4.2- t=12896 -> @@7
```

(a) Draw the graph corresponding to this listing. Be sure to put the feasible breakpoints in the vertices, number the vertices, and put the demerits on each edge.

(b) Find all of the paths from the start of the graph to the last feasible breakpoint and compute the total demerits for each path.

Problem 5.

(a) Try this little experiment. Draw two vertical parallel lines about 3 inches apart. For each of the paths you found through the graph in the previous problem, "typeset" the text by hand by printing the letters, breaking the lines at the indicated feasible breakpoints, and leaving the proper amount of space between each word.

(b) Compare the results of the work you just did. Can you see why the text from the shortest path through the graph is considered the best choice?

Solutions

1. Drofnats is Stanford, backwards.

2. There are 2^n possible ways to break the paragraph into lines. If $n = 30$, this amounts to 1,073,741,824 possibilities.

3. If the tolerance is increased, more possible breakpoints will qualify as feasible.

4. (a)

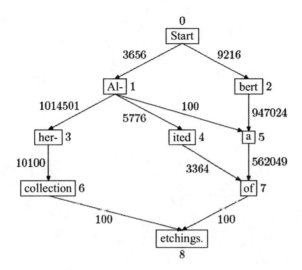

Figure 2. Problem 4(a).

(b) 0-1-3-6-8 has 1,028,357 demerits
    0-1-4-7-8 has 12,896 demerits
    0-1-5-7-8 has 565,905 demerits
    0-2-5-7-8 had 1,518,389 demerits

5. Path 0-1-3-6-8 has total demerits 1,028,357 and corresponds to the following way to set the text:

> Mrs. Johnson, the last living relative of Al-
> bert Johnson, was surprised when she inher-
> ited a small—but in the end adequate—collection
> of etchings.

Path 0-1-4-7-8 has total demerits 12,896 and produces:

> Mrs. Johnson, the last living relative of Al-
> bert Johnson, was surprised when she inherited
> a small—but in the end adequate—collection of
> etchings.

Path 0-1-5-7-8 has total demerits 565,905:

> Mrs. Johnson, the last living relative of Al-
> bert Johnson, was surprised when she inherited a
> small—but in the end adequate—collection of
> etchings.

Path 0-2-5-7-8 has total demerits 1,518,389, producing:

> Mrs. Johnson, the last living relative of Albert
> Johnson, was surprised when she inherited a
> small—but in the end adequate—collection of
> etchings.

Graph Complexity

Michael Orrison
Harvey Mudd College

Summary

This project asks students to define, motivate, and explore an objective measure of the complexity of a graph.

Notes for the instructor

This is an open-ended, capstone-like project designed to come at the end of the graph theory portion of a discrete mathematics course. I like this project because it provides a natural opportunity to discuss the general nature of mathematical research. Based on student feedback, I believe it is particularly successful because it gives students a genuine (and often unexpected) opportunity to express themselves in a mathematics course. Moreover, their insights have led, in some cases, to intriguing, worthwhile, and enjoyable independent student research projects.

At least two one-hour class periods should be set aside for this project. The first hour can be devoted to allowing students to work individually or in groups to devise and experiment with a measure of the complexity of a graph. The second hour can then be devoted to the presentation and discussion of the results. If you have the time and interest, this project may easily be turned into a more substantial research project (which I have done in a graph theory course) by asking the students, at each stage, to explore more fully their measure of complexity.

To get the students going, I think it is worth mentioning that mathematicians and computer scientists have defined the complexity of a graph in several different ways (and for several different purposes). After that, however, I suggest stepping aside to give the students an opportunity to grapple with idea of graph complexity on their own. That said, as the instructor, it may be helpful to know that the complexity of a graph has been defined, for example, as

- the number of spanning trees of a graph [1],

- the value of a certain formula involving the number of vertices, edges, and proper paths in a graph [2],

- the minimum number of additions, subtractions, and scalar multiplications required to multiply an adjacency matrix of a graph and an arbitrary vector [3], and

- the number of Boolean operations, based on a pre-determined set of Boolean operators (usually union and intersection), necessary to construct a graph from a fixed generating set of graphs [4].

Examples of straightforward measures of complexity that have come up in my class include

- the genus of a graph,

- the crossing number of a graph,

- the chromatic number of a graph,

- the sample variance of the degrees of the vertices of a graph,

- the average runtime of a fixed algorithm that takes a graph as input, and

- the minimum number of complete graphs required to construct a graph.

Note, for example, that if a student uses the chromatic number of a graph to measure complexity, then the graph on the left in Figure 1 is more complex than the graph on the right (6 vs. 3). On the other hand, if the sample variance of the degrees of the vertices is used to measure complexity, then the graph on the right in Figure 1 is more complex than the graph on the left (1.07 vs. 1.36).

Editor's note: This project was adapted by the author and his colleague Darryl Yong for the 2006 Imagine Math Day for high school students and their teachers; it was very successful.

Bibliography

[1] Biggs, N. *Algebraic graph theory*, Cambridge University Press, London, 1974, Cambridge Tracts in Mathematics, No. 67.

[2] Minoli, D. "Combinatorial graph complexity," *Atti Accad. Naz. Lincei Rend. Cl. Sci. Fis. Mat. Natur.* 59 (1975) 651–661.

[3] Neel, D. and M. Orrison, "The linear complexity of a graph," *Electronic Journal of Combinatorics* 13 (2006) R9, 19pp.

[4] Pudlák, P., V. Rödl, and P. Savický, "Graph complexity," *Acta Inform.* 25 (1988) 515–535.

Graph Complexity

Complexity is in the eye of the beholder. For example, suppose you are given the graphs in Figure 1. Which would you say is more complex? Do you think your classmates would agree with you? Can you imagine situations in which one, both, or neither would be considered complex?

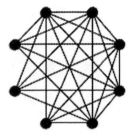 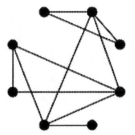

Figure 1. Which graph is more complex?

For this project, you will define a measure of the complexity of a graph. You are encouraged to use any of the graph theoretic concepts you have encountered so far. You may even want to define and use some new concepts along the way!

Project

1. Define a measure of the complexity of a graph. For simplicity, your measure should assign to each graph a nonnegative real number such that if a graph G is assigned a higher number than that of G', then we will say that G is "more complex" than G'.

2. Motivate your measure of the complexity of a graph. Give several examples of graphs and compute (or approximate) their complexity according to your measure.

3. Describe some properties of your measure. For example, according to your measure, is the complexity of a graph always greater than or equal to the complexity of any one of its subgraphs? State and attempt to prove at least one conjecture about your measure.

Part II

Historical Projects in
Discrete Mathematics
and
Computer Science

Introduction

Janet Barnett, Guram Bezhanishvili, Hing Leung,
Jerry Lodder, David Pengelley, Desh Ranjan

Introduction

A course in discrete mathematics is a relatively recent addition, within the last 30 or 40 years, to the modern American undergraduate curriculum, born out of a need to instruct computer science majors in algorithmic thought. The roots of discrete mathematics, however, are as old as mathematics itself, with the notion of counting a discrete operation, usually cited as the first mathematical development in ancient cultures. By contrast, a course in finite mathematics is sometimes presented as a fast-paced news reel of facts and formulae, often memorized by the students, with the text offering only passing mention of the motivating problems and original work that eventually found resolution in the modern concepts of induction, recursion and algorithm. This chapter focuses on the pedagogy of historical projects, which offer excerpts from original sources, place the material in context, and provide direction to the subject matter.

Each historical project is centered around a publication of mathematical significance, such as Blaise Pascal's "Treatise on the Arithmetical Triangle" [2, vol. 30] from the 1650s or Alan Turing's 1936 paper "On Computable Numbers with an Application to the Entscheidungsproblem" [3]. The projects are designed to introduce or provide supplementary material for topics in the curriculum, such as induction in a discrete mathematics course, or compilers and computability for a computer science course. Each project provides a discussion of the historical exigency of the piece, a few biographical comments about the author, excerpts from the original work, and a sequence of questions to help the student appreciate the source. The main pedagogical idea is to teach and learn certain course topics from the primary historical source, thus recovering motivation for studying the material.

For use in the classroom, allow several weeks per project with one or two projects per course. The time spent working on a project may vary from two to six weeks, depending on the depth and breadth in which the material is covered. Some instructors may wish not to assign all parts of a project, while others may wish to add further questions. The source files for the projects may be downloaded from the web resource [1] and edited. Before assignment, the instructor should work through all details of a project, not only for acquaintance with the historical source, but to determine which questions best suit the course material. Pay particular attention to the language of the historical author, and its mathematical content. Be prepared to examine a quote from the source in class and discuss its meaning, either verbally or with present-day formulas. Student solutions may be collected in installments or only once upon completing the project. In either event, monitor student progress on what for them is a lengthy assignment. Each project should count for a significant portion of the course grade (about 20%) and may take the place of an in-class examination. Begin early in the course with a discussion of the relevance of the historical piece, its relation to the course curriculum, and how modern textbook techniques owe their development to problems often posed centuries earlier. Each project offers a "Notes for the instructor" for more detailed information about the level, content and implementation of the project.

The topics covered by the projects include naive set theory, mathematical induction, binary arithmetic, computability, graph theory, and the combinatorics of the Catalan numbers. They range in level from beginning undergraduate courses in discrete mathematics to advanced undergraduate courses in logic, graph theory, and computer science. Most are independent of each other, although a few projects build on the same source or present closely related topics via different sources. Each historical source is one which either solved an outstanding problem, inaugurated a mathematical technique, or offered a novel point of view on existing material.

The next page is a list of instructions to students concerning "How to Work on Your Project." A good student solution should reflect the benefits afforded by close attention to these points. The instructor may wish to photocopy the following instructions for distribution to the entire class.

After completion of a course using historical projects, students write the following about the benefits of history:

"See how the concepts developed and understand the process."

"Learn the roots of what you've come to believe in."

"Appropriate question building."

"Helps with English-math conversion."

"It leads me to my own discoveries."

Acknowledgment

The development of curricular materials for discrete mathematics and computer science has been partially supported by the National Science Foundation's Course, Curriculum and Laboratory Improvement Program under grant DUE-0231113, for which the authors are most appreciative. Any opinions, findings, and conclusions or recommendations expressed in this material are those of the authors and do not necessarily reflect the views of the National Science Foundation.

Bibliography

[1] Bezhanishvili, G., Leung, H., Lodder, J., Pengelley, D., Ranjan, D., "Teaching Discrete Mathematics via Primary Historical Sources," `www.math.nmsu.edu/hist_projects/`

[2] Pascal, B., "Treatise on the Arithmetical Triangle," in *Great Books of the Western World*, Mortimer Adler (editor), Encyclopædia Britannica, Inc., Chicago, 1991.

[3] Turing, A. M., "On Computable Numbers with an Application to the Entscheidungsproblem," *Proceedings of the London Mathematical Society* **42** (1936), 230–265.

How to Work on Your Project:
What is Expected

Here are some words of advice and encouragement for completing the historical project. This is a significant course requirement, in which you will examine a key historical episode in the development of a mathematical topic. The historical presentation is often more verbal than modern textbook formulations, which carries the advantage that no technical knowledge is assumed a priori. On the other hand, you must carefully read the historical author's words and be prepared to experiment with your own calculations to verify or amplify the historical source. Here are some specific instructions.

1. Start today. This is a major assignment, requiring more work than most homework problems from a textbook. Work on the project every day before it is due.

2. Read the entire project to see what is involved. From what time period is the historical piece? What modern topics have evolved from the source?

3. Next read the project very carefully and make a list of any unfamiliar words or concepts. Do you understand the language of the historical author? If not, ask your instructor.

4. Feel free to seek additional historical or mathematical information about the source from either the library or the Internet. Always question the validity of Internet sources.

5. Begin answering questions in the project. Use complete sentences along with equations, tables, or diagrams to support your work. Some questions will be rather simple, while others require detailed solutions. Identify the difficult questions early and seek help from your instructor if you don't know how to proceed.

6. Seek feedback from your instructor even if you feel that you have answered the questions correctly, since you may have misinterpreted a passage.

7. The final solution to the project should be a written paper in which you address all the issues raised in the project. Use complete sentences along with modern equations to support your claims. Discuss any connections from the historical source with present-day techniques that you may have encountered. How does the historical source differ from the textbook presentation?

Binary Arithmetic:
From Leibniz to von Neumann

Jerry M. Lodder
New Mexico State University

The Era of Leibniz

Gottfried Wilhelm Leibniz (1646–1716) is often described as the last universalist, having contributed to virtually all fields of scholarly interest of his time, including law, history, theology, politics, engineering, geology, physics, and perhaps most importantly, philosophy, mathematics and logic [1, 9, 11]. The young Leibniz began to teach himself Latin at the age of 8, and Greek a few years later, in order to read classics not written in his native language, German. Later in life, he wrote:

> Before I reached the school-class in which logic was taught, I was deep into the historians and poets, for I began to read the historians almost as soon as I was able to read at all, and I found great pleasure and ease in verse. But as soon as I began to learn logic, I was greatly excited by the division and order in it. I immediately noticed, to the extent that a boy of 13 could, that there must be a great deal in it [5, p. 516].

His study of logic and intellectual quest for order continued throughout his life and became a basic principle to his method of inquiry. At the age of 20 he published *Dissertatio de arte combinatoria* (Dissertation on the Art of Combinatorics) in which he sought a *characteristica generalis* (general characteristic) or a *lingua generalis* (general language) that would serve as a universal symbolic language and reduce all debate to calculation. Leibniz maintained:

> If controversies were to arise, there would be no more need of disputation between two philosophers than between two accountants. For it would suffice to take their pencils in their hands, to sit down to their slates, and to say to each other: ... Let us calculate [14, p. 170].

The Leipzig-born scholar traveled extensively with diplomatic visits to Paris and London, and extended trips to Austria and Italy to research the history of the House of Brunswick. The years 1672–1676 were spent in Paris in an attempt to persuade King Louis XIV (1638–1715) not to invade Germany, but Egypt instead, although Leibniz was never granted an audience with the French king. During this time in Paris, however, the young German became acquainted with several of the leading philosophers of the day, acquired access to unpublished manuscripts of Blaise Pascal (1623–1662) and René Descartes (1596–1650), and met the renowned Christiaan Huygens (1629–1695), from whom he learned much about contemporary mathematics. During these years he laid the foundation of his calculus and the core of what would become his philosophical legacy.

Leibniz's invention of the differential and integral calculus is, in part, a product of his search for a universal language. Questions in the calculus can be reduced to the rules of calculation which the symbols for derivative, d, and integral, $\int$, satisfy. Sadly a priority dispute with Isaac Newton (1642–1727) over the invention of calculus cast a pall over Leibniz's later years. Moreover, he became a subject of ridicule with his philosophy that this is the best of all possible worlds bitterly satirized in Voltaire's (1694–1778) play *Candide*.

Let's turn to the universal genius's 1703 publication "Explication de l'arithmétique binaire, qui se sert des seuls caractères 0 et 1, avec des remarques sur son utilité, et sur ce qu'elle donne le sens des anciennes figures Chinoises de Fohy" [6, p. 223–227] (An Explanation of Binary Arithmetic Using only the Characters 0 and 1, with Remarks about its Utility and the Meaning it Gives to the Ancient Chinese Figures of Fuxi), which originally appeared in the journal *Memoires de l'Académie Royale des Sciences* [13]. Here again, with the reduction of arithmetic to expressions involving only zeroes and ones, we see a possible candidate for Leibniz's *characteristica generalis*. Of binary numeration, he writes "it permits new discoveries [in] . . . arithmetic . . . in geometry, because when the numbers are reduced to the simplest principles, like 0 and 1, a wonderful order appears everywhere." Concerning the binary calculations themselves " . . . these operations are so easy that we shall never have to guess or apply trial and error, as we must do in ordinary division. Nor do we need to learn anything by rote." Certainly Leibniz was not the first to experiment with binary numbers or the general concept of a number base [7]. However, with base 2 numeration, Leibniz witnessed the confluence of several intellectual ideas of his world view, not just the *characteristica generalis*, but also theological and mystical ideas of order, harmony and creation [16]. Additionally his 1703 paper [13] contains a striking application of binary numeration to the ancient Chinese text of divination, the *Yijing* (*I-Ching* or *Book of Changes*).

Early in life Leibniz developed an interest in China, corresponded with Catholic missionaries there, and wrote on questions of theology concerning the Chinese. Surprisingly he believed that he had found an historical precedent for his binary arithmetic in the ancient Chinese lineations or 64 hexagrams of the *Yijing*. This, he thought, might be the origin of a universal symbolic language. A hexagram consists of six lines atop one another, each of which is either solid or broken, forming a total of 64 possibilities, while a grouping of only three such lines is called a trigram (or cova). Leibniz lists the eight possible trigrams in his exposition on binary arithmetic, juxtaposed with their binary equivalents.

He had been in possession of his ideas concerning binary arithmetic well before his 1703 publication. In 1679 Leibniz outlined plans for a binary digital calculating machine, and in 1697 he sent a congratulatory birthday letter to his patron Duke Rudolph August of Brunswick, in which he discusses binary numeration and the related creation theme with 0 denoting nothing and 1 denoting God [16]. Furthermore, Leibniz sent the French Jesuit Joachim Bouvet (1656–1730) an account of his binary system while Bouvet was working in China. The Jesuits are an educational order of Catholic priests, who, while in China, sought the conversion of the Chinese to Christianity, hopefully by the identification of an ancient theology common to both religions. Bouvet began a study of the *Yijing*, viewing this text as the possible missing link between the two religions [16]. It was from this Jesuit priest that Leibniz received the hexagrams attributed to Fuxi, the mythical first Emperor of China and legendary inventor of Chinese writing. In actuality, the hexagrams are derived from the philosopher Shao Yong's (1011–1077) *Huangji jingshi shu* (Book of Sublime Principle Which Governs All Things Within the World). Shortly after receiving Bouvet's letter containing the hexagrams and Bouvet's identification of a relation between them and binary numeration, Leibniz submitted for publication his 1703 paper "Explanation of Binary Arithmetic" [3, p. 44].

Here is the first question of the student module:

1. Concerning the utility of the binary system, Leibniz cites an application to weighing masses. Suppose that a two-pan balance is used for weighing stones. A stone of unknown (integral) weight is placed on the left pan, while standard weights are placed only on the right pan until both sides balance. For example, if standard weights of 1, 4, 6 are used, then a stone of weight 7 on the left pan would balance the standard weights 1 and 6 on the right. Two standard weights with the same value cannot be used. Leibniz claims that all stones of integral weight between 1 and 15 inclusive can be weighed with just four standard weights. What are these four standard weights? Explain how each stone of weight between 1 and 15 inclusive can be weighed with the four standard weights. Make a table with one column for each of the four standard weights and another column for the stone of unknown weight. For each of the 15 stones, place an "X" in the columns for the standard weights used to weigh the stone.

Let's now read from an "Explanation of Binary Arithmetic," using a modified version of the Ching-Oxtoby translation [3, p. 81–86].

An Explanation of Binary Arithmetic
Using only the Characters 0 and 1, with Remarks
about its Utility and the Meaning it Gives to the
Ancient Chinese Figures of Fuxi

By G.W. Leibniz

Ordinary arithmetical calculations are performed according to a progression of tens. We use ten characters, which are 0, 1, 2, 3, 4, 5, 6, 7, 8, 9, that signify zero, one and the following numbers up to nine, inclusive. After reaching ten, we begin again, writing *ten* with 10, and ten times ten or *one hundred* with 100, and ten times one hundred or *one thousand* with 1000, and ten times one thousand with 10000, and so on.

2. Write the numbers 1, 10, 100, 1000 and 10000 as powers of ten. Express your answer in complete sentences or with equations. What pattern do you notice in the exponents?

But instead of the progression by tens, I have already used for several years the simplest of all progressions, that by twos, having found that this contributes to the perfection of the science of numbers. Thus I use no characters other than 0 and 1, and then, reaching two, I begin again. This is why *two* is written here as 10, and two times two or *four* as 100, and two times four or *eight* as 1000, and two times eight or *sixteen* as 1000, and so on.

3. Write the numbers 1, 2, 4, 8 and 16 as powers of two. Express your answer in complete sentences or with equations. What pattern do you notice in the exponents? How do the exponents compare with those in question 2? How does the progression by twos compare with the standard weights in question 1?

Here is the *Table of Numbers* according to this pattern, which we can continue as far as we wish.

4. Compare the entries from 1 to 15 in Leibniz's "Table of Numbers" on the following page with the table for weighing stones that you constructed previously in question 1. Today a number written only with the characters 0 and 1 according to Leibniz's "progression of twos" is said to be written in binary notation, or base 2. Find the binary equivalents of the (base 10) numbers

$$34, \quad 64, \quad 100, \quad 1015.$$

Be sure to explain your work.

At a glance we see the reason for the *famous property of the double geometric progression* in whole numbers, which states that given only one of these numbers in each degree, we can form all other whole numbers below the double of the highest degree. Since, as we would say, for example, 111 or 7 is the sum of four, two and one, and that 1101 or 13 is the sum of eight, four and one.

1	0	0	4		1	0	0	0	8
	1	0	2			1	0	0	4
		1	1					1	1
1	1	1	7		1	1	0	1	13

This property is useful to investigators for weighing all kinds of masses with just a few weights or could be used in monetary systems to provide a range of change with just a few coins.

Table of Numbers

o	o	o	o	o	0	0
o	o	o	o	o	1	1
o	o	o	o	1	0	2
o	o	o	o	1	1	3
o	o	o	1	0	0	4
o	o	o	1	0	1	5
o	o	o	1	1	0	6
o	o	o	1	1	1	7
o	o	1	0	0	0	8
o	o	1	0	0	1	9
o	o	1	0	1	0	10
o	o	1	0	1	1	11
o	o	1	1	0	0	12
o	o	1	1	0	1	13
o	o	1	1	1	0	14
o	o	1	1	1	1	15
o	1	0	0	0	0	16
o	1	0	0	0	1	17
o	1	0	0	1	0	18
o	1	0	0	1	1	19
o	1	0	1	0	0	20
o	1	0	1	0	1	21
o	1	0	1	1	0	22
o	1	0	1	1	1	23
o	1	1	0	0	0	24
o	1	1	0	0	1	25
o	1	1	0	1	0	26
o	1	1	0	1	1	27
o	1	1	1	0	0	28
o	1	1	1	0	1	29
o	1	1	1	1	0	30
o	1	1	1	1	1	31
1	0	0	0	0	0	32

etc.

5. Using modern notation, Leibniz's "double geometric progression" or "progression by twos" would be written 2^0, $2^1, 2^2, 2^3, \ldots, 2^n$. Guess a simple formula for

$$2^0 + 2^1 + 2^2 + 2^3 + \cdots + 2^n$$

based on Leibniz's verbal description "that given only one of these numbers in each degree," we should be able to achieve all whole numbers "below the double of the highest degree." Prove that your guess is correct using only algebra (addition and multiplication). Hint: Multiply $(2^0 + 2^1 + 2^2 + 2^3 + \cdots + 2^n)$ by 1, where 1 is written as $2 - 1$.

Leibniz continues:

This expression of numbers, once established, facilitates all kinds of operations.

For example, addition (1)

```
  1  1  0 ‖ 6              1  0  1 ‖ 5           1  1  1  0 ‖ 14
  1  1  1 ‖ 7           1  0  1  1 ‖ 11       1  0  0  0  1 ‖ 17
  .  .                  .  .  .  .           
 ───────────           ──────────────        ──────────────────
  1  1  0  1 ‖ 13       1  0  0  0  0 ‖ 16    1  1  1  1  1 ‖ 31
```

For subtraction

```
  1  1  0  1 ‖ 13    1  0  0  0  0 ‖ 16    1  1  1  1  1 ‖ 31
     1  1  1 ‖ 7        1  0  1  1 ‖ 11    1  0  0  0  1 ‖ 17
 ──────────────      ─────────────────    ─────────────────
     1  1  0 ‖ 6          1  0  1 ‖ 5       1  1  1  0 ‖ 14
```

For multiplication (2)

```
       1  1 ‖ 3           1  0  1 ‖ 5           1  0  1 ‖ 5
       1  1 ‖ 3              1  1 ‖ 3           1  0  1 ‖ 5
    ──────────             ──────────          ──────────
       1  1                 1  0  1               1  0  1
    1  1                 1  0  1             1  0  1  0
    .                      .                   .
 ──────────────       ──────────────      ──────────────────
 1  0  0  1 ‖ 9       1  1  1  1 ‖ 15      1  1  0  0  1 ‖ 25
```

For division

```
 15 ‖ 1̸  1̸  1  1         1  0  1 ‖ 5
  3 ‖ 1̸  1̸  1̸  1
    ‖    1̸  1
```

All these operations are so easy that we shall never have to guess or apply trial and error, as we must do in ordinary division. Nor do we need to learn anything by rote here, as must be done in ordinary calculation, where, for example, it is necessary to know that 6 and 7 taken together makes 13, and that 5 multiplied by 3 gives 15, following the so-called Pythagorean table[1] that one times one is one. But here everything is found and proven from the source, just as we see in the preceding examples under the signs (1) and (2).

6. Using your knowledge of base 10 addition, explain the examples of base 2 addition given above by Leibniz. What is the likely meaning of the dot Leibniz includes in certain columns for addition? Using binary arithmetic, compute $1101 + 1110$, without converting these numbers to base 10. Explain the examples for binary subtraction, multiplication and division above. Keep in mind that these should be base 2 analogues of base 10 procedures. Since Leibniz's example for division may be incomplete by today's standards, you may wish to supplement his work with additional steps, indicating clearly what multiples of 3 are subtracted from 15 in base 2. Finally, using binary arithmetic, compute the following.

$$11010 - 1101, \quad (1101) \cdot (11), \quad 1101 \div 101 \,.$$

Be sure to explain your work. For the division problem, you may state what the remainder is in terms of a binary whole number, without writing it as a fraction.

However, I am not at all recommending this manner of counting as a replacement for the ordinary practice of tens. For aside from the fact that we are accustomed to this, there is no need to learn what we have already memorized; the practice of tens is shorter, the numbers not as long. If we were accustomed to proceed by twelves or by sixteens, there would be even more of an

---

[1] This is likely a reference to the multiplication table.

advantage. As compensation for its length, however, calculation by twos, that is by 0 and by 1, is most basic for science; it permits new discoveries which become useful even in the practice of arithmetic, and especially in geometry, because when the numbers are reduced to the simplest principles, like 0 and 1, a wonderful order appears everywhere. For example, even in the Table of Numbers, we see in each column those periods which always reappear. In the first column it is 01, in the second 0011, in the third 00001111, in the fourth 0000000011111111, and so on. Small zeroes are put into the table to fill the void at the beginning of the column, and to mark these periods better. Lines are also traced in the table indicating that what these lines enclose always reoccurs below them. The square numbers, cubes and other powers, as well as the triangular numbers,[2] pyramidal numbers,[3] and other figurate numbers, also have similar periods, so that one can immediately write the tables without even calculating. A certain tedium at the beginning, which later serves to spare us calculation and to allow us to go by rule infinitely far, is extremely advantageous.

What is surprising in this calculation is that this arithmetic of 0 and 1 contains the mystery of lines of an ancient king and philosopher named Fuxi, who is believed to have lived more than four thousand years ago and whom the Chinese regard as the founder of their empire and of their sciences. There are several figures of lines that are attributed to him; they all go back to this arithmetic. But it is enough to place here the so-called figures of the eight Cova [trigrams], which are basic, and to add to these an explanation which is manifest, so that it is understood that a whole line — signifies unity or one, and that a broken line – – signifies zero or 0.

0	⊢	0	⊢	0	⊢	0	⊢
0	0	⊢	⊢	0	0	⊢	⊢
0	0	0	0	⊢	⊢	⊢	⊢
0	1	10	11	100	101	110	111
0	1	2	3	4	5	6	7

... [S]carcely two years ago I sent to the Reverend Father Bouvet, the famous French Jesuit living in Peking, my manner of counting by 0 and 1, and it was all he needed to recognize that this holds the key to Fuxi's[4] figures. So he wrote to me on November 14, 1701, sending me the great figure of this princely philosopher which goes to 64. ...

Although striking to Leibniz, the link between the 64 hexagrams of the *Yijing* and binary numeration appears today as only an intellectual curiosity. The base 2 system did not provide a common origin to Christianity and Confucianism, as Leibniz and Bouvet had sought.

The Electronic Age

John von Neumann (1903–1957) was a leading mathematician, physicist and engineer of the twentieth century, having contributed significantly to the foundations of quantum mechanics, the development of the atomic bomb, and the logical structure of the electronic digital computer [8, 10]. Born in Budapest Hungary, the young von Neumann showed a gift for mathematics, received a doctorate in the subject from the University of Budapest and a degree in chemical engineering from the *Eidgenössische Technische Hochschule* (Swiss Federal Polytechnic) in Zurich. He met the renowned David Hilbert (1862–1943) on a visit to Göttingen in 1926, after which he was offered the position of a

---

[2]The sequence 1, 3, 6, 10, 15, . . . , giving the number of dots in certain triangles [12, p. 49] forms the triangular numbers.

[3]The sequence 1, 4, 10, 20, 35, . . . , giving the number of dots in certain pyramids [19, p. 76] forms the pyramidal numbers.

[4]The mythical first Emperor of China.

Privatdozent (an un-salaried lecturer) at the University of Berlin and then at the University of Hamburg. In 1930 he visited the United States, accepting a salaried lectureship at Princeton University, a move which would shape the rest of his life.

Becoming a Professor of Mathematics at the prestigious Institute for Advanced Study (Princeton, New Jersey) in 1933, von Neumann was able to devote his time to the study of analysis, continuous geometry, fluid dynamics, wave propagation and differential equations. In 1943 he became a member of the Los Alamos Laboratory and helped develop the atomic bomb. The particular problem he faced, the implosion problem, was how to produce an extremely fast reaction in a small amount of the uranium isotope U^{235} in order to cause a great amount of energy to be released. In conjunction with Seth Neddermeyer, Edward Teller and James Tuck, this problem was solved with a high explosive lens designed to produce a spherical shock wave to cause the implosion necessary to detonate the bomb. Von Neumann's strength was his ability to model theoretical phenomena mathematically and solve the resulting equations numerically [8, p. 181], which required adroit skills in calculation.

Meanwhile in 1941 John William Mauchly (1907–1980), as a newly appointed Assistant Professor at the University of Pennsylvania's Moore School of Electrical Engineering, began discussions with graduate student John Presper Eckert (1919–1995) and others about the possibility of an electronic digital computing device that would be faster and more accurate than any existing machine, designed in part to meet the computing needs of the Ballistics Research Laboratory (BRL) of the Army Ordnance Department in Aberdeen, Maryland. With the help of mathematician and First Lieutenant Herman Goldstine (1913–), Mauchly's proposal for a high-speed vacuum-tube computer received funding from the BRL in 1943. The device was dubbed the Electronic Numerical Integrator and Computer (ENIAC). Tested in late 1945, and unveiled in 1946, the ENIAC was a behemoth containing 18,000 vacuum tubes and requiring 1,800 square feet of floor space for the computer alone [15, p. 133]. Arithmetic on the ENIAC was performed using the base 10 decimal system, requiring the ability to store ten different values for each digit of a numerical quantity. The multiplication table for all digits between zero and nine was also stored on the machine. The ENIAC was not programmable in the modern sense of a coded program, and contained no sub-unit similar to a present-day compiler. To alter its function, i.e., to implement a different numerical algorithm, external switches and cables had to be repositioned. Designs for a more robust machine may have been in place as the ENIAC went into production, but the rush to meet the needs of the war effort took precedence.

By serendipity, in the summer of 1944 Goldstine met von Neumann at a railway station in Aberdeen, both working on separate highly classified projects. Goldstine writes: "Prior to that time I had never met this great mathematician, but I knew much about him of course and had heard him lecture on several occasions" [8, p. 182]. After a discussion of the computing power of the ENIAC, von Neumann became keenly interested in this machine, and in late 1945 tested it on computations needed for the design of the hydrogen bomb. Von Neumann quickly became involved with the logical structure of the next generation of computing machinery, the Electronic Discrete Variable Automatic Computer (EDVAC), designed around the "stored program" concept. The instructions of an algorithm could be stored electronically on the EDVAC and then executed in sequential order. In this way, von Neumann had outlined a "universal computing machine" in the sense of Alan Turing (1912–1954), with the universal character referring to the machine's ability to execute any algorithmic procedure that could be reduced to simple logical steps. Turing first introduced a logical description of his universal computing machine in 1936 [17] as the solution to a problem posed by David Hilbert. Von Neumann, who had studied logic early in his career, was certainly aware of Turing's work, and in 1938 had offered Turing an assistantship at the Institute for Advanced Study [8, p. 271]. In 1945 von Neumann issued his white paper "First Draft of a Report on the EDVAC" [18] under the auspices of the University of Pennsylvania and the United States Army Ordnance Department. Although this draft was never revised, the ideas therein soon became known as von Neumann architecture in computer design. Let's read a few excerpts from this paper [18] related to binary arithmetic.

First Draft of a Report on the EDVAC

2.2 First: Since the device is primarily a computer, it will have to perform the elementary operations of arithmetic most frequently. There are addition, subtraction, multiplication and division: $+$, $-$, $\times$, $\div$. It is therefore reasonable that it should contain specialized organs for just these operations. At any rate a *central arithmetical* part of the device will probably have to exist, and this constitutes *the first specific part: CA.*

4.3 It is clear that a very high speed computing device should ideally have vacuum tube elements. Vacuum tube aggregates like counters and scalers have been used and found reliable at reaction times (synaptic delays) as short as a microsecond ($= 10^{-6}$ seconds).

5.1 Let us now consider certain functions of the first specific part: the central arithmetical part CA.

The element in the sense of 4.3, the vacuum tube used as a current valve or *gate*, is an all-or-none device, or at least it approximates one: According to whether the grid bias is above or below cut-off; it will pass current or not. It is true that it needs definite potentials on all its electrodes in order to maintain either state, but there are combinations of vacuum tubes which have perfect equilibria: Several states in each of which the combination can exist indefinitely, without any outside support, while appropriate outside stimuli (electric pulses) will transfer it from one equilibrium into another. These are the so called *trigger circuits*, the basic one having two equilibria. The trigger circuits with more than two equilibria are disproportionately more involved.

Thus, whether the tubes are used as gates or as triggers, the all-or-none, two equilibrium arrangements are the simplest ones. Since these tube arrangements are to handle numbers by means of their digits, it is natural to use a system of arithmetic in which the digits are also two valued. This suggests the use of the binary system.

5.2 A consistent use of the binary system is also likely to simplify the operations of multiplication and division considerably. Specifically it does away with the decimal multiplication table. In other words: Binary arithmetic has a simpler and more one-piece logical structure than any other, particularly than the decimal[5] one.

7. Let a and b denote binary variables with one digit each. Using only the logical connectives $\wedge$ (and), $\vee$ (or) and $\sim$ (not), find a logical expression which gives the digit in the ones place (the right-hand digit) of $a + b$. Find a logical expression which gives the digit in the twos place (the left-hand digit) of $a + b$. Explain your answer.

Extra Credit A: Find a pattern in the binary representation of the square numbers $1, 4, 9, 16, 25, \ldots$. Leibniz claims to have found such patterns.

Extra Credit B: If in question 1 above, standard weights can be placed on both sides of the balance, what four standard weights should be used in order to weigh all stones of integral weight between 1 and 40 inclusive?

---

[5]base 10

Notes for the instructor

The project is ideal for a beginning-level discrete mathematics or computer programming course. It could be assigned independently or in conjunction with the project "Turing Machines and Binary Addition," available from the web resource [2]. After completion of the project, the instructor may wish to lead the class in the discovery of divisibility properties of integers that become apparent after studying their expression in binary form. The pattern of zeroes in the base 2 expansion of the perfect squares leads to the conjecture that $8|(n^2 - 1)$, n an odd integer, where the vertical bar denotes "divides." In this way, the class is following in Leibniz's footsteps when he claims that "calculation by twos ... is most basic for science; it permits new discoveries, ..., because when the numbers are reduced to the simplest principles, like 0 and 1, a wonderful order appears everywhere" [13].

Bibliography

[1] Aiton, E. J., *Leibniz: A Biography*, Adam Hilger, Boston, 1985.

[2] Bezhanishvili, G., Leung, H., Lodder, J., Pengelley, D., Ranjan, D., "Teaching Discrete Mathematics via Primary Historical Sources," www.math.nmsu.edu/hist_projects/

[3] Ching, J., Oxtoby, W. G., *Moral Enlightenment: Leibniz and Wolff on China*, Steyler Verlag, Nettetal, 1992.

[4] Davis, M., *The Undecidable. Basic Papers on Undecidable Propositions, Unsolvable Problems and Computable Functions*, Martin Davis (editor), Raven Press, Hewlett, N.Y., 1965.

[5] Gerhardt, C. I., (editor) *Die Philosophischen Schriften von Leibniz*, vol. VII, Olms, Hildesheim, 1965.

[6] Gerhardt, C. I., (editor) *G. W. Leibniz Mathematische Schriften*, Vol. VII, Olms, Hildesheim, 1962.

[7] Glaser, A., *History of Binary and Other Nondecimal Numeration*, Anton Glaser, Southampton, PA, 1971.

[8] Goldstine, H. H., *The Computer from Pascal to von Neumann*, Princeton University Press, Princeton, New Jersey, 1972.

[9] Hollingdale, S., *Makers of Mathematics*, Penguin Books, New York, 1994.

[10] James, I., *Remarkable Mathematicians: From Euler to von Neumann*, Cambridge University Press, Cambridge, 2002.

[11] Jolley, N., (editor) *The Cambridge Companion to Leibniz*, Cambridge University Press, Cambridge, 1995.

[12] Katz, V., *A History of Mathematics: An Introduction*, Second Edition, Addison-Wesley, New York, 1998.

[13] Leibniz, G. W., "Explication de l'arithmétique binaire, qui se sert des seuls caractères 0 et 1, avec des remarques sur son utilité, et sur ce qu'elle donne le sens des anciennes figures Chinoises de Fohy," *Memoires de l'Académie Royale des Sciences*, **3** (1703), 85–89.

[14] Russell, B., *A Critical Exposition of the Philosophy of Leibniz*, second ed., Allen and Unwin, London, 1937.

[15] Stern, N., *From ENIAC to UNIVAC: An Appraisal of the Eckert-Mauchly Computers*, Digital Press, Bedford, Massachusetts, 1981.

[16] Swetz, F. J., "Leibniz, the *Yijing*, and the Religious Conversion of the Chinese," *Mathematics Magazine*, **76**, No. 4 (2003), 276–291.

[17] Turing, A. M., "On Computable Numbers with an Application to the Entscheidungsproblem," *Proceedings of the London Mathematical Society* **42** (1936), 230–265. A correction, **43** (1937), 544–546. This paper with a short foreword by Davis was reprinted on pages 115–154 of [4].

[18] von Neumann, J., "First Draft of a Report on the EDVAC," in *From ENIAC to UNIVAC: An Appraisal of the Eckert-Mauchly Computers*, N. Stern, Digital Press, Bedford, Massachusetts, 1981, 177–246.

[19] Young, R. M., *Excursions in Calculus: An Interplay of the Continuous and the Discrete*, Mathematical Association of America, Washington, D.C., 1992.

Arithmetic Backwards from Shannon to the Chinese Abacus

Jerry M. Lodder

New Mexico State University

Recall that in the 1945 white paper "First Draft of a Report on the EDVAC" (Electronic Discrete Variable Automatic Computer), John von Neumann (1903–1957) advocated the use of binary arithmetic for the digital computers of his day. Vacuum tubes afforded these machines a speed of computation unmatched by other calculational devices, with von Neumann writing: "Vacuum tube aggregates ...have been found reliable at reaction times as short as a microsecond ..." [7, p. 188].

Predating this, in 1938 Claude Shannon (1916–2001) published a ground-breaking paper "A Symbolic Analysis of Relay and Switching Circuits" [4] in which he demonstrated how electronic circuits can be used for binary arithmetic, and more generally for computations in Boolean algebra and logic. These relay contacts and switches performed at speeds slower than vacuum tubes. Shannon identified an economy of representing numbers electronically in binary notation as well as an ease for arithmetic operations, such as addition. These advantages of base two arithmetic are nearly identical to those cited by von Neumann. Shannon [4] writes:

> A circuit is to be designed that will automatically add two numbers, using only relays and switches. Although any numbering base could be used the circuit is greatly simplified by using the scale of two. Each digit is thus either 0 or 1; the number whose digits in order are a_k, $a_{k-1}, a_{k-2}, \ldots, a_2, a_1, a_0$ has the value $\sum_{j=0}^{k} a_j 2^j$.

1. Explain how the base 10 number 95 can be written in base 2 using the above formula. In particular, compute a_k, $a_{k-1}, a_{k-2}, \ldots, a_2, a_1, a_0$ for the number 95. What is k in this case? Write $\sum_{j=0}^{k} a_j 2^j$ in terms of addition symbols using the above value for k and each value of a_j.

Claude Elwood Shannon was a pioneer in electrical engineering, mathematics and computer science, having founded the field of information theory, and discovered key relationships between Boolean algebra and computer circuits [6]. Born in the state of Michigan in 1916, he showed an interest in mechanical devices, and studied both electrical engineering and mathematics at the University of Michigan. Having received Bachelor of Science degrees in both of these subjects, he then accepted a research assistantship in the Department of Electrical Engineering at the Massachusetts Institute of Technology. The fundamental relation between Boolean logic and electrical circuits formed the topic of his master's thesis at MIT, and became his first published paper [4], for which he received an award from the combined engineering societies of the United States. In 1940 he earned his doctorate in mathematics from MIT with the dissertation "An Algebra for Theoretical Genetics." He spent the academic year 1940–41 visiting the Institute for Advanced Study in Princeton, New Jersey, where he began to formulate his ideas about information theory and efficient communication systems. The next fifteen years were productively spent at Bell Laboratories, and in 1948 he launched a new field of study, information theory, with the paper "A Mathematical Theory of Communication" [5]. He published extensively in communication theory, cryptography, game theory and computer science. In 1956 Dr. Shannon accepted a professorship at MIT, and retired in 1978.

Let's read a few excerpts from "A Symbolic Analysis of Relay and Switching Circuits" [4] with an eye toward understanding the circuitry behind binary arithmetic.

A Symbolic Analysis of Relay and Switching Circuits

Claude E. Shannon

I. Introduction

In the control and protective circuits of complex electrical systems it is frequently necessary to make intricate interconnections of relay contacts and switches. Examples of these circuits occur in automatic telephone exchanges, industrial motor-control equipment, and in almost any circuits designed to perform complex operations automatically. In this paper a mathematical analysis of certain of the properties of such networks will be made. . . .

II. Series-Parallel Two-Terminal Circuits

Fundamental Definitions and Postulates

We shall limit our treatment of circuits containing only relay contacts and switches, and therefore at any given time the circuit between two terminals must be either open (infinite impedance) or closed (zero impedance). Let us associate a symbol X_{ab} or more simply X with the terminals a and b. This variable, a function of time, will be called the hindrance of the two-terminal circuit a–b. The symbol 0 (zero) will be used to represent the hindrance of a closed circuit and the symbol 1 (unity) to represent the hindrance of an open circuit. Thus when the circuit a–b is open $X_{ab} = 1$ and when closed $X_{ab} = 0$. . . . Now let the symbol + (plus) be defined to mean the series connection of the two-terminal circuits whose hindrances are added together. Thus $X_{ab} + X_{cd}$ is the hindrance of the circuit a–d when b and c are connected together. Similarly the product of two hindrances $X_{ab} \cdot X_{cd}$ or more briefly $X_{ab}X_{cd}$ will be defined to mean the hindrance of the circuit formed by connecting the circuits a–b and c–d in parallel. A relay contact or switch will be represented in a circuit by the symbol in Figure 1, the letter being the corresponding hindrance function. Figure 2 shows the interpretation of the plus sign and Figure 3 the multiplication sign.

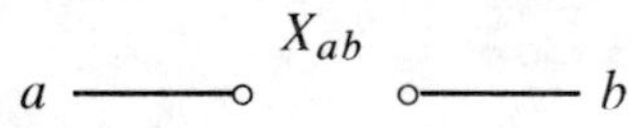

Figure 1. Figure 1. Symbol for hindrance function.

Figure 2. Figure 2. Interpretation of addition.

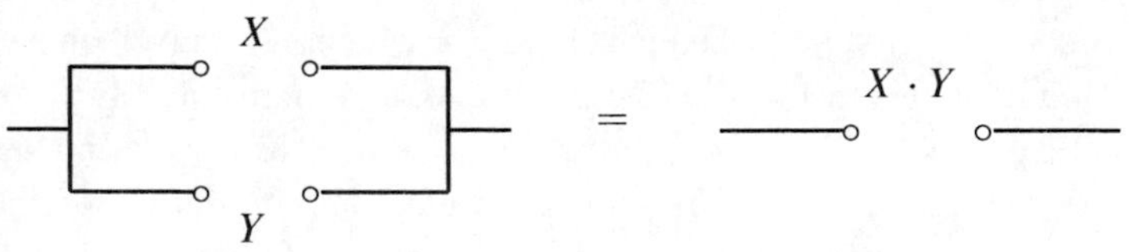

Figure 3. Figure 3. Interpretation of multiplication.

2. Consider X_{ab} as a Boolean variable with only two possible values 0 or 1. If $X_{ab} = 0$, then the switch in Figure 1 is closed, and current flows from a to b. If $X_{ab} = 1$, then the switch is open, and current does not flow from a to b. Now let X and Y be two Boolean variables. In a table listing all possible values for X and Y, record the results for X and Y joined in series, i.e., $X + Y$. Be sure to justify your answer by discussing whether current flows in the series circuit. Note that if current flows from left to right in Figure 2, then $X + Y = 0$, while if current does not flow, then $X + Y = 1$. To what extent does $X + Y$ represent a usual notation of addition? To what extent does $X + Y$ represent a construction in logic?

3. In a table listing all possible values for X and Y, record the results for X and Y joined in parallel, i.e., $X \cdot Y$. Justify your answer by discussing whether current flows in the parallel circuit. To what extent does $X \cdot Y$ represent a usual notion of multiplication? To what extent does $X \cdot Y$ represent a construction in logic?

4. These basic series and parallel circuits may be combined in any combination, using any finite number of switches. let X, Y, Z and W be Boolean variables, used as switches in the picture below. In a table listing all possible values for X, Y, Z and W, compute the value (0 or 1) of the circuit:

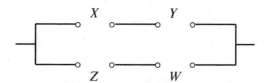

Explain your result in terms of a simple logical construction involving the results for the basic circuits $X + Y$ and $Z + W$:

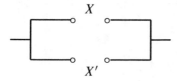

Shannon continues:

> We shall now define a new operation to be called negation. The negative of a hindrance X will be written X' and is defined to be a variable which is equal to 1 when X equals 0 and equal to 0 when X equals 1.

5. In a table listing both values for X, compute the value of the circuit:

$$\underline{\qquad X \qquad X' \qquad}$$

In another table, compute the value of the circuit:

$$\begin{array}{c} X \\ X' \end{array}$$

Can you interpret these tables in terms of simple logical constructions?

6. Now let a and b be binary variables with one digit. Recall from the project "Binary Arithmetic: From Leibniz to von Neumann" that the digit in the ones place (the right-hand digit) of the *arithmetic sum $a + b$* can be expressed using the "exclusive or" operation. Find a circuit which gives this digit. Justify your answer with a table that lists all possible values of a and b. Note that the arithmetic sum $a + b$ is the result of adding the binary values of a and b, not a and b combined in series.

7. Let a and b be binary variables with one digit as in question 6. Find a circuit which gives the digit in the twos place (the left-hand digit) of the arithmetic sum $a + b$. Justify your answer using a table listing all possible values of a and b.

8. Let a and b be binary variables with two possible digits. In Shannon's notation,

$$a = a_1 2^1 + a_0 2^0, \quad b = b_1 2^1 + b_0 2^0.$$

The digits of a are a_1, a_0, and the digits of b are b_1, b_0. Let c_0 be the result of the carried digit from $a_0 + b_0$. Either $c_0 = 0$ or $c_0 = 1$. In terms of the values for a_1, b_1 and c_0, when is the digit in the twos place for the arithmetic sum $a + b$ equal to zero? equal to one? Using a_1, b_1 and c_0 as switches, find a circuit which gives the digit in the twos place for $a + b$. Justify your answer using a table that lists all possible values for a_1, b_1 and c_0.

9. Let a and b be binary variables with two digits as in question 8. Using a_1, b_1 and c_0 as switches, find a circuit which gives the digit in the fours place (the left-most digit) of $a + b$. Justify your answer using a table listing all possible values for a_1, b_1 and c_0.

Recall that of binary numeration, Leibniz [1, p. 225] writes:

> However, I am not at all recommending this manner of counting as a replacement for the ordinary practice of tens. . . . [The] practice of tens is shorter, the numbers not as long. If we were accustomed to proceed by twelves or by sixteens, there would be even more of an advantage.

Writing large numbers by hand in binary notation easily results in transcription errors, since there are often many digits in a number base 2. However, entering base 10 numbers on a computer requires a conversion to base 2 at some level, a conversion which is not readily made, since 10 is not an integral power of 2. Let's now examine the Chinese abacus, and remember that Leibniz's intellectual curiosity had led him to the study of Chinese culture and religion, with an interpretation of the ancient text *Yijing* (the *I-ching* or *Book of Changes*) in terms of binary numeration.

The Chinese abacus (*suan pan*) consists of bars set in a rectangular frame, with the number of bars being 9, 11, 13, 17 or more [2, p. 211]. Each bar contains two upper beads and five lower beads, separated by a crossbar. Each upper bead counts as five units, and each lower bead counts as one unit. Traditionally numbers are represented positionally using base 10. A decimal point can be arbitrarily chosen between two bars of the abacus, and the digits are then arranged from left to right in decreasing powers of ten, so that the ones place is to the right of the tens place, the tens place is to the right of the hundreds place [3, p. 74–75]. Before representing a number on the abacus, all beads are moved away from the central crossbar so that they rest against the frame. Placing the decimal point at the far right of the frame, the base 10 number 197 would be displayed by moving one upper bead and two lower beads against the crossbar of the right-most bar, one upper bead and four lower beads against the crossbar of the bar immediately to the left of that, and one lower bead against the crossbar of the next bar.

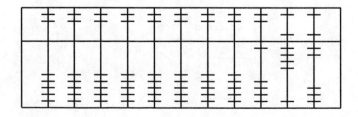

The number 197 set on a Chinese abacus.

Go to the library or the world wide web and research how (base 10) addition is performed on an abacus. Pay particular attention to the operation known today as "carrying." Notice that representing a number in base 10 requires a minimum of ten distinct values for bead arrangements along a given bar, which includes the value zero.

10. On a Chinese abacus, how many distinct numerical values can be represented along a given bar, including the value zero? Note that certain numbers greater than ten can be constructed by using two five-beads on the same bar. Let N denote this number of distinct values. If the full range of values for bead arrangements is employed on each bar, what number base is represented on a Chinese abacus?

11. Using N as in question 10, write the base 10 numbers from 1 to 20 inclusive in base N. Although you may invent any symbols that you wish for additional digits in this new base, be sure to explain what your symbols mean. For now, call the new digits n_1, n_2, n_3, etc., where $n_1 = 10, n_2 = 11, n_3 = 12$, etc.

12. In a table list the base 10 numbers 1 through 32 inclusive, their binary equivalents and their base N equivalents. Is there a pattern between the base 2 and base N representations? Explain.

13. Consider the number $9n_25n_4$ in base N, where n_4 is in the ones place, 5 in the N's place, n_2 in the N^2's place, and 9 in the N^3's place. Explain how to perform the addition

$$9n_25n_4 + n_172n_2$$

in base N on a Chinese abacus. What is the value of this sum in base N? Convert the final sum to base 10, and explain the conversion process.

Extra Credit: What is the sum $9n_25n_4 + n_172n_2$ in base 2? Justify your answer.

Notes for the instructor

The project builds naturally on the previous offering "Binary Arithmetic: From Leibniz to von Neumann," and is well suited for a first course in discrete mathematics or computer science. The project also includes an examination of arithmetic on a Chinese abacus, and, in a departure from the historical record, explores base 16 (hexadecimal) operations on the abacus. This use results from the full potential of all numerical values that can be represented on one bar of the Chinese abacus, and provides as an enrichment exercise in two-power base arithmetic. Today base 16 is often used by computer scientists as a shorthand for base 2, since, as observed by Leibniz, larger bases afford shorter lengths of notation in place value numeration.

Bibliography

[1] Gerhardt, C. I., (editor) *G. W. Leibniz Mathematische Schriften*, Vol. VII, Olms, Hildesheim, 1962.

[2] Martzloff, J.-L., *A History of Chinese Mathematics*, Wilson, S.S. (translator), Springer Verlag, Berlin, 1997.

[3] Needham, J., *Science and Civilisation in China*, vol. 3, Cambridge University Press, Cambridge, 1959.

[4] Shannon, C., "A Symbolic Analysis of Relay and Switching Circuits," *Transactions American Institute of Electrical Engineers*, **57** (1938), 713–723.

[5] Shannon, C.E., "A Mathematical Theory of Communication," *Bell System Technical Journal*, **27** (1948), 379–423 and 623–656.

[6] Sloane, N.J.A., Wyner, A.D. (editors), *Claude Elwood Shannon: Collected Papers*, The Institute of Electrical and Electronics Engineers, Inc., New York, 1993.

[7] von Neumann, J., "First Draft of a Report on the EDVAC," in *From ENIAC to UNIVAC: An Appraisal of the Eckert-Mauchly Computers*, N. Stern, Digital Press, Bedford, Massachusetts, 1981, 177–246.

Pascal's Treatise on the Arithmetical Triangle:
Mathematical Induction, Combinations, the Binomial Theorem and Fermat's Theorem[1]

David Pengelley
New Mexico State University

Introduction

Blaise Pascal (1623–1662) was born in Clermont-Ferrand in central France. Even as a teenager his father introduced him to meetings for mathematical discussion in Paris run by Marin Mersenne, who served as a primary conduit for transmitting mathematical ideas widely at that time, before the existence of any research journals. He quickly became involved in the development of projective geometry, the first in a sequence of highly creative mathematical and scientific episodes in his life, punctuated by periods of religious fervor. Around age twenty-one he spent several years developing a mechanical addition and subtraction machine, in part to help his father in tax computations as a local administrator. It was the first of its kind ever to be marketed. Then for several years he was at the center of efforts to understand vacuum, which led to an understanding of barometric pressure. In fact the scientific unit of pressure is named the *pascal*. He is also known for Pascal's Law on the behavior of fluid pressure.

Around 1654 Pascal conducted his studies on the Arithmetical Triangle ("Pascal's Triangle") and its relationship to probabilities. His correspondence with Pierre de Fermat (1601–1665) in that year marks the beginning of probability theory. Several years later, Pascal refined his ideas on area problems via the method of indivisibles already being developed by others, and solved various problems of areas, volumes, centers of gravity, and lengths of curves. Later in the seventeenth century, Gottfried Leibniz, one of the two inventors of the infinitesimal calculus which supplanted the method of indivisibles, explicitly credited Pascal's approach as stimulating his own ideas on the so-called characteristic triangle of infinitesimals in his fundamental theorem of calculus. After only two years of work on the calculus of indivisibles, Pascal fell gravely ill, abandoned almost all intellectual work to devote himself to prayer and charitable work, and died three years later at age thirty-nine. In addition to his work in mathematics and physics, Pascal is prominent for his *Provincial Letters* defending Christianity, which gave rise to his posthumously published *Pensées* (Thoughts) on religious philosophy [1, 2]. Pascal was an extremely complex person, and one of the outstanding scientists of the mid-seventeenth century, but we will never know how much more he might have accomplished with more sustained efforts and a longer life.

Pascal's *Traité du Triangle Arithmetique* (in English translation in [5, vol. 30]) makes a systematic study of the numbers in his triangle. They have simultaneous roles in mathematics as figurate numbers[2], combination numbers, and binomial coefficients, and he elaborates on all these. Given their multifaceted nature, it is no wonder that these ubiquitous numbers had already been in use for over 500 years, in places ranging from China to the Islamic world [3]. Pascal, however, was the first to connect binomial coefficients with combinatorial coefficients in probability. In fact, a

---

[1] With thanks to Joel Lucero-Bryan and Jerry Lodder.

[2] Figurate numbers count the number of equally spaced dots in geometric figures. Especially important to Pascal were the numbers of dots in equilateral triangles, triangular pyramids, and so forth in higher dimensions.

major motivation for Pascal was a question from the beginnings of probability theory, about the equitable division of stakes in an interrupted game of chance. The question had been posed to Pascal around 1652 by Antoine Gombaud, the Chevalier de Méré, who wanted to improve his chances at gambling: Suppose two players are playing a fair game, to continue until one player wins a certain number of rounds, but the game is interrupted before either player reaches the winning number. How should the stakes be divided equitably, based on the number of rounds each player has won [3, p. 431, 451ff]? The solution requires the combinatorial properties inherent in the numbers in the Arithmetical Triangle, as Pascal demonstrated in his treatise, since they count the number of ways various occurrences can combine to produce a given result. The Arithmetical Triangle overflows with fascinating patterns and applications, and we will see several of these in reading his treatise. We will study parts of Pascal's explanation of the connections between the numbers in his triangle and combination numbers. The reader is encouraged to read his entire treatise to see its many other aspects and connections.

From Pascal's treatise we will also learn the principle of mathematical induction. Pascal explains this in the specific context of proofs about the numbers in the triangle. The basic idea of mathematical induction had occurred in the mathematics of the Islamic world during the Middle Ages, and in southern Europe in the fourteenth century [3], but Pascal's was perhaps the first text to make a complete explicit statement and justification of this extremely powerful method of proof in modern mathematics. Mathematical induction is an astonishingly clever technique that allows us to prove claims about infinitely many interlinked phenomena all at once, even when proving just a single one of them in isolation would be very difficult! It will be a challenging technique to master, but will provide tremendous power for future mathematical work.

Learning about the connections of the Arithmetical Triangle to the binomial theorem in algebra will also allow an application to proving a famous and extremely important theorem on prime numbers discovered by Pascal's correspondent Pierre de Fermat (1601–1665) of Toulouse, on congruence remainders and prime numbers. This prepares one to understand the RSA cryptosystem, which today is at the heart of securing electronic transactions. We'll see how all these things are interconnected, and along the way we'll also acquire important mathematical tools, like notations for general indexing, summations, and products, and learn how to work with recurrence relations.

Part One: The Arithmetical Triangle and Mathematical Induction

Let us begin reading Blaise Pascal's

TREATISE ON THE ARITHMETICAL TRIANGLE

Definitions

I call *arithmetical triangle* a figure constructed as follows:

From any point, G, I draw two lines perpendicular to each other, GV, Gζ in each of which I take as many equal and contiguous parts as I please, beginning with G, which I number 1, 2, 3, 4, etc., and these numbers are the *exponents* of the sections of the lines.

Next I connect the points of the first section in each of the two lines by another line, which is the base of the resulting triangle.

In the same way I connect the two points of the second section by another line, making a second triangle of which it is the base.

And in this way connecting all the points of section with the same exponent, I construct as many triangles and bases as there are exponents.

Through each of the points of section and parallel to the sides I draw lines whose intersections make little squares which I call *cells*.

Cells between two parallels drawn from left to right are called *cells of the same parallel row*, as, for example, cells G, σ, π, etc., or φ, ψ, θ, etc.

Those between two lines drawn from top to bottom are called *cells of the same perpendicular row*, as, for example, cells G, φ, A, D, etc., or σ, ψ, B, etc.

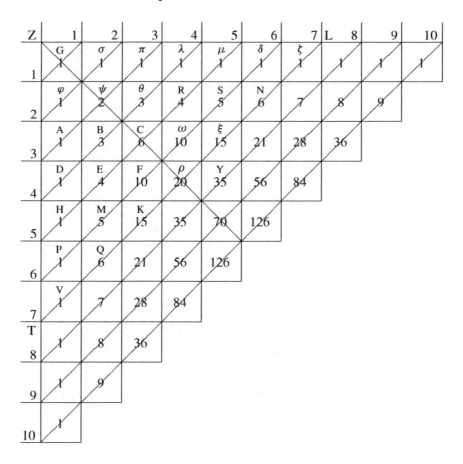

Those cut diagonally by the same base are called *cells of the same base*, as, for example, D, B, θ, λ, or A, ψ, π.

Cells of the same base equidistant from its extremities are called *reciprocals*, as, for example, E, R and B, θ, because the parallel exponent of one is the same as the perpendicular exponent of the other, as is apparent in the above example, where E is in the second perpendicular row and in the fourth parallel row and its reciprocal, R, is in the second parallel row and in the fourth perpendicular row, reciprocally. It is very easy to demonstrate that cells with exponents reciprocally the same are in the same base and are equidistant from its extremities.

It is also very easy to demonstrate that the perpendicular exponent of any cell when added to is parallel exponent exceeds by unity the exponent of its base.

For example, cell F is in the third perpendicular row and in the fourth parallel row and in the sixth base, and the exponents of rows 3 and 4, added together, exceed by unity the exponent of base 6, a property which follows from the fact that the two sides of the triangle have the same number of parts; but this is understood rather than demonstrated.

Of the same kind is the observation that each base has one more cell than the preceding base, and that each has as many cells as its exponent has units; thus the second base, $\varphi\sigma$, has two cells, the third, $A\psi\pi$, has three, etc.

Now the numbers assigned to each cell are found by the following method:

The number of the first cell, which is at the right angle, is arbitrary; but that number having been assigned, all the rest are determined, and for this reason it is called the *generator* of the triangle. Each of the others is specified by a single rule as follows:

The number of each cell is equal to the sum of the numbers of the perpendicular and parallel cells immediately preceding. Thus cell F, that is, the number of cell F, equals the sum of cell C and cell E, and similarly with the rest.

Whence several consequences are drawn. The most important follow, wherein I consider triangles generated by unity, but what is said of them will hold for all others.

First Consequence

In every arithmetical triangle all the cells of the first parallel row and of the first perpendicular row are the same as the generating cell.

For by definition each cell of the triangle is equal to the sum of the immediately preceding perpendicular and parallel cells. But the cells of the first parallel row have no preceding perpendicular cells, and those of the first perpendicular row have no preceding parallel cells; therefore they are all equal to each other and consequently to the generating number.

Thus $\varphi = G + 0$,   that is, $\varphi = G$,

$A = \varphi + 0$,   that is, φ,

$\sigma = G + 0$, $\pi = \sigma + 0$,

And similarly of the rest.

Second Consequence

In every arithmetical triangle each cell is equal to the sum of all the cells of the preceding parallel row from its own perpendicular row to the first, inclusive.

Let any cell, ω, be taken. I say that it is equal to $R + \theta + \psi + \varphi$, which are the cells of the next higher parallel row from the perpendicular row of ω to the first perpendicular row.

This is evident if we simply consider a cell as the sum of its component cells.

For ω equals $R + C$

$\underbrace{\hphantom{R + C}}$

$\theta + B$

$\underbrace{\hphantom{\theta + B}}$

$\psi + A$

$\underbrace{\hphantom{\psi + A}}$

φ,

for A and φ are equal to each other by the preceding consequence.

Therefore $\omega = R + \theta + \psi + \varphi$.

Third Consequence

In every arithmetical triangle each cell is equal to the sum of all the cells of the preceding perpendicular row from its own parallel row to the first, inclusive.

Let any cell, C, be taken. I say that it is equal to $B + \psi + \sigma$, which are the cells of the preceding perpendicular row from the parallel row of cell C to the first parallel row.

This is also apparent, as above, simply by the interpretation of cells.

For $C = B + \theta$

$\underbrace{\hphantom{B + \theta}}$

$\psi + \pi$

$\underbrace{\hphantom{\psi + \pi}}$

σ,

for $\pi = \sigma$ by the first consequence.

Therefore $C = B + \psi + \sigma$.

Fourth Consequence

In every arithmetical triangle each cell exceeds by unity the sum of all the cells within its parallel and perpendicular rows, exclusive.

Let any cell, ξ, be taken. I say that $\xi - G = R + \theta + \psi + \varphi + \lambda + \pi + \sigma + G$, which are all the numbers between row $\xi\omega CBA$ and row $\xi S\mu$ exclusive.

This is also apparent by interpretation.

For $\xi = \lambda + R + \omega$
$$\underbrace{\pi + \theta + C}$$
$$\underbrace{\sigma + \psi + B}$$
$$\underbrace{G + \varphi + A}$$
$$\underbrace{G.}$$

Therefore $\xi = \lambda + R + \pi + \theta + \sigma + \psi + G + \varphi + G$.

N.B. I have written in the enunciation *each cell exceeds by unity* because the generator is unity. If it were some other number, the enunciation should read: *each cell exceeds by the generating number.*

1. Pascal's Triangle and its numbers

   (a) Let us use the notation $T_{i,j}$ to denote what Pascal calls the number assigned to the cell in *parallel row i* (which we today call just *row i*) and *perpendicular row j* (which we today call *column j*). We call the i and j by the name *indices* (plural of *index*) in our notation. Using this notation, explain exactly what Pascal's rule is for determining all the numbers in all the cells. Be sure to give full details. This should include explaining for exactly which values of the indices he defines the numbers.

   (b) In terms of our notation $T_{i,j}$, explain his terms *exponent, base, reciprocal, parallel row, perpendicular row,* and *generator.*

   (c) Rewrite Pascal's first two "Consequences" entirely in the $T_{i,j}$ notation.

   (d) Rewrite his proofs of these word for word in our notation also.

   (e) Do you find his proofs entirely satisfactory? Explain why or why not.

   (f) Improve on his proofs using our notation. In other words, make them apply for arbitrary prescribed situations, not just the particular examples he lays out.

2. Modern mathematical notation

   Read in a modern textbook about index, summation, and product notations, and recurrence relations. Do some exercises.

Fifth Consequence

In every arithmetical triangle each cell is equal to its reciprocal.

For in the second base, $\varphi\sigma$, it is evident that the two reciprocal cells, φ, σ, are equal to each other and to G.

In the third base, A, ψ, π, it is also obvious that the reciprocals, π, A, are equal to each other and to G.

In the fourth base it is obvious that the extremes, D, λ, are again equal to each other and to G.

And those between, B, θ, are obviously equal since $B = A + \psi$ and $\theta = \pi + \psi$. But $\pi + \psi = A + \psi$ by what has just been shown. Therefore, etc.

Similarly it can be shown for all the other bases that reciprocals are equal, because the extremes are always equal to G and the rest can always be considered as the sum of cells in the preceding base which are themselves reciprocals.

3. Symmetry in the triangle: first contact with mathematical induction

   Write the Fifth Consequence using our index notation. Use index notation and the ideas in Pascal's proof to prove the Consequence in full generality, not just for the example he gives. Explain the conceptual ideas behind the general proof.

4. Mathematical induction: gaining more familiarity

   (a) Read in a modern textbook about mathematical induction.

   (b) Prove Pascal's First Consequence by mathematical induction. (Hint: for a proof by mathematical induction, always first state very clearly exactly what the n-th mathematical statement $P(n)$ says. Then state and prove the base step. Then state the inductive step very clearly before you prove it.)

   (c) Write the general form of Pascal's Second Consequence, and give a general proof using summation notation, but following his approach.

   (d) Now prove the Second Consequence by mathematical induction, i.e., a different proof.

   (e) **Optional**: More patterns.

      i. Write the Fourth Consequence using summation notation. Hint: You can write it using a sum of sums. Try writing Pascal's proof in full generality, using summation notation to help. If you don't complete it his way, explain why it is difficult.

      ii. Prove the Fourth Consequence by mathematical induction.

Seventh Consequence

In every arithmetical triangle the sum of the cells of each base is double that of the preceding base.

Let any base, $DB\theta\lambda$, be taken. I say that the sum of its cells is double the sum of the cells of the preceding base, $A\psi\pi$.

For the extremes	$D,$	$\lambda,$
are equal to the extremes	$A,$	$\pi,$
and each of the rest	$B,$	$\theta,$
is equal to two cells of the other base ...	$A + \psi,$	$\psi + \pi.$

Therefore $D + \lambda + B + \theta = 2A + 2\psi + 2\pi$.

The same thing is demonstrated in the same way of all other bases.

Eighth Consequence

In every arithmetical triangle the sum of the cells of each base is a number of the double progression beginning with unity whose exponent is the same as that of the base.

For the first base is unity.

The second is double the first; therefore it is 2.

The third is double the second; therefore it is 4.

And so on to infinity.

N.B. If the generator were not unity but some other number, such as 3, the same thing would be true. But we should have to take not the numbers of the double progression beginning with unity, that is, 1, 2, 4, 8, 16, etc., but those of the double progression beginning with the generator 3, that is, 3, 6, 12, 24, 48, etc.

5. Sums of bases in the triangle: a geometric progression

   (a) Use our index notation $T_{i,j}$ to explain exactly which are the numbers in the n-th base.

   (b) In full generality, write the Seventh Consequence and its proof, using our $T_{i,j}$ notation.

   (c) Write the statement of the Eighth Consequence in our notation, using modern exponential notation to describe his double progression. Use summation notation as needed, and introduce additional new notation if helpful. Then prove the Eighth Consequence by mathematical induction.

The next consequence is the most important and famous in the whole treatise. Pascal derives a formula for the ratio of consecutive numbers in a base. From this he will obtain an elegant and efficient formula for all the numbers in the triangle.

Twelfth Consequence

In every arithmetical triangle, of two contiguous cells in the same base the upper is to the lower as the number of cells from the upper to the top of the base is to the number of cells from the lower to the bottom of the base, inclusive.

Let any two contiguous cells of the same base, E, C, be taken. I say that

E	:	C	::	2	:	3
the		the		because there are two		because there are three
lower		upper		cells from E to the		cells from C to the top,
				bottom, namely E, H,		namely C, R, μ.

Although this proposition has an infinity of cases, I shall demonstrate it very briefly by supposing two lemmas:

The first, which is self-evident, that this proportion is found in the second base, for it is perfectly obvious that $\varphi : \sigma :: 1 : 1$;

The second, that if this proportion is found in any base, it will necessarily be found in the following base.

Whence it is apparent that it is necessarily in all the bases. For it is in the second base by the first lemma; therefore by the second lemma it is in the third base, therefore in the fourth, and to infinity.

It is only necessary therefore to demonstrate the second lemma as follows: If this proportion is found in any base, as, for example, in the fourth, $D\lambda$, that is, if $D : B :: 1 : 3$, and $B : \theta :: 2 : 2$, and $\theta : \lambda :: 3 : 1$, etc., I say the same proportion will be found in the following base, $H\mu$, and that, for example, $E : C :: 2 : 3$.

For $D : B :: 1 : 3$, by hypothesis.

Therefore $\underbrace{D + B}_{E}$: B :: $\underbrace{1 + 3}_{4}$: 3

E : B :: 4 : 3

Similarly $B : \theta :: 2 : 2$, by hypothesis

Therefore $\underbrace{B + \theta}_{C}$: B :: $\underbrace{2 + 2}_{4}$: 2

C : B :: 4 : 2

But B : E :: 3 : 4

Therefore, by compounding the ratios, $C : E :: 3 : 2$.   Q.E.D.

The proof is the same for all other bases, since it requires only that the proportion be found in the preceding base, and that each cell be equal to the cell before it together with the cell above it, which is everywhere the case.

6. Pascal's Twelfth Consequence: the key to our modern factorial formula

(a) Rewrite Pascal's Twelfth Consequence as a generalized modern formula, entirely in our $T_{i,j}$ terminology. Also verify its correctness in a couple of examples taken from his table in the initial definitions section.

(b) Adapt Pascal's proof by example of his Twelfth Consequence into modern generalized form to prove the formula you obtained above. Use the principle of mathematical induction to create your proof.

Now Pascal is ready to describe a formula for an arbitrary number in the triangle.

Problem

Given the perpendicular and parallel exponents of a cell, to find its number without making use of the arithmetical triangle.

Let it be proposed, for example, to find the number of cell ξ of the fifth perpendicular and of the third parallel row.

All the numbers which precede the perpendicular exponent, 5, having been taken, namely 1, 2, 3, 4, let there be taken the same number of natural numbers, beginning with the parallel exponent, 3, namely 3, 4, 5, 6.

Let the first numbers be multiplied together and let the product be 24. Let the second numbers be multiplied together and let the product be 360, which, divided by the first product, 24, gives as quotient 15, which is the number sought.

For ξ is to the first cell of its base, V, in the ratio compounded of all the ratios of the cells between, that is to say, $\xi : V$

in the ratio compounded of $\xi : \rho, \quad \rho : K, \quad K : Q, \quad Q : V$
or by the twelfth consequence $3 : 4 \qquad 4 : 3 \qquad 5 : 2 \qquad 6 : 1$

Therefore $\xi : V :: 3 \cdot 4 \cdot 5 \cdot 6 : 4 \cdot 3 \cdot 2 \cdot 1$.

But V is unity; therefore ξ is the quotient of the division of the product of $3 \cdot 4 \cdot 5 \cdot 6$ by the product of $4 \cdot 3 \cdot 2 \cdot 1$.

N.B. If the generator were not unity, we should have had to multiply the quotient by the generator.

7. Pascal's formula for the numbers in the Arithmetical Triangle

   (a) Write down the general formula Pascal claims in solving his "Problem." Your formula should read $T_{i,j} =$ "some formula in terms of i and j." Also write your formula entirely in terms of factorials.

   (b) Look at the reason Pascal indicates for his formula for a cell, and use it to make a general proof for your formula above for an arbitrary $T_{i,j}$. You may try to make your proof just like Pascal is indicating, or you may prove it by mathematical induction.

VARIOUS USES OF THE ARITHMETICAL TRIANGLE
WHOSE GENERATOR IS UNITY

Having given the proportions obtaining between the cells and the rows of arithmetical triangles, I turn in the following treatises to various uses of those triangles whose generator is unity. But I leave out many more than I include; it is extraordinary how fertile in properties this triangle is. Everyone can try his hand. I only call your attention here to the fact that in everything that follows I am speaking exclusively of arithmetical triangles whose generator is unity.

Part Two: Combinations and the Arithmetical Triangle

We continue reading Pascal's *Treatise on the Arithmetical Triangle*:

USE OF THE ARITHMETICAL TRIANGLE FOR COMBINATIONS

The word *combination* has been used in several different senses, so that to avoid ambiguity I am obliged to say how I understand it.

When of many things we may choose a certain number, all the ways of taking as many as we are allowed out of all those offered to our choice are here called the *different combinations*.

For example, if of four things expressed by the four letters, A, B, C, D, we are permitted to take, say any two, all the different ways of taking two out of the four put before us are called *combinations*.

Thus we shall find by experience that there are six different ways of choosing two out of four; for we can take A and B, or A and C, or A and D, or B and C, or B and D, or C and D.

I do not count A and A as one of the ways of taking two; for they are not different things, they are only one thing repeated.

Nor do I count A and B and B and A as two different ways; for in both ways we take only the same two things but in a different order, and I am not concerned with the order; so that I could make myself understood at once by those who are used to considering combinations, simply by saying that I speak only of combinations made without changing the order.

We shall also find by experience that there are four ways of taking three things out of four; for we can take ABC or ABD or ACD or BCD.

Finally we shall find that we can take four out of four in one way only, $ABCD$.

I shall speak therefore in the following terms:

> 1 in 4 can be combined 4 times.
> 2 in 4 can be combined 6 times.
> 3 in 4 can be combined 4 times.
> 4 in 4 can be combined 1 time.

Or:

> the number of combinations of 1 in 4 is 4.
> the number of combinations of 2 in 4 is 6.
> the number of combinations of 3 in 4 is 4.
> the number of combinations of 4 in 4 is 1.

But the sum of all the combinations in general that can be made in 4 is 15, because the number of combinations of 1 in 4, of 2 in 4, of 3 in 4, of 4 in 4, when joined together, is 15.

After this explanation I shall give the following consequences in the form of lemmas:

Lemma 1

There are no combinations of a number in a smaller number; for example, 4 *cannot be combined in* 2.

...

Proposition 2

The number of any cell is equal to the number of combinations of a number less by unity than its parallel exponent in a number less by unity than the exponent of its base.

Let any cell be taken, say F in the fourth parallel row and in the sixth base. I say that it is equal to the number of combinations of 3 in 5, less by unity than 4 and 6, for it is equal to the cells $A + B + C$. Therefore by the preceding proposition, etc.

1. Combinations according to Pascal

   (a) Explain in your own words what Pascal says about how many combinations there are for choosing two things out of four things.

   (b) Write Pascal's Proposition 2 using our $T_{i,j}$ notation for numbers in the triangle. In other words, fill in a sentence saying "$T_{i,j}$ is the number of combinations of choosing ____ things from ____ things." Pascal's justification for his Proposition 2 is based on his Lemma 4 and Proposition 1, which are not included in this project. However, the reader is encouraged to study and understand them, to wit:

(c) **Optional:** From Pascal's treatise [5, vol. 30], rewrite his statements and explanations of his Lemma 4 and Proposition 1 in your own words. State and prove Lemma 4 in the general case; that is, show that the number of combinations of k in n is the sum of the combinations of $k-1$ in $n-1$ and the combinations of k in $n-1$. Also explain why Proposition 2 follows from Proposition 1.

2. Combinations and Pascal's recursion formula

(a) The modern symbol $\binom{n}{r}$ means the number of ways ("combinations") of choosing r things from amongst n things. Explain how this is related to what we have been learning about the Arithmetical Triangle from reading Pascal. In particular, explain how the numbers $T_{i,j}$ are related to the numbers $\binom{n}{r}$. Do this by writing an equation expressing $T_{i,j}$ in $\binom{n}{r}$ notation, and also writing an equation expressing $\binom{n}{r}$ in $T_{i,j}$ notation. Now use the formula we learned earlier, from Pascal's solution to his *Problem*,[3] to write a formula for the combination number $\binom{n}{r}$, and manipulate it to express it entirely in terms of factorials.

(b) Now read in a modern textbook about the multiplication rule for counting possibilities, about permutations, and about combinations. Explain how a combination is different from a permutation.

(c) Read in a modern textbook about the algebra of combinations, Pascal's recursion formula, and how the text presents Pascal's Triangle. How is it different from Pascal's presentation?

Part Three: The Binomial Theorem and Fermat's Theorem

Now we can put together all of what we have learned from Pascal to prove an extremely important result in number theory, called *Fermat's Little Theorem*, which is at the heart of today's encryption methods in digital communications. The ingredients will be the binomial theorem, proof by mathematical induction as learned from Pascal, Pascal's formula for the numbers in his triangle (solved in his *Problem*), and uniqueness of prime factorization.

1. The binomial theorem and combinations

Read in a modern textbook about the binomial theorem. Write an explanation of the proof of the binomial theorem using the idea of counting combinations.

2. Discovering Fermat's "Little" Theorem: prime numbers and congruence remainders

(a) Make a table of the remainders of a^n upon division by n for positive integer values of both a and n ranging up to 14. To do this you should learn about congruence arithmetic, and figure out how to do these calculations quickly and easily without a calculator.

(b) Based on your table, make a conjecture of the form $a^p \equiv ?$ (mod p) for p a prime number and a any integer. This is called Fermat's "Little" Theorem; it is one of the most important phenomena in number theory. Also make some other interesting conjectures from patterns in your table, and try to prove them, perhaps using the binomial theorem.

(c) Write up the details of proving Fermat's Theorem by mathematical induction on a, with p held fixed. Use the binomial theorem, our knowledge of Pascal's "factorial" formula for binomial coefficients, and the Fundamental Theorem of Arithmetic (uniqueness of prime factorization) to analyze divisibility of the binomial coefficients by a prime p.

(d) **Optional:** Read about what Fermat was trying to do when he discovered his Theorem [4, p. 159ff]. Describe what you find in your own words.

3. **Optional:** The RSA cryptosystem

Read and study the RSA cryptosystem and its applications to digital security, including how it works, which follows from Fermat's Theorem. Write up the details in your own words, with some example calculations.

---

[3] *Given the perpendicular and parallel exponents of a cell, to find its number without making use of the arithmetical triangle.*

Notes for the instructor

The project's primary aim is for students of introductory discrete mathematics to learn the concept of mathematical induction and its application directly from reading the pioneering work *Treatise on the Arithmetical Triangle* of Blaise Pascal in the 1650s. There are three project parts, covering several standard topics: mathematical induction, combinations, and the binomial theorem and Fermat's Theorem. The project even ends with optional application to the RSA cryptosystem. With the entirety of its applications the project can constitute as much as 20% of a semester course. It works well to have students complete and submit small pieces of the project as one goes along.

In some places the instructor should give students guidance on reading and exercises from their textbook recommended by the project. There are also some places, especially in the third part, where the project expects either substantial independent learning from the student or more substantial instructor guidance.

In his treatise, Pascal, after arranging the figurate numbers in a defining triangular table, notices several patterns in the table, which he would like to claim continue indefinitely. Exhibiting unusual rigor for his day, Pascal offers a condition for the persistence of a pattern, stated verbally in his Twelfth Consequence, a condition known today as mathematical induction. This is perhaps the first complete enunciation and justification in the literature of the logical principle of mathematical induction, all provided in the context of a particular application. Moreover, this Twelfth Consequence immediately results in the modern formula for the combination numbers or binomial coefficients.

When Pascal's original writing becomes a student's initial contact for learning the idea of mathematical induction, a textbook is then merely a supplement. We have found that students come to grips with mathematical induction better by first seeing how Pascal eases into the idea through the proof of several patterns in the triangle, and then formalizes the principle and applies it further. Have students first read and work quite a bit with Pascal's verbal description, and then hold an instructor-moderated class discussion comparing this to the axiomatic formulation of induction students can read in the textbook. We intentionally wait on having students read the textbook approach until they have become comfortable with Pascal's. In fact, many students become and remain more comfortable with Pascal's more verbal way of handling mathematical induction than with their textbooks. Students can become so comfortable with Pascal's treatise that, on a final exam, many will voluntarily choose a question requiring new analysis of a part of Pascal's treatise using mathematical induction that they have never seen, over an analogous exercise from their modern textbook.

Bibliography

[1] *Encyclopædia Britannica*, Chicago, 1986.

[2] Gillispie, C. C., Holmes, F. L., (editors) *Dictionary of Scientific Biography*, Scribner, New York, 1970.

[3] Katz, V., *A History of Mathematics: An Introduction*, Second Edition, Addison-Wesley, New York, 1998.

[4] Laubenbacher, R., Pengelley, D., *Mathematical Expeditions: Chronicles by the Explorers*, Springer Verlag, New York, 1999.

[5] Pascal, B., "Treatise on the Arithmetical Triangle," in *Great Books of the Western World*, Mortimer Adler (editor), Encyclopædia Britannica, Inc., Chicago, 1991.

Early Writings on Graph Theory:
Euler Circuits and The Königsberg Bridge Problem

Janet Heine Barnett

Colorado State University - Pueblo

Figure 1. The Königsberg Bridges.

In a 1670 letter to Christian Huygens (1629–1695), the celebrated philosopher and mathematician Gottfried W. Leibniz (1646–1716) wrote as follows:

> I am not content with algebra, in that it yields neither the shortest proofs nor the most beautiful constructions of geometry. Consequently, in view of this, I consider that we need yet another kind of analysis, geometric or linear, which deals directly with position, as algebra deals with magnitude. [1, p. 30]

Known today as the field of 'topology,' Leibniz's study of position was slow to develop as a mathematical field. As C. F. Gauss (1777–1855) noted in 1833,

> Of the geometry of position, which Leibniz initiated and to which only two geometers, Euler and Vandermonde, have given a feeble glance, we know and possess, after a century and a half, very little more than nothing. [1, p. 30]

The 'feeble glance' which Leonhard Euler (1707–1783) directed towards the geometry of position consists of a single paper now considered to be the starting point of modern graph theory. Within the history of mathematics, the eighteenth century itself is commonly known as 'The Age of Euler' in recognition of the tremendous contributions that Euler made to mathematics during this period. Born in Basel, Switzerland, Euler studied mathematics under Johann Bernoulli (1667–1748), then one of the leading European mathematicians of the time and among the first — along with his brother Jakob Bernoulli (1654–1705) — to apply the new calculus techniques developed by Leibniz in the late seventeenth century to the study of curves. Euler soon surpassed his early teacher, and made important contributions to an astounding variety of subjects, ranging from number theory and analysis to astronomy and optics to mapmaking, in addition to graph theory and topology. His work was particularly important in re-defining calculus as the study of analytic functions, in contrast to the seventeenth century view of calculus as the study of curves. Amazingly, nearly half of Euler's nearly 900 books, papers and other works were written after he became almost totally blind in 1771.

The paper we examine in this project appeared in *Commentarii Academiae Scientiarum Imperialis Petropolitanae* in 1736. In it, Euler undertakes a mathematical formulation of the now-famous Königsberg Bridge Problem: is it possible to plan a stroll through the town of Königsberg which crosses each of the town's seven bridges once and only once? Like other early graph theory work, the Königsberg Bridge Problem has the appearance of being little more than an interesting puzzle. Yet from such deceptively frivolous origins, graph theory has grown into a powerful and deep mathematical theory with applications in the physical, biological, and social sciences. The resolution of the Four Color Problem — one of graph theory's most famous historical problems — even raised new questions about the notion of mathematical proof itself. First formulated by Augustus De Morgan in an 1852 letter to Hamilton, the Four Color Problem asks whether four colors are sufficient to color every planar map in such a way that regions sharing a boundary are colored in different colors. In 1976, after a long history of failed attempts to prove this is the case, Kenneth Appel (1932–) and Wolfgang Haken (1928–) published a computer-assisted proof which many mathematicians were unwilling to accept as valid. At the heart of the issue is a question that could be asked of any computer-assisted proof: should an argument that can not be directly checked by any member of the mathematical community be considered a valid proof?

This modern controversy highlights the historical fact that standards of proof have always varied from century to century, and from culture to culture. This project will highlight one part of this historical story by examining the differences in precision between an eighteenth century proof and a modern treatment of the same result. In particular, we wish to contrast Euler's approach to the problem of finding necessary and sufficient conditions for the existence of what is now known as an 'Euler circuit' to a modern proof of the main result of the paper.

In what follows, we take our translation from [1, pp. 3–8], with some portions eliminated in order to focus only on those most relevant to Euler's reformulation of the 'bridge crossing problem' as a purely mathematical problem. Definitions of modern terminology are introduced as we proceed through Euler's paper; modern proofs of two lemmas used in the proof of the main result are also included in an appendix.

SOLUTIO PROBLEMATIS AD GEOMETRIAM SITUS PERTINENTIS

1   In addition to that branch of geometry which is concerned with magnitudes, and which has always received the greatest attention, there is another branch, previously almost unknown, which Leibniz first mentioned, calling it the *geometry of position*. This branch is concerned only with the determination of position and its properties; it does not involve measurements, nor calculations made with them. It has not yet been satisfactorily determined what kind of problems are relevant to this geometry of position, or what methods should be used in solving them. Hence, when a problem was recently mentioned, which seemed geometrical but was so constructed that it did not require the measurement of distances, nor did calculation help at all, I had no doubt that it was concerned with the geometry of position — especially as its solution involved only position, and no calculation was of any use. I have therefore decided to give here the method which I have found for solving this kind of problem, as an example of the geometry of position.

2   The problem, which I am told is widely known, is as follows: in Königsberg in Prussia, there is an island *A*, called *the Kneiphof*; the river which surrounds it is divided into two branches, as can be seen in Fig. [1.2], and these branches are crossed by seven bridges, *a, b, c, d, e, f* and *g*. Concerning these bridges, it was asked whether anyone could arrange a route in such a way that he would cross each bridge once and only once. I was told that some people asserted that this was impossible, while others were in doubt: but nobody would actually assert that it could be done. From this, I have formulated the general problem: whatever be the arrangement and division of the river into branches, and however many bridges there be, can one find out whether or not it is possible to cross each bridge exactly once?

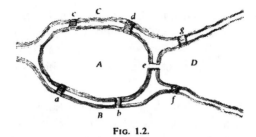

FIG. 1.2.

Notice that Euler begins his analysis of the 'bridge crossing' problem by first replacing the map of the city by a simpler diagram showing only the main feature. In modern graph theory, we simplify this diagram even further to include only points (representing land masses) and line segments (representing bridges). These points and line segments are referred to as *vertices* (singular: vertex) and *edges* respectively. The collection of vertices and edges together with the relationships between them is called a *graph*. More precisely, a graph consists of a set of vertices and a set of edges, where each edge may be viewed as an ordered pair of two (usually distinct) vertices. In the case where an edge connects a vertex to itself, we refer to that edge as a *loop*.

1. Sketch the diagram of a graph with five vertices and eight edges to represent the following bridge problem.

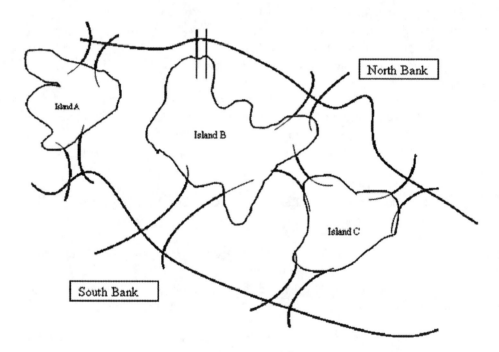

3 As far as the problem of the seven bridges of Königsberg is concerned, it can be solved by making an exhaustive list of all possible routes, and then finding whether or not any route satisfies the conditions of the problem. Because of the number of possibilities, this method of solution would be too difficult and laborious, and in other problems with more bridges it would be impossible. Moreover, if this method is followed to its conclusion, many irrelevant routes will be found, which is the reason for the difficulty of this method. Hence I rejected it, and looked for another method concerned only with the problem of whether or not the specified route could be found; I considered that such a method would be much simpler.

4 My whole method relies on the particularly convenient way in which the crossing of a bridge can be represented. For this I use the capital letters *A, B, C, D,* for each of the land areas separated by the river. If a traveller goes from *A* to *B* over bridge *a* or *b*, I write this as *AB* — where the first letter refers to the area the traveller is leaving, and the second refers to

the area he arrives at after crossing the bridge. Thus, if the traveller leaves *B* and crosses into *D* over bridge *f*, this crossing is represented by *BD*, and the two crossing *AB* and *BD* combined I shall denote by the three letters *ABD*, where the middle letter *B* refers to both the area which is entered in the first crossing and to the one which is left in the second crossing.

5   Similarly, if the traveller goes on from *D* to *C* over the bridge *g*, I shall represent these three successive crossings by the four letters *ABDC*, which should be taken to mean that the traveller, starting in *A*, crosses to *B*, goes on to *D*, and finally arrives in *C*. Since each land area is separated from every other by a branch of the river, the traveller must have crossed three bridges. Similarly, the successive crossing of four bridges would be represented by five letters, and in general, however many bridges the traveller crosses, his journey is denoted by a number of letters one greater than the number of bridges. Thus the crossing of seven bridges requires eight letters to represent it.

After rejecting the impractical strategy of solving the bridge-crossing problem by making an exhaustive list of all possible routes, Euler again reformulates the problem in terms of sequences of letters (vertices) representing land masses, thereby making the diagram itself unnecessary to the solution of the problem. Today, we say that two vertices joined by an edge in the graph are *adjacent*, and refer to a sequence of adjacent vertices as a *walk*. Technically, a walk is a sequence of alternating (adjacent) vertices and edges $v_0 e_1 v_1 e_1 \ldots e_n v_n$ in which both the order of the vertices and the order of the edges used between adjacent vertices are specified. In the case where no edge of the graph is repeated (as required in a bridge-crossing route), the walk is known as a *path*. If the initial and terminal vertex are equal, the path is said to be a *circuit*.

There is one more definition we need: a graph is said to be *connected* if for every pair of vertices u, v in the graph, there is a walk from u to v. Notice that a graph which is not connected will consist of several components, or subgraphs, each of which is connected. We can now give the modern definition of the concept from Euler's article. If *every* edge of a connected graph is used *exactly once* (as desired in a bridge-crossing route), the path (circuit) is said to be an *Euler path (circuit)*.

2. For the bridge problem shown in Question 1 above, how many capital letters (representing graph vertices) will be needed to represent an Euler path?

Having reformulated the bridge crossing problem in terms of sequences of letters (vertices) alone, Euler now turns to the question of determining *whether* a given bridge crossing problem admits of a solution. As you read through Euler's development of a procedure for deciding this question in paragraphs 7–13 below (we omit paragraph 6), pay attention to the style of argument employed, and how this differs from that used in a modern textbook.

7   The problem is therefore reduced to finding a sequence of eight letters, formed from the four letters *A*, *B*, *C* and *D*, in which the various pairs of letters occur the required number of times. Before I turn to the problem of finding such a sequence, it would be useful to find out whether or not it is even possible to arrange the letters in this way, for if it were possible to show that there is no such arrangement, then any work directed toward finding it would be wasted. I have therefore tried to find a rule which will be useful in this case, and in others, for determining whether or not such an arrangement can exist.

8   In order to try to find such a rule, I consider a single area *A*, into which there lead any number of bridges *a, b, c, d*, etc. (Fig. [1.3]). Let us take first the single bridge *a* which leads into *A*: if a traveller crosses this bridge, he must either have been in *A* before crossing, or have come into *A* after crossing, so that in either case the letter *A* will occur once in the representation described above. If three bridges (*a, b* and *c*, say) lead to *A*, and if the traveller crosses all three, then in the representation of his journey the letter *A* will occur twice, whether he starts his journey from *A* or not. Similarly, if five bridges lead to *A*, the

FIG. 1.3.

representation of a journey across all of them would have three occurrences of the letter *A*. And in general, if the number of bridges is any odd number, and if it is increased by one, then the number of occurrences of *A* is half of the result.

3. In paragraph 8, Euler deduces a rule for determining how many times a vertex must appear in the representation of the route for a given bridge problem for the case where an odd number of bridges leads to the land mass represented by that vertex. **Before reading further**, use this rule to determine how many times each of the vertices *A*, *B*, *C* and *D* would appear in the representation of a route for the Königsberg Bridge Problem. Given Euler's earlier conclusion (paragraph 5) that a solution to this problem requires a sequence of eight vertices, is such a sequence possible? Explain.

9   In the case of the Königsberg bridges, therefore, there must be three occurrences of the letter *A* in the representation of the route, since five bridges (*a, b, c, d, e*) lead to the area *A*. Next, since three bridges lead to *B*, the letter *B* must occur twice; similarly, *D* must occur twice, and *C* also. So in a series of eight letters, representing the crossing of seven bridges, the letter *A* must occur three times, and the letters *B*, *C* and *D* twice each — but this cannot happen in a sequence of eight letters. It follows that such a journey cannot be undertaken across the seven bridges of Königsberg.

10   It is similarly possible to tell whether a journey can be made crossing each bridge once, for any arrangement of bridges, whenever the number of bridges leading to each area is odd. For if the sum of the number of times each letter must occur is one more than the number of bridges, then the journey can be made; if, however, as happened in our example, the number of occurrences is greater than one more than the number of bridges, then such a journey can never be accomplished. The rule which I gave for finding the number of occurrences of the letter *A* from the number of bridges leading to the area *A* holds equally whether all of the bridges come from another area *B*, as shown in Fig. [1.3], or whether they come from different areas, since I was considering the area *A* alone, and trying to find out how many times the letter *A* must occur.

11   If, however, the number of bridges leading to *A* is even, then in describing the journey one must consider whether or not the traveller starts his journey from *A*; for if two bridges lead to *A*, and the traveller starts from *A*, then the letter *A* must occur twice, once to represent his leaving *A* by one bridge, and once to represent his returning to *A* by the other. If, however, the traveller starts his journey from another area, then the letter *A* will only occur once; for this one occurrence will represent both his arrival in *A* and his departure from there, according to my method of representation.

12   If there are four bridges leading to *A*, and if the traveller starts from *A*, then in the representation of the whole journey, the letter *A* must occur three times if he is to cross each bridge once; if he begins his walk in another area, then the letter *A* will occur twice. If there are six bridges leading to *A*, then the letter *A* will occur four times if the journey starts from *A*, and if the traveller does not start by leaving *A*, then it must occur three times. So, in general, if the number of bridges is even, then the number of occurrences of *A* will be half of this number if the journey is not started from *A*, and the number of occurrences will be one greater than half the number of bridges if the journey does start at *A*.

13   Since one can start from only one area in any journey, I shall define, corresponding to the number of bridges leading to each area, the number of occurrences of the letter denoting

that area to be half the number of bridges plus one, if the number of bridges is odd, and if the number of bridges is even, to be half of it. Then, if the total of all the occurrences is equal to the number of bridges plus one, the required journey will be possible, and will have to start from an area with an odd number of bridges leading to it. If, however, the total number of letters is one less than the number of bridges plus one, then the journey is possible starting from an area with an even number of bridges leading to it, since the number of letters will therefore be increased by one.

Notice that Euler's definition concerning 'the number of occurrences of the letter denoting that area' depends on whether the number of bridges (edges) leading to each area (vertex) is even or odd. In contemporary terminology, the number of edges incident on a vertex v is referred to as the *degree of vertex v*.

4. Let $\deg(v)$ denote the degree of vertex v in a graph G. Euler's definition of 'the number of occurrences of v' can then be re-stated as follows:

   - If $\deg(v)$ is even, then v occurs $\frac{1}{2}\deg(v)$ times.
   - If $\deg(v)$ is odd, then v occurs $\frac{1}{2}[\deg(v)+1]$ times.

   Based on Euler's discussion in paragraphs 9–12, how convinced are you that this definition gives a correct description of the Königsberg Bridge Problem? How convincing do you find Euler's claim (in paragraph 13) that the required route can be found in the case where 'the total of all the occurrences is equal to the number of bridges plus one'? Comment on how a proof of this claim in a modern textbook might differ from the argument which Euler presents for it in paragraphs 9–12.

14   So, whatever arrangement of water and bridges is given, the following method will determine whether or not it is possible to cross each of the bridges: I first denote by the letters *A, B, C,* etc. the various areas which are separated from one another by the water. I then take the total number of bridges, add one, and write the result above the working which follows. Thirdly, I write the letters *A, B, C,* etc. in a column, and write next to each one the number of bridges leading to it. Fourthly, I indicate with an asterisk those letters which have an even number next to them. Fifthly, next to each even one I write half the number, and next to each odd one I write half the number increased by one. Sixthly, I add together these last numbers, and if this sum is one less than, or equal to, the number written above, which is the number of bridges plus one, I conclude that the required journey is possible. It must be remembered that if the sum is one less than the number written above, then the journey must begin from one of the areas marked with an asterisk, and it must begin from an unmarked one if the sum is equal. Thus in the Königsberg problem, I set out the working as follows:

Number of bridges 7, which gives 8

	Bridges	
A,	5	3
B,	3	2
C,	3	2
D,	3	2

Since this gives more than 8, such a journey can never be made.

15   Suppose that there are two islands *A* and *B* surrounded by water which leads to four rivers as shown in Fig. [1.4]. Fifteen bridges (*a, b, c, d,* etc.) cross the rivers and the water surrounding the islands, and it is required to determine whether one can arrange a journey which crosses each bridge exactly once. First, therefore, I name all the areas separated by water as *A, B, C, D, E, F,* so that there are six of them. Next, I increase the number of bridges (15) by one, and write the result (16) above the working which follows.

		16
A*,	8	4
B*,	4	2
C*,	4	2
D,	3	2
E,	5	3
F*,	6	3
		16

Thirdly, I write the letters *A, B, C,* etc. in a column, and write next to each one the number of bridges which lead to the corresponding area, so that eight bridges lead to *A,* four to *B,* and so on. Fourthly, I indicate with an asterisk those letters which have an even number next to them. Fifthly, I write in the third column half the even numbers in the second column, and then I add one to the odd numbers and write down half the result in each case. Sixthly, I add up all the numbers in the third column in turn, and I get the sum 16; since this is equal to the number (16) written above, it follows that the required journey can be made if it starts from area *D* or *E,* since these are not marked with an asterisk. The journey can be done as follows:

EaFbBcFdAeFfCgAhCiDkAmEnApBoEID,

where I have written the bridges which are crossed between the corresponding capital letters.

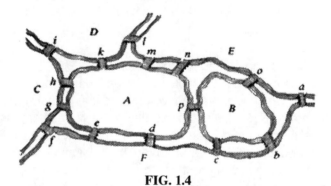

FIG. 1.4

5. Apply Euler's procedure to determine whether the graph representing the 'bridge-crossing' problem in question 1 above contains an Euler path. If so, find one.

In paragraphs 16 and 17, Euler makes some observations intended to simplify the procedure for determining whether a given bridge-crossing problem has a solution. As you read these paragraphs, consider how to reformulate these observations in terms of degree.

16   In this way it will be easy, even in the most complicated cases, to determine whether or not a journey can be made crossing each bridge once and once only. I shall, however, describe a much simpler method for determining this which is not difficult to derive from the present method, after I have first made a few preliminary observations. First, I observe that the numbers of bridges written next to the letters *A, B, C*, etc. together add up to twice the total number of bridges. The reason for this is that, in the calculation where every bridge leading to a given area is counted, each bridge is counted twice, once for each of the two areas which it joins.

17   It follows that the total of the numbers of bridges leading to each area must be an even number, since half of it is equal to the number of bridges. This is impossible if only one of these numbers is odd, or if three are odd, or five, and so on. Hence if some of the numbers of bridges attached to the letters *A, B, C*, etc. are odd, then there must be an even number of these. Thus, in the Königsberg problem, there were odd numbers attached to the letters *A, B, C* and *D*, as can be seen from Paragraph 14, and in the last example (in Paragraph 15), only two numbers were odd, namely those attached to *D* and *E*.

6. The result described in Paragraph 16 is sometimes referred to as 'The Handshake Theorem,' based on the equivalent problem of counting the number of handshakes that occur during a social gathering at which every person present shakes hands with every other person present exactly once. A modern statement of the Handshake Theorem would be: *The sum of the degree of all vertices in a finite graph equals twice the number of edges in the graph.* Locate this theorem in a modern textbook, and comment on how the proof given there compares to Euler's discussion in paragraph 16.

7. The result described in Paragraph 17 can be re-stated as follows: *Every finite graph contains an even number of vertices with odd degree.* Locate this theorem in a modern textbook, and comment on how the proof given there compares to Euler's discussion in paragraph 17.

Euler now uses the above observations to develop simplified rules for determining whether a given bridge-crossing problem has a solution. Again, consider how you might reformulate this argument in modern graph theoretic terms; we will consider a modern proof of the main results below.

18   Since the total of the numbers attached to the letters *A, B, C,* etc. is equal to twice the number of bridges, it is clear that if this sum is increased by 2 and then divided by 2, then it will give the number which is written above the working. If, therefore, all of the numbers attached to the letters *A, B, C, D,* etc. are even, and half of each of them is taken to obtain the numbers in the third column, then the sum of these numbers will be one less than the number written above. Whatever area marks the beginning of the journey, it will have an even number of bridges leading to it, as required. This will happen in the Königsberg problem if the traveller crosses each bridge twice, since each bridge can be treated as if it were split in two, and the number of bridges leading into each area will therefore be even.

19   Furthermore, if only two of the numbers attached to the letters *A, B, C,* etc. are odd, and the rest are even, then the journey specified will always be possible if the journey starts from an area with an odd number of bridges leading to it. For, if the even numbers are halved, and the odd ones are increased by one, as required, the sum of their halves will be one greater than the number of bridges, and hence equal to the number written above. It can further be seen from this that if four, or six, or eight. . . odd numbers appear in the second column, then the sum of the numbers in the third column will be greater by one, two, three. . . than the number written above, and the journey will be impossible.

20   So whatever arrangement may be proposed, one can easily determine whether or not a journey can be made, crossing each bridge once, by the following rules:

> *If there are more than two areas to which an odd number of bridges lead, then such a journey is impossible.*

If, however, the number of bridges is odd for exactly two areas, then the journey is possible if it starts in either of these areas.

If, finally, there are no areas to which an odd number of bridges leads, then the required journey can be accomplished starting from any area.

With these rules, the given problem can always be solved.

21   When it has been determined that such a journey can be made, one still has to find how it should be arranged. For this I use the following rule: let those pairs of bridges which lead from one area to another be mentally removed, thereby considerably reducing the number of bridges; it is then an easy task to construct the required route across the remaining bridges, and the bridges which have been removed will not significantly alter the route found, as will become clear after a little thought. I do not therefore think it worthwhile to give any further details concerning the finding of the routes.

The main results of Euler's paper can be stated in modern terminology as follows:

Theorem. *A finite graph G contains an Euler circuit if and only if G is connected and contains no vertices of odd degree.*

Corollary. *A finite graph G contains an Euler path if and only if G is connected and contains at most two vertices of odd degree.*

8. Illustrate why the modern statement specifies that G is connected by giving an example of a disconnected graph which has vertices of even degree only and contains no Euler circuit. Explain how you know that your example contains no Euler circuit.

9. Comment on Euler's proof of this theorem and corollary as they appear in paragraphs 16–19. How convincing do you find his proof? Where and how does he make use of the assumption that the graph is connected in his proof?

10. Below is the sketch of a modern proof of the 'if' direction of the main theorem. The first published proof of this direction is due to the German mathematician Carl Hierholzer (1840–1871); following Hierholzer's premature death, this proof was prepared for publication by a colleague and appeared in 1873 [3]. **Complete the proof sketch below** by filling in the missing details. *(Specific questions that you will need to address in your completed proof are indicated in italics.)*

Note: You may make use of the lemmas that are provided (with proofs) in the appendix of this project to do so.

Claim. If G is connected and has no vertices of odd degree, then G contains an Euler circuit.

Proof. Suppose G is connected and has no vertices of odd degree.

We show that G contains an Euler circuit as follows:

Case I: Consider the case where every edge in G is a loop.

- Since every edge in G is a loop, G must contain only one vertex.
  How do we know a connected graph in which every edge in G is a loop contains only one vertex?
- Since every edge in G is a loop on the single vertex v, the graph G must contain an Euler circuit.
  What will an Euler circuit in a connected graph on the single vertex v look like as a sequence of alternating vertices and edges?

Case II: Consider the case where at least one edge in G is not a loop.

- Choose any vertex v in G that is incident on at least one edge that is not a loop.

- Let $\{vu\}$ and $\{vw\}$ be two distinct edges on which v is incident which are not loops, where the vertices u and w may be equal.

  How do we know two such edges exist?

- Let W be a **simple path** (i.e., a path with no repeated edges and no repeated vertices) from v to w that does not use the edge $\{vw\}$.

  How do we know there is a walk from v to w that does not use this edge?

  (You may wish to consider what happens in the case where every walk from v to w uses the edge $\{vw\}$; what happens to the graph when the edge $\{vw\}$ is removed? The graph property discussed above in Question 7 of this project may be useful here.)

  Why can we assume that this walk is, in fact, a simple path?

- Use W to obtain a circuit C starting and ending at v.

  How is this done?

- Consider the two cases:

  - C uses every edge of G.

    Why are we now done?

  - C does not use every edge in G.

    * Consider the graph G' obtained by removing the edges of C from the graph G along with any vertices that become isolated when these edges are removed. Note that G' consists of finitely many connected components and that each of these connected components has vertices of even degree only. *How do we know that each connected component of G' has only vertices of even degree?*

    * Select a vertex v' in G' which appears in C.

      How do we know that such a vertex exists?

    * Apply the process outlined above (beginning with CASE I) to the connected component of G' that contains v' in order to obtain a circuit C' in G' with no repeated edges, and combine C with C' to obtain a new circuit C_1 in G that does not repeat any edges. *How do we know that we can apply this process to the connected component in question? How do we combine the circuits C and C' from our construction into a single circuit? How do we know that the combined walk C_1 is a circuit? How do we know that the combined circuit C_1 does not contain any repeated edges?*

    * Repeat this process as required until a circuit is obtained that includes every edge of G.

      How do we know this process will eventually terminate?

11. Now write a careful (modern) proof of the 'only if' direction. Begin by assuming that G is a connected graph which contains an Euler circuit. Then prove that G has no vertices of odd degree.

12. Finally, give a careful (modern) proof of the corollary.

Notes for the instructor

The project is suitable for a beginning-level discrete mathematics course, or for a 'transition to proof' course. It could be assigned independently or in conjunction with one or both of the projects "Early Writings on Graph Theory: Hamiltonian Circuits and The Icosian Game" and "Early Writings on Graph Theory: Topological Connections," both of which appear in this volume. By introducing modern graph theory terminology alongside Euler's original writing, the project assumes no prior background in graph theory. Thus, the project can be assigned prior to, concurrently with, or immediately following the introduction of basic graph theory concepts. The first part of the project in which students are required to read and understand Euler's analysis of the 'bridge problem' is well suited for small group discussion. Questions 6 and 7 ask students to compare Euler's treatment of key results to the treatment of these same results in a modern textbook, with the objective of drawing students' attention to current standards regarding formal proof. The instructor may wish to provide access to several different modern textbooks for students to reference on these questions. The project culminates with exercises which require students to 'fill in the gaps' in a modern proof of

Euler's main theorem (Questions 10–12). These questions are ideally suited for individual practice in proof writing, but could also be completed in small groups.

Appendix: Lemmas Used in Proving Euler's Theorem

Lemma I. *For every graph G, if W is a walk in G that has repeated edges, then W has repeated vertices.*

Proof. Let G be a graph and W a walk in G that has a repeated edge e. Let v and w be the endpoint vertices of e. If e is a loop, note that $v = w$, and v is a repeated vertex of W since the sequence 'vev' must appear somewhere in W. Thus, we need only consider the case where e is not a loop and $v \neq w$. In this case, one of the following must occur:

1. The edge e is immediately repeated in the walk W. That is, W includes a segment of the form '$vewev$' a segment of the form '$wevew$'.

2. The edge e is not immediately repeated, but occurs later in the walk W and in the same order. That is, either W includes a segment of the form '$vew \ldots vew$' or W includes a segment of the form '$wev \ldots wev$'.

3. The edge e is not immediately repeated, but occurs later in the walk W in the reverse order. That is, either W includes a segment of the form '$vew \ldots wev$' or W includes a segment of the form '$wev \ldots vew$'.

Since one of the vertices v or w is repeated in the first case, while both the vertices v and w are repeated in the latter two cases, this completes the proof.

Corollary. *For every graph G, if W is a walk in G that has no repeated vertices, then W has no repeated edges.*

Proof. This is the contrapositive of Lemma I.

Lemma II. *If G is a connected graph, then every pair of vertices of G is connected by a 'simple path' which repeats neither edges nor vertices.*

Proof. Let G be a connected graph. Let u and w be any arbitrary vertices in G. Since G is connected, we know G contains a walk W from u to w. Denote this walk by the sequence '$v_0 e_0 v_1 e_1 \ldots v_n e_n v_{n+1}$', where $e_0, e_1, \ldots e_n$ denote edges, $v_0, \ldots, v_{n+1}$ denote vertices with $v_0 = u$ the starting vertex and $v_{n+1} = w$ the ending vertex.

Note that W may include repeated vertices. If so, construct a new walk W' from u to w as follows:

- Let v be the first repeated vertex in the walk W. Then $v = v_i$ and $v = v_j$ for some $i < j$. To construct the new walk W', delete the segment of the original walk between the first occurrence of v and its next occurrence, including the second occurrence of v. That is, replace

$$\underbrace{v_0}_{u} \, e_1 v_1 e_2 \ldots v_{i-1} e_{i-1} \, \overbrace{v_i}^{v} \, \underbrace{e_i v_{i+1} e_{i+1} \ldots e_{j-1} v_{j-1} e_j \, \overbrace{v_j}^{v}}_{\text{delete}} \, e_{j+1} v_{j+1} \ldots v_{n-1} e_{n-1} v_n e_n \, \underbrace{v_{n+1}}_{w}$$

by

$$u e_1 v_1 e_2 \ldots v_{i-1} e_{i-1} \, \overbrace{v_i}^{v} \, e_{j+1} v_{j+1} \ldots v_{n-1} e_{n-1} v_n e_n w.$$

Since '$v e_{j+1}$' appeared in the original walk W, we know the edge e_{j+1} is incident on the vertex $v = v_i$. Thus, the new sequence of alternating edges and vertices is also a walk from $u = v_0$ to $w = v_{n+1}$.

(Also note that if $j = n + 1$, then the repeated vertex was $w = v_{n+1}$ and the walk now ends at v_i, where we know that $v_i = v_j = v_{n+1} = w$; thus, the new walk also ends at w.)

- If the new walk W' contains a repeated vertex, we repeat the above process. Since the sequence is finite, we know that we will obtain a walk with no repeated vertices after a finite number of deletions.

In this way, we obtain a new walk S from u to w that contains no repeated vertices. By the corollary to Lemma I, it follows that S contains no repeated edges. Thus, by definition of simple path, S is a simple path from u to w. Since u and w were arbitrary, this completes the proof.

Bibliography

[1] Biggs, N., Lloyd, E., Wilson, R., *Graph Theory: 1736–1936*, Clarendon Press, Oxford, 1976.

[2] Euler, L., *Novi Commentarii Academiae Scientarium Imperialis Petropolitanque* **7** (1758–59), p. 9–28.

[3] Hierholzer, C. , "Ueber die Möglichkeit, einen Linienzug ohne Wiederholung und ohne Unterbrechnung zu um-fahren," *Math. Ann.* **6** (1873), 30–32.

[4] James, I., *Remarkable Mathematicians: From Euler to von Neumann*, Cambridge University Press, Cambridge, 2002.

[5] Katz, V., *A History of Mathematics: An Introduction*, Second Edition, Addison-Wesley, New York, 1998.

Counting Triangulations of a Convex Polygon[1]

Desh Ranjan
New Mexico State University

Introduction

In a 1751 letter to Christian Goldbach (1690–1764), Leonhard Euler (1707–1783) discusses the problem of counting the number of triangulations of a convex polygon. Euler, one of the most prolific mathematicians of all times, and Goldbach, who was a Professor of Mathematics and historian at St. Petersburg and later served as a tutor for Tsar Peter II, carried out extensive correspondence, mostly on mathematical matters. In his letter, Euler provides a "guessed" method for computing the number of triangulations of a polygon that has n sides but does not provide a proof of his method. The method, if correct, leads to a formula for calculating the number of triangulations of an n-sided polygon which can be used to quickly calculate this number [1, p. 339–350] [2]. Later, Euler communicated this problem to the Hungarian mathematician Jan Andrej Segner (1704–1777). Segner, who spent most of his professional career in Germany (under the German name Johann Andreas von Segner), was the first Professor of Mathematics at the University of Göttingen, becoming the chair in 1735. Segner "solved" the problem by providing a proven correct method for computing the number of triangulations of a convex n-sided polygon using the number of triangulations for polygons with fewer than n sides [5]. However, this method did not establish the validity (or invalidity) of Euler's guessed method. Segner communicated his result to Euler in 1756 and in his communication he also calculated the number of triangulations for the n-sided polygons for $n = 1, \ldots, 20$ [5]. Interestingly enough, he made simple arithmetical errors in calculating the number of triangulations for polygons with 15 and 20 sides. Euler corrected these mistakes and also calculated the number of triangulations for polygons with up to 25 sides. It turns out that with the corrections, Euler's guessed method gives the right number of triangulations of polygons with up to 25 sides.

Was Euler's guessed method correct? It looked as if it were but there was no proof. The problem was posed as an open challenge to mathematicians by Joseph Liouville (1809–1882) in the late 1830s. He received solutions or purported solutions to the problem by many mathematicians (including one by Belgian mathematician Catalan which was correct but not so elegant), some of which were later published in the Liouville journal, one of the primary journals of mathematics at that time and for many decades. The most elegant of these solutions was communicated to him in a paper by Gabriel Lamé (1795–1870) in 1838. French mathematician, engineer and physicist Lamé was educated at the prestigious École Polytechnique and later at the École des Mines [3, p. 601–602]. From 1832 to 1844 he served as the chair of physics at the École Polytechnique, and in 1843 joined the Paris Academy of Sciences in the geometry section. He contributed to the fields of differential geometry, number theory, thermodynamics and applied mathematics. Among his publications are textbooks in physics and papers on heat transfer, where he introduced the rather useful technique of curvilinear coordinates. In 1851 he was appointed Professor of Mathematical Physics and Probability at the University of Paris, and resigned eleven years later after becoming deaf. Gauss considered Lamé the foremost French mathematician of his day [3, p. 601–602].

A translated version of this paper by Lamé is the key historical source for this project. This translation is presented in its entirety before the description of the major project tasks in the next section. Interestingly, and perhaps somewhat ironically, today these numbers (the number of triangulations of an n sided polygon for $n = 1, 2, 3, \ldots$) are called

---

[1] With thanks to David Pengelley, Inna Pivkina and Karen Villaverde.

Catalan numbers. Although none of Segner, Euler or Catalan provided any insight into why they were interested in calculating the number of triangulations of convex polygons, the related problem of computing optimal polygon triangulations has become a well-studied problem in the realm of algorithm design and is fairly commonly used to illustrate the effectiveness of dynamic programming paradigm for designing efficient algorithms.

Understanding Triangulations

A *diagonal* in a (convex) polygon is a straight line that connects two non-adjacent vertices of the polygon. Two diagonals are different if they have at least one different endpoint. A *triangulation* of a polygon is a division of the polygon into triangles by drawing *non-intersecting* diagonals. For example, the 6-sided polygon $ABCDEF$ below is triangulated into four triangles by using the diagonals AD, AE, BD.

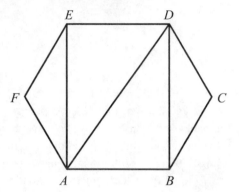

Figure 1. A Triangulation of $ABCDEF$.

Two triangulations are different if at least one of the diagonals in a triangulation is different from all diagonals in the other triangulation.

TASKS:

1.1 Draw a triangulation of $ABCDEF$ that is different from the triangulation in Figure 1. How many diagonals does your triangulation have? How many triangles does it divide $ABCDEF$ into?

1.2 Consider an n-sided polygon $A_1 A_2 \ldots A_n$. How many different possible diagonals does this polygon have? **Note:** We are talking about all possible diagonals, not just diagonals in a triangulation.

1.3 Use mathematical induction to prove that any triangulation of an n-sided polygon has $n - 2$ triangles and $n - 3$ diagonals.

Let's now read from Lamé's letter to Liouville [4].

Excerpt from a letter of Monsieur Lamé to Monsieur Liouville on the question:
Given a convex polygon, in how many ways can one
partition it into triangles by means of diagonals?[2]

The formula that you communicated to me yesterday is easily deduced from the comparison of two methods leading to the same goal.

Indeed, with the help of two different methods, one can evaluate the number of decompositions of a polygon into triangles: by consideration of the sides, or of the vertices.

---

[2]See a Memoir of Segner (*Novi Commentarii Acad. Petrop.*, vol. VII, p. 203). The author found equation (1) of M. Lamé; but formula (3) presents a much simpler solution. Formula (3) is no doubt due to Euler. It is pointed out without proof on page 14 of the volume cited above. The equivalence of equations (1) and (3) is not easy to establish. M. Terquem proposed this problem to me, achieving it with the help of some properties of factorials. I then communicated it to various geometers: none of them solved it; M. Lamé has been very successful: I am unaware of whether others before him have obtained such an elegant solution.   J. LIOUVILLE

I.

Let $ABCDEF\ldots$ be a convex polygon of $n+1$ sides, and denote by the symbol P_k the total number of decompositions of a polygon of k sides into triangles. An arbitrary side AB of $ABCDEF\ldots$ serves as the base of a triangle, in each of the P_{n+1} decompositions of the polygon, and the triangle will have its vertex at C, or D, or $F\ldots$; to the triangle CBA there will correspond P_n different decompositions; to DBA another group of decompositions, represented by the product $P_3 P_{n-1}$; to EBA the group $P_4 P_{n-2}$; to FBA, $P_5 P_{n-3}$; and so forth, until the triangle ZAB, which will belong to a final group P_n. Now, all these groups are completely distinct: their sum therefore gives P_{n+1}. Thus one has

$$P_{n+1} = P_n + P_3 P_{n-1} + P_4 P_{n-2} + P_5 P_{n-3} + \cdots + P_{n-3} P_5 + P_{n-2} P_4 + P_{n-1} P_3 + P_n. \quad (1)$$

II.

Let $abcde\ldots$ be a polygon of n sides. To each of the $n-3$ diagonals, which end at one of the vertices a, there will correspond a group of decompositions, for which this diagonal will serve as the side of two adjacent triangles: to the first diagonal ac corresponds the group $P_3 P_{n-1}$; to the second ad corresponds $P_4 P_{n-2}$; to the third ae, $P_5 P_{n-3}$, and so forth until the last ax, which will occur in the group $P_3 P_{n-1}$. These groups are not totally different, because it is easy to see that some of the partial decompositions, belonging to one of them, is also found in the preceding ones. Moreover they do not include the partial decompositions of P_n in which none of the diagonals ending in a occurs.

But if one does the same for each of the other vertices of the polygon, and combines all the sums of the groups of these vertices, by their total sum

$$n\left(P_3 P_{n-1} + P_4 P_{n-2} + \cdots + P_{n-2} P_4 + P_{n-1} P_3\right)$$

one will be certain to include all the partial decompositions of P_n; each of these is itself repeated therein a certain number of times.

Indeed, if one imagines an arbitrary such decomposition, it contains $n-2$ triangles, having altogether $3n-6$ sides; if one removes from this number the n sides of the polygon, and takes half of the remainder, which is $n-3$, one will have the number of diagonals appearing in the given decomposition. Now, it is clear that this partial decomposition is repeated, in the preceding total sum, as many times as these $n-3$ diagonals have ends, that is $2n-6$ times: since each end is a vertex of the polygon, and in evaluating the groups of this vertex, the diagonal furnished a group including the particular partial decomposition under consideration.

Thus, since each of the partial decompositions of the total group P_n is repeated $2n-6$ times in $n\left(P_3 P_{n-1} + P_4 P_{n-2} + \cdots + P_{n-2} P_4 + P_{n-1} P_3\right)$, one obtains P_n upon dividing this sum by $2n-6$. Therefore one has

$$P_n = \frac{n\left(P_3 P_{n-1} + P_4 P_{n-2} + \cdots + P_{n-2} P_4 + P_{n-1} P_3\right)}{2n-6}. \quad (2)$$

III.

The first formula (1) gives

$$P_3 P_{n-1} + P_4 P_{n-2} + \cdots + P_{n-2} P_4 + P_{n-1} P_3 = P_{n+1} - 2P_n,$$

and the second (2) gives

$$P_3 P_{n-1} + P_4 P_{n-2} + \cdots + P_{n-2} P_4 + P_{n-1} P_3 = \frac{2n-6}{n} P_n;$$

so finally

$$P_{n+1} - 2P_n = \frac{2n - 6}{n} P_n,$$

or

$$P_{n+1} = \frac{4n - 6}{n} P_n. \tag{3}$$

This is what was to be proven.

Paris, 25 August, 1838

Optimal Triangulation and Counting Triangulations

The *Optimal Polygon Triangulation* Problem is the following: Given an n-sided polygon $A_1 A_2 \ldots A_n$ and a weight $w_{i,j}$ for each diagonal $A_i A_j$, find a triangulation of the polygon such that the sum of the weights of the diagonals in the triangulations is minimized. A naïve way to solve the problem is to generate all possible triangulations one by one, calculate their weight (i.e. sum of weights of all the diagonals in the triangulation) and keep the best. The efficiency of this naïve method depends on the number of possible triangulations of a polygon with n sides. Thus, we would like to count how many different triangulations an n-sided polygon has. As mentioned in the introduction, the problem of counting the number of triangulations of an n-sided convex polygon was already being discussed in the mid-eighteenth century by well-known figures in mathematics like Euler and Segner and an elegant solution was provided by Lamé in 1838. As noted before, the numbers P_n in Lamé's paper are now called Catalan numbers.

TASKS:

2.1 Read Section I in Lamé's paper. Explain what Lamé is saying in your own words and derive the general recursive formula for P_{n+1}, i.e., formula (1) in Lamé's paper.

2.2 Use the recursive formula to calculate P_i for $i = 2, 3, 4, 5, 6, 7, 8$ by hand and display it as a table.

2.3 Draw all triangulations of polygons with n sides for $n = 4, 5$.

2.4 Lamé's recurrence relation in his section 1 for $n = 5$ yields

$$P_6 = P_5 + P_3 P_4 + P_4 P_3 + P_5.$$

Draw all triangulations of a 6-sided polygon classified into groups according to the idea of the recurrence relation, i.e., the triangulations should be classified into four groups with each group corresponding to a term on the right-hand side of the recurrence above.

2.5 Write a simple recursive function $SRCAT(n)$ (for "Simple Recurrence CATalan") in Java that given an input n calculates P_n using the recurrence relation (1) in Lamé's paper directly.

2.6 Write another Java program that repeatedly uses $SRCAT$ to calculate P_i for $i = 3, 4, 5, \ldots$ Restrict the total time your program uses to 10 minutes. What is the largest value N_0 of i for which your program calculates P_i? Print out a table with i and the time required in seconds by $SRCAT$ to calculate each of the P_i values. Your table should have a row for each $i = 3, 4, 5, \ldots, N_0$.

2.7 From your calculations you may observe that it seems that for all n, if $n \geq 3$ then $P_{n+1} \geq 2 * P_n$. Give a simple mathematical argument that establishes the truth of this statement.

2.8 Prove that for all n, if $n \geq 3$ then $P_n \geq 2^n/8$.

2.9 What does this tell you about the efficiency of the naïve algorithm for solving the optimal polygon triangulation problem?

2.10 Write a Java program that repeatedly uses the recurrence given in formula (1) in Lamé's paper to calculate P_i for $i = 3, 4, 5 \ldots$ but that stores the computed values in an array systematically and uses them as needed. Restrict the total time your program uses to 10 minutes. What is the largest value M_0 of i for which your program calculates P_i?

2.11 Extend your program to print out a table of values of i and time required in seconds to compute P_i for $i = 3, 4, \ldots, M_0$.

2.12 Graph the tables obtained in 2.6 and 2.11. Analyse these graphs and write down your observations.

Lame's Method for deriving a formula for P_n

Section II of Lamé's paper gives an alternative way of counting triangulations of a polygon. Read this section carefully.

TASKS:

Consider a 6-sided polygon $ABCDEF$.

3.1 Draw all triangulations of the polygon where:

- AC is one of the diagonals in the triangulation.
- AD is one of the diagonals in the triangulation.
- AE is one of the diagonals in the triangulation.

How many total triangulations did you draw?

3.2 Repeat the same with vertex B as the "special" vertex, i.e., draw all triangulations where:

- BD is one of the diagonals in the triangulation.
- BE is one of the diagonals in the triangulation.
- BF is one of the diagonals in the triangulation.

How many total triangulations did you draw?

3.3 Do the same with vertices C, D, E, F being "special".

3.4 Consider the triangulation of $ABCDEF$ in figure 1 (of section 1). How many times is that triangulation repeated in all the triangulations that you drew for $ABCDEF$ in this section? Identify the diagonals in whose group it was drawn.

3.5 Do the same for the different triangulations of $ABCDEF$ that you drew in section 1.

3.6 What would you guess about the number of times any triangulation of $ABCDEF$ is repeated? Argue why your guess is correct.

3.7 Consider the n-sided polygon $A_1 A_2 \ldots A_n$. Let P_i denote the number of different triangulations of a polygon with i sides.

(a) Calculate, in terms of P_i's, the number of triangulations of this polygon that have $A_1 A_3$ as a diagonal, that have $A_1 A_4$ as a diagonal, that have $A_1 A_j$ as a diagonal.

(b) Consider drawing triangulations treating A_1 as the "special" vertex. That is, draw all triangulations where $A_1 A_3$ is a diagonal, then draw all triangulations where $A_1 A_4$ is a diagonal, etc. all the way up to where $A_1 A_{n-1}$ is a diagonal. What is the number of triangulations you draw (in terms of P_i's) when A_1 is treated as a special vertex?

(c) Suppose we repeat the above process with another vertex (say A_2) being the special vertex instead of A_1. What can you say about the number of triangulations drawn as compared to the number of triangulations drawn when A_1 was chosen as the special vertex? Explain in your own words why this is true.

(d) Consider doing what you did for A_1 in (b) successively for each vertex. That is, enumerate all triangulations treating A_1 as a special vertex, treating A_2 as a special vertex, ... treating A_n as a special vertex. Now consider the specific triangulation of $A_1, A_2, \ldots, A_n$ obtained by drawing the diagonals $A_1 A_3, A_1 A_4, \ldots, A_1 A_{n-1}$. How many times is this triangulation enumerated? What about the triangulation obtained by drawing the diagonals $A_1 A_4, A_1 A_5, \ldots, A_1 A_{n-2}$ and the two diagonals $A_2 A_4, A_{n-2} A_n$? Justify your answer.

(e) What is your guess as to how many times any specific triangulation is enumerated? Explain in your own words why this is the case.

3.8 Combine (b) and (e) to derive the formula (2) in Lamé's paper. Explain in your own words how this formula is obtained.

3.9 Combine formulas (1) and (2) in Lamé's paper to obtain the formula (3) in Lamé's paper. Show all the steps in your calculation. Explain why this formula is better for calculating P_n.

3.10 Using formula (3) in Lamé's paper, show that $P_{n+2} = \frac{1}{n+1}\binom{2n}{n}$ where $\binom{2n}{n} = \frac{(2n)!}{n!n!}$.

3.11 Write a simple recursive function $ASRCAT(n)$ (for "Another Simple Recurvise CATalan") in Java that given an input n calculates P_n using the recurrence relation (3) in Lamé's paper directly.

3.12 Write another Java program that repeatedly uses $ASRCAT$ to calculate P_i for $i = 3, 4, 5, \ldots$ Restrict the total time your program uses to 10 minutes. What is the largest value L_0 of i for which your program calculates P_i?

3.13 Extend your program to print out a table of values of i and time required in seconds by $ASRCAT(n)$ to compute each of the P_i values for $i = 3, 4, \ldots, L_0$. Your table should have a row for each $i = 3, 4, \ldots, L_0$.

3.14 Write a better Java program using the ideas from dynamic programming ("store and re-use") that repeatedly calculates P_i for $i = 3, 4, \ldots$ Restrict the total time your program uses to 10 minutes. What is the largest value L_1 of i for which your program calculates P_i.

3.15 Extend your program to print out a table of values of i and the time required in seconds to calculate each of the P_i values. Your table should have a row for each $i = 3, 4, \ldots, L_1$.

3.16 Graph the tables obtained in 3.13 and 3.15. Analyse all four graphs obtained and write down your observations. How do the results for the second two programs compare with your first two programs? How fast does the running time of the last two programs grow?

3.17 Discuss how the choice of Lamé's formulas (1) or (3), or using dynamic versus naïve recursive programming influences the effectiveness of computation.

Notes for the instructor

This project is most suitable for use in an upper-division undergraduate algorithm design and analysis course. In particular, it is best used at the time when the students are learning the *dynamic programming* paradigm for algorithm design. It allows them to see why a naïve solution is infeasible for solving the Optimal Polygon Triangulation Problem for a polygon with a large number of sides and how the dynamic programming technique allows one to do the same calculation much more efficiently.

Some of the tasks, as stated in the project, ask the students to write JAVA programs. Use of JAVA is not critical. Any programming language that allows for recursion and use of arrays (e.g., C, C++) can be substituted for JAVA without affecting the project.

Bibliography

[1] Euler, L., *Leonhard Euler und Christian Goldbach, Briefwechsel 1729–1764*, Juskevic, A. P., Winter, E. (editors), Akademie Verlag, Berlin, 1965.

[2] Euler, L., *Novi Commentarii Academiae Scientarium Imperialis Petropolitanque* **7** (1758–59), p. 9–28.

[3] Gillispie, C. C., Holmes, F. L., (editors) *Dictionary of Scientific Biography*, Scribner, New York, 1970.

[4] Lamé, G., "Un polygone convexe étant donné, de combien de manières peut-on le partager en triangles au moyen de diagonales?" *Journal de Mathématiques Pures et Appliquées*, **3** (1838), 505–507.

[5] Segner, A., "Enumeratio Modorum Quibus Figurae Planae Rectilinae per Diagonales Dividuntur in Triangula," *Novi Commentarii Academiae Scientarium Imperialis Petropolitanque* **7** (1758-59), 203–209.

Early Writings on Graph Theory: Hamiltonian Circuits and The Icosian Game

Janet Heine Barnett
Colorado State University - Pueblo

Introduction

Problems that are today considered to be part of modern graph theory originally appeared in a variety of different connections and contexts. Some of these original questions appear little more than games or puzzles. In the instance of the 'Icosian Game', this observation seems quite literally true. Yet for the game's inventor, the Icosian Game encapsulated deep mathematical ideas which we will explore in this project.

Sir William Rowan Hamilton (1805–1865) was a child prodigy with a gift for both languages and mathematics. His academic talents were fostered by his uncle James Hamilton, an Anglican clergyman with whom he lived from the age of 3. Under his uncle's tutelage, Hamilton mastered a large number of languages — including Latin, Greek, Hebrew, Persian, Arabic and Sanskrit — by the age of 10. His early interest in languages was soon eclipsed by his interests in mathematics and physics, spurred in part by his contact with an American calculating prodigy. Hamilton entered Trinity College in Dublin in 1823, and quickly distinguished himself. He was appointed Astronomer Royal of Ireland at the age of 22 based on his early work in optics and dynamics. Highly regarded not only by his nineteenth century colleagues, Hamilton is today recognized as a leading mathematician and physicist of the nineteenth century.

In mathematics, Hamilton is best remembered for his creation of a new algebraic system known as the 'quaternions' in 1843. The system of quaternions consists of 'numbers' of the form $Q = a + bi + cj + dk$ subject to certain basic 'arithmetic' rules. The project that led Hamilton to the discovery of quaternions was the search for an algebraic system that could be reasonably interpreted in the three-dimensional space of physics, in a manner analogous to the interpretation of the algebra of complex numbers $a + bi$ in a two-dimensional plane. Although this geometrical interpretation of the complex numbers is now standard, it was discovered by mathematicians only in the early 1800s and thus was relatively new in Hamilton's time. Hamilton was one of several nineteenth century British mathematicians interested in developing a purely *algebraic* foundation for complex numbers that would capture the essence of this geometrical interpretation. His algebraic development of the complex numbers as ordered pairs of real numbers (a, b) subject to certain operations appeared in a landmark 1837 essay entitled *Theory of Conjugate Functions, or Algebraic Couples; with a Preliminary and Elementary Essay on Algebra as the Science of Pure Time*.

Hamilton concluded his 1837 essay with a statement concerning his hope that he would soon publish a similar work on the algebra of triplets. After years of unsuccessful work on this problem, Hamilton was able to solve it in 1843 only by abandoning the property of commutativity. For example, two of the basic multiplication rules of the quaternion system are $ij = k$ and $ji = -k$, so that $ij \neq ji$. Hamilton also replaced 'triplets' by the 'four-dimensional' quaternion $a + bi + cj + dk$. Soon after Hamilton's discovery, physicists realized that only the 'vector part' $bi + cj + dk$ of a quaternion was needed to represent three-dimensional space. Although vectors replaced the use of quaternions in physics by the end of the nineteenth century, the algebraic system of vectors retains the non-commutativity of quaternions.

Today's students of mathematics are familiar with a variety of non-commutative algebraic operations, including vec-

217

tor cross-product and matrix multiplication. In Hamilton's day, however, quaternions constituted a major breakthrough comparable to the discovery of non-Euclidean geometry. Immediately following Hamilton's 1843 announcement of his discovery, at least seven other "new numbers systems" were discovered by several other British algebraists. In the 'Icosian Game', Hamilton himself developed yet another example of a non-commutative algebraic system. In this project, we explore both the algebra of that system and the graph theoretical notion of 'Hamiltonian circuit' on which Hamilton's interpretation of this algebra is based. The idea for the game was first exhibited by Hamilton at an 1857 meeting of the British Association in Dublin [3], and later sold for 25 pounds to 'John Jacques and Son,' a wholesale dealer in games. We begin with the preface to the instructions pamphlet which Hamilton prepared for marketing of the game in 1859 [1, pp. 32–35].

The Icosian Game and Hamiltonian Circuits

THE ICOSIAN GAME

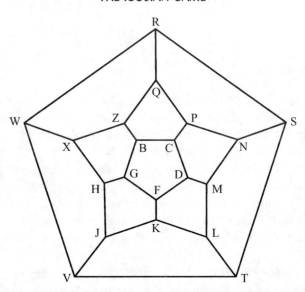

In this new Game (invented by Sir WILLIAM ROWAN HAMILTON, LL.D., &c., of Dublin, and by him named *Icosian* from a Greek word signifying 'twenty') a player is to place the whole or part of a set of twenty numbered pieces or men upon the points or in the holes of a board, represented by the diagram above drawn, in such a manner as always to proceed *along the lines* of the figure, and also to fulfill certain *other* conditions, which may in various ways be assigned by another player. Ingenuity and skill may thus be exercised in *proposing* as well as in *resolving* problems of the game. For example, the first of the two players may place the first five pieces in any five consecutive holes, and then require the second player to place the remaining fifteen men consecutively in such a manner that the succession may be *cyclical*, that is, so that No. 20 may be adjacent to No. 1; and it is always possible to answer any question of this kind. Thus, if *B C D F G* be the five given initial points, it is allowed to complete the succession by following the alphabetical order of the twenty consonants, as suggested by the diagram itself; but after placing the piece No. 6 in hole *H*, as before, it is *also* allowed (by the supposed conditions) to put No. 7 in *X* instead of *J*, and then to conclude with the succession, *W R S T V J K L M N P Q Z*. Other Examples of Icosian Problems, with solutions of some of them, will be found in the following page.

In graph theoretic terminology, the holes of the game board are referred to as *vertices* (singular: *vertex*) and the lines that join two holes (vertices) are called *edges*. The collection of vertices and edges in a given relationship (as represented by a diagram such as the game board) is called a *graph*. Two vertices that are joined by an edge in the graph are said to be *adjacent*. Thus, the instruction "always to proceed *along the lines* of the figure" requires the player to find a sequence of adjacent vertices; such a sequence is known as a *path*. In the case where no vertex is repeated in the sequence, the path is said to be a *simple path*. In the case where every vertex of the graph is used exactly once in the sequence, the path is said to be a *Hamiltonian path*. The term *cycle* is now used to describe what Hamilton referred to as a 'cyclical' path.

1. Explain why the rules of the Icosian Game require players to always find a simple path.

2. Use modern terminology to formally define the terms *cycle* and *Hamiltonian cycle*.

Following the preface, Hamilton includes several examples of Icosian Problems in the Instruction Pamphlet. We consider only the first two problems as a means of familiarizing ourselves with the concepts of Hamiltonian cycle and Hamiltonian path.

EXAMPLES OF ICOSIAN PROBLEMS
FIRST PROBLEM

Five initial points are given; cover the board, and finish *cyclically*. (As hinted in the preceding page, a succession is said to be *cyclical* when the *last* piece is adjacent to the *first*.)

[This problem is always possible in at least two, and sometimes in four, different ways. Two examples have been assigned: the following are a few others.]

Example 3. Given *B C P N M* as initial: two solutions exist; one is the succession, *D F K L T S R Q Z X W V J H G*; the other is *D F G H X W V J K L T S R Q Z*.

Example 4. Five initials, *L T S R Q*. Four solutions.

Example 5. Five initials, *J V T S R*. Two solutions.

3. Explain why Hamilton's first problem is equivalent to the problem of finding a Hamiltonian circuit beginning from a given initial sequence of five vertices.

4. In *Example 3*, Hamilton specifies *B C P N M* as the first five vertices in the desired circuit. He then claims that the two solutions listed in the example are the *only* two solutions of this particular problem. Prove that these are in fact the only two solutions by completing the details of the following argument. Include copies of the diagram illustrating each step of the argument in a different color as part of your proof.

   (a) Explain why the initial conditions for this example imply that the solution to the problem must include either the sequence *R S T* or the sequence *T S R*.

   (b) Explain why the initial conditions for this example imply that the solution to the problem must include either the sequence *R Q Z* or the sequence *Z Q R*.

   (c) Explain why we can now conclude that the solution to this problem must include either the sequence *X W V* or the sequence *V W X*.

   (d) Explain why the initial conditions for this example imply that the solution to the problem must include either the sequence *F D M* or the sequence *M D F*.

   (e) Explain why we can now conclude that the solution to this problem must include either the sequence *K L T* or the sequence *T L K*.

   (f) Use the information from above concerning which edges and vertices we know must be part of the solution to prove that the two circuits Hamilton lists are the only solutions to the problem.

5. In *Example 4*, Hamilton claims there are four Hamiltonian circuits that begin with the vertices *L T S R Q*. Find them. (You do not need to prove these are the only four.)

Although Hamilton claims that every initial sequence of five vertices will lead to at least two solutions (and possibly four) within the Icosian Game, he does not offer a proof of this claim. Nor does he claim that this is true of all graphs.

6. Show that it is not true of every graph that any initial sequence of five vertices will lead to at least one Hamiltonian circuit by finding an example of a graph with at least five vertices that has no Hamiltonian circuit. Prove that your graph does not contain a Hamiltonian circuit.

The next question pertains to Hamilton's second problem, which Hamilton describes in the pamphlet as follows:

EXAMPLES OF ICOSIAN PROBLEMS (continued)

SECOND PROBLEM

Three initial points are given; cover the board *non-cyclically*. (A succession is said to be *non-cyclical* when the *last* piece is *not* adjacent to the first.)

[This problem is sometimes soluble in only *one* way; sometimes in only *two* ways; sometimes in *four* ways; and sometimes it is not soluble at all, as will be seen in the following examples.]

Example 6. Three initial points, *B C D*; cover, and end with *T*. There is in this case only one solution, namely, *F G H X Z Q P N M L K J V W R S T*.

Example 7. Same initials; cover, and end with *W*. Two solutions.

Example 8. Same initials; cover, and end with *J*. Two solutions.

[The same number of solutions exists, if it be required, having the same three initials, to end with *K*, or *L*, or *N*, or *V*.]

Example 9. Same initials; cover, and end with *R*. Four solutions.

Example 10. Same initials; cover, and end with *M*. Impossible.

[The same result, if it be required to end with *F*, or *H*, or *P*, or *Q*, or *S*, or *X*.]

7. In *Example 10*, Hamilton claims the problem of finding a 'non-cyclical' path that uses all vertices beginning with *B C D* and ending with *M* is impossible. Prove that he is correct.

The Icosian Game and Non-Commutative Algebra

We now turn to the portion of Hamilton's pamphlet which links the Icosian Game to a non-commutative algebra.

HINTS ON THE ICOSIAN CALCULUS, OF WHICH THE ICOSIAN GAME IS DESIGNED TO BE AN ILLUSTRATION.

I. In a "MEMORANDUM respecting a New System of Roots of Unity," which appeared in the *Philosophical magazine* for December 1856, Sir W. R. Hamilton expressed himself nearly as follows (a few words only being here omitted):

'I have lately been led to the conception of a new system, or rather *family of systems,* of *non-commutative roots of unity*, which are entirely distinct from the *i j k* of quaternions, though having some general analogy thereto; and which admit, even more easily than the quaternion symbols do, of geometrical interpretation. In the system which seems at present to be the most interesting one among those included in this new family, I assume three symbols, ι, κ, λ, such that $\iota^2 = 1, \kappa^3 = 1, \lambda^5 = 1, \lambda = \iota\kappa$; where $\iota\kappa$ must be *distinguished* from $\kappa\iota$, since otherwise we should have $\lambda^6 = 1, \lambda = 1$. As a very simple *specimen* of the symbolical conclusions deduced from these fundamental assumptions I may mention that if we make $\mu = \iota\kappa^2 = \lambda\iota\lambda$, we shall have also $\mu^5 = 1, \lambda = \mu\iota\mu$; so that μ is a new fifth root of reciprocity. A long train of such symbolical deductions is found to follow; and every one of the results may be *interpreted* as having reference to the passage from *face to face* (or from corner to corner) of the *icosahedron* (or of the dodecahedron): on which account, I am at present disposed to give the name of 'Icosian Calculus' to this new system of symbols, and of rules for their operations.'

The system of *"non-commutative roots of unity"* described above employs three symbols ι, κ, λ subject to the following (non-commutative) rules:

$$\iota^2 = 1 , \kappa^3 = 1 , \lambda^5 = 1 , \lambda = \iota\kappa , \iota\kappa \neq \kappa\iota.$$

The symbol '1' represents the identity, so that $1\iota = \iota1 = \iota$, $1\kappa = \kappa1 = \kappa$, and $1\lambda = \lambda1 = \lambda$. In Part II of *Hints on the Icosian Calculus*, Hamilton describes in detail how to interpret his system of 'non-commutative roots of unity' within the Icosian Game. First, consider only the symbolic action of ι, κ, λ as defined by the above multiplication rules to complete Questions 8 and 9.

8. Prove symbolically that $\kappa = \iota\lambda$.

9. Prove symbolically that $\iota\kappa^2 = \lambda\iota\lambda$.

   (This shows that it makes sense to define the new symbol μ by $\mu = \iota\kappa^2 = \lambda\iota\lambda$.)

Extra Credit Question. Show symbolically that $\mu^5 = 1$.

We now consider Hamilton's interpretation of this algebraic system within the Icosian Game. This interpretation provides a concrete method for deriving new symbolic equations such as those mentioned in the following excerpt.

HINTS ON THE ICOSIAN CALCULUS (continued)

II. In a LITHOGRAPH, which was distributed in Section A of the British Association, during its Meeting at Dublin in 1857, Sir W. R. H. pointed out a few other symbolical results of the same kind: especially the equations $\lambda\mu^2\lambda = \mu\lambda\mu$, $\mu\lambda^2\mu = \lambda\mu\lambda$, $\lambda\mu^3\lambda = \mu^2$, $\mu\lambda^3\mu = \lambda^2$; and the formula $(\lambda^3\mu^3(\lambda\mu)^2)^2 = 1$, which serves as a *common mathematical type* for the solution of *all cases* of the First Problem of the Game. He also gave at the same time an oral (and hitherto unprinted) account of his rules of *interpretation* of the principal symbols; which rules, with reference to the present Icosian Diagram (or ICOSIAN), may be briefly stated as follows:

1. The operation ι *reverses* (or reads backwards) a *line* of the figure; changing, for example, BC to CB.

2. The operation κ causes a line to *turn* in a particular direction round its final point; changing, for instance, BC to DC.

3. The operation λ changes a line considered as a *side* of a pentagon to the *following side* thereof, proceeding always *right-handedly* for every pentagon except the large or outer one; thus λ changes BC to CD, but SR to RW.

4. The operation μ is *contrasted* with λ, and changes a line considered as a side of a *different pentagon*, and in the *opposite order* or rotation, to the consecutive side of that *other* pentagon; thus μ changes BC to CP, and SR to RQ; but it changes also RS to ST, whereas λ would change RS to SN.

5. The only operations employed in the *game* are those marked λ and μ; but another operation, $\omega = \lambda\mu\lambda\mu\lambda = \mu\lambda\mu\lambda\mu$, having the property that $\omega^2 = 1$, was also mentioned in the Lithograph above referred to; and to complete the present statement of *interpretations*, it may be added that the effect of this operation ω is to change an *edge* of a pentagonal *dodecahedron* to the *opposite edge* of that *solid*; for example, in the diagram, BC to TV.

Note that proceeding "right handedly" may be described as moving clockwise around the appropriate pentagon, so that action which is "in the opposite order" of proceeding "right handedly" may be described as moving counterclockwise.

10. Use the interpretation of ι in (1) to explain why $\iota^2 = 1$.

    Begin by looking at the effect of applying the operation ι *twice in succession*, beginning with the edge BC. Then explain in general.

11. Use the interpretation of κ in (2) to explain why $\kappa^3 = 1$.

    Begin by looking at the effect of applying the operation κ *three times in succession*, beginning with the edge BC. Then look at the effect of applying the operation κ *three times in succession*, beginning with the edge PN. Finally, explain in general.

12. Use the interpretation of λ in (3) to explain why $\lambda^5 = 1$.

    Begin by looking at the effect of applying the operation λ *five times in succession*, beginning with the edge BC. Then look at the effect of applying the operation λ *five times in succession*, beginning with the edge SR. Finally, explain in general.

13. Use the interpretation of μ in (4) to explain why $\mu^5 = 1$.

    Begin by looking at the effect of applying the operation μ *five times in succession*, beginning with the edge BC. Then look at the effect of applying the operation μ *five times in succession*, beginning with the edge RS. Finally, explain in general. (Note that this provides a geometric solution of the extra credit question stated above, immediately following question 9.)

14. Beginning with the edge BC, use the interpretations given for the four symbols $\iota, \kappa, \lambda, \mu$ to illustrate that $\mu = \iota\kappa^2 = \lambda\iota\lambda$.

Extra Credit Question.   Establish the equation $\lambda\mu^2\lambda = \mu\lambda\mu$ symbolically; then illustrate this equation within the Icosian Game, first beginning with the edge BC and then beginning with the edge RS.

Some Closing Remarks

Notice Hamilton's claim that an *algebraic proof* using equations of the type $(\lambda^3\mu^3(\lambda\mu)^2)^2 = 1$ can be used to find all Hamiltonian cycles beginning with a specified initial sequence of five vertices. Notice also the contrast between the graph theoretical proof that you completed in question 4 above, and the proof that we would get of this same result by interpreting this degree 20 equation within the context of the Icosian Game Board. In general, however, it is not viable to associate an algebraic system with an arbitrary graph as a means to find all Hamiltonian circuits within that graph. In fact, as you demonstrated in question 6 above, a graph may contain no Hamiltonian circuits at all. Unlike the known situation for other kinds of circuits (e.g., Euler circuits), there is no known simple condition on a graph which allows one to determine in all cases whether a Hamiltonian circuit exists or not. In the case that a graph does contain a Hamiltonian circuit, we say the graph is *Hamiltonian*.

The more general question of determining a condition under which a graph is Hamiltonian was first studied by Thomas Penyngton Kirkman (1840–1892). Unlike Hamilton, who was primarily interested in the algebraic connections of one specific graph, Kirkman was interested in the general study of 'Hamiltonian circuits' in arbitrary graphs. The rector of a small and isolated English parish, Kirkman presented a paper on this subject to the Royal Society on 6 August 1855. Regrettably, his solution of the problem was incorrect. He did, however, present a second paper in 1856 in which he described a general class of graphs which do not contain such a circuit. Kirkman also studied the existence of Hamiltonian circuits on the dodecahedron, a variation of the Icosian Game which Hamilton also studied. In fact, the two men met once in 1861 when Hamilton visited Kirkman at his rectory. That Hamilton's name became associated with the circuits, and not Kirkman's, appears to be one of the accidents of history, or perhaps a credit to the fame of Hamilton's quaternions and work in mathematical physics.

Notes for the instructor

This project contains two sub-sections "The Icosian Game and Hamiltonian Circuits" and "The Icosian Game and Non-Commutative Algebra," both of which were developed specifically for use in an introductory undergraduate course in discrete mathematics. Because no prior background in graph theory is assumed, the connection to symbolic algebra makes the project suitable for use in a junior-level abstract algebra course as well. In a discrete mathematics course, the project could be assigned independently or in conjunction with one or both of the projects "Early Writings on Graph Theory: Euler Circuits and The Königsberg Bridge Problem" and "Early Writings on Graph Theory: Topological Connections," both of which appear in this volume. For students with no prior knowledge of non-commutative algebras, the instructor may wish to provide more explicit directions for Questions 8 and 9, or work these together as a whole class. Otherwise, the project may be completed by students working in small groups over 2–3 in-class days, or assigned as a week-long individual project outside of class. Multiple copies of the Icosian Game diagram will be needed for each student; use of color pencils or markers is also highly recommended.

Bibliography

[1] Biggs, N., Lloyd, E., Wilson, R., *Graph Theory: 1736–1936*, Clarendon Press, Oxford, 1976.

[2] Crowe, M. J., *A History of Vector Analysis: The Evolution of the Idea of a Vectorial System*, Dover Publications, New York, 1994.

[3] Hamilton, W. R., "Account of the Icosian Game," *Proc. Roy. Irish. Acad.* **6** (1853-7), 415–416.

[4] Katz, V., *A History of Mathematics: An Introduction*, Second Edition, Addison-Wesley, New York, 1998.

Are All Infinities Created Equal?[1]

Guram Bezhanishvili
New Mexico State University

Georg Ferdinand Ludwig Philip Cantor (1845–1918), the founder of set theory, and considered by many as one of the most original minds in the history of mathematics, was born in St. Petersburg, Russia in 1845. His parents, who were of Jewish descent, moved the family to Frankfurt, Germany in 1856. Georg entered the Wiesbaden Gymnasium at the age of 15, and two years later began his university career at Zürich. In 1863 he moved to the University of Berlin, which during Cantor's time was considered the world's center of mathematical research. Four years later Cantor received his doctorate from the great Karl Weierstrass (1815–1897). In 1869 Cantor obtained an unpaid lecturing post, which ten years later flourished into a full professorship, at the minor University of Halle. However, he never achieved his dream of holding a Chair of Mathematics at Berlin. It is believed that one of the main reasons for this was the rejection of his theories of infinite sets by the leading mathematicians of that time, most noticeably by Leopold Kronecker (1823–1891), a professor at the University of Berlin and a very influential figure in German mathematics, both mathematically and politically.

Cantor married in 1874 and had two sons and four daughters. Ten years later Georg suffered the first of the mental breakdowns that were to plague him for the rest of his life. He died in 1918 in a mental hospital at Halle. By that time his revolutionary ideas were becoming accepted by some of the leading figures of the new century. For example, one of the greatest mathematicians of the twentieth century, David Hilbert (1862–1943), described Cantor's new mathematics as "the most astonishing product of mathematical thought" [5, p. 359], and claimed that "no one shall ever expel us from the paradise which Cantor has created for us" [5, p. 353].

In this project we will learn about Cantor's treatment of infinite sets. We will discuss the cardinality of a set, the notion of equivalence of two sets, and study how to compare infinite sets with each other. We will introduce countable sets and show that many sets are countable, including the set of integers and the set of rational numbers. We will also discuss Cantor's diagonalization method which allows us to show that not every infinite set is countable. In particular, we will show that the set of real numbers is not countable. We will also examine the cardinal number $\aleph_0$, the first in the hierarchy of transfinite cardinal numbers, and obtain a method that allows us to create infinitely many transfinite cardinal numbers.

We will learn much of this by studying and working with the historical source [3], which is an English translation of two papers by Cantor [1, 2] that appeared in 1895 and 1897. More on Georg Cantor can be found in [4, 5, 6] and in the literature cited therein.

We begin by reading Cantor's definition of the cardinal number of a given set. Note that in this translation Jourdain uses "aggregate" instead of the more familiar "set."

1. Read carefully the following quote from Cantor.

> Every aggregate M has a definite "power," which we also call its "cardinal number."
>
> We will call by the name "power" or "cardinal number" of M the general concept which, by means of our active faculty of thought, arises from the aggregate M when we make abstraction of the

[1] Thanks are due to Joel Lucero-Bryan for team-teaching this project in an undergraduate course in discrete mathematics at New Mexico State University.

nature of its various elements m and of the order in which they are given.

We denote the result of this double act of abstraction, the cardinal number or power of M, by

$$\overline{\overline{M}}.$$

What do you think Cantor means by "cardinal number"? Why? Given a set M consisting of ten round marbles, each of a different color, what is $\overline{\overline{M}}$?

2. Read the following quote from Cantor.

> We say that two aggregates M and N are "equivalent," in signs
>
> $$M \sim N \quad \text{or} \quad N \sim M,$$
>
> if it is possible to put them, by some law, in such a relation to one another that to every element of each one of them corresponds one and only one element of the other.

In modern terminology describe what it means for two sets to be equivalent.

3. Prove the following claim of Cantor.

> Every aggregate is equivalent to itself:
> $$M \sim M.$$

4. Prove the following claim of Cantor.

> If two aggregates are equivalent to a third, they are equivalent to one another, that is to say:
>
> $$\text{from} \quad M \sim P \quad \text{and} \quad N \sim P \quad \text{follows} \quad M \sim N.$$

5. Read carefully the following quote from Cantor.

> Of fundamental importance is the theorem that two aggregates M and N have the same cardinal number if, and only if, they are equivalent: thus,
>
> $$\text{from } M \sim N, \text{ we get } \overline{\overline{M}} = \overline{\overline{N}},$$
>
> and
>
> $$\text{from } \overline{\overline{M}} = \overline{\overline{N}}, \text{ we get } M \sim N.$$
>
> Thus the equivalence of aggregates forms the necessary and sufficient condition for the equality of their cardinal numbers.

Explain in your own words what Cantor means in the above.

6. Let **P** be the set of all perfect squares

$$\{0, 1, 4, 9, 16, 25, \ldots\},$$

and let **N** denote the set of all natural numbers

$$\{0, 1, 2, 3, 4, 5, \ldots\}.$$

From Cantor's statement above, do **P** and **N** have the same cardinality? Justify your answer.

7. Let **Z** denote the set of all integers. Do **N** and **Z** have the same cardinality? Justify your answer.

8. Let **N** × **N** denote the Cartesian product of **N** with itself; that is

$$\mathbf{N} \times \mathbf{N} = \{(n, m) : n, m \in \mathbf{N}\}.$$

Do **N** and **N** × **N** have the same cardinality? Justify your answer. Hint: Draw a picture of **N** × **N**. Can you label each element of **N** × **N** by a unique natural number?

9. Let **Q** denote the set of all rational numbers; that is

$$\mathbf{Q} = \left\{ \frac{a}{b} : a \in \mathbf{Z} \text{ and } b \in \mathbf{N} - \{0\} \right\}.$$

What is the cardinality of **Q**? Justify your answer. Hint: Establish a 1-1 correspondence between **Q** and (a subset of) **Z** × (**N** − {0}) and modify your solution to (8).

10. Read carefully the following quote from Cantor.

> If for two aggregates M and N with the cardinal numbers $\mathfrak{a} = \overline{\overline{M}}$ and $\mathfrak{b} = \overline{\overline{N}}$, both the conditions:
>
> (a) There is no part[2] of M which is equivalent to N,
> (b) There is a part N_1 of N, such that $N_1 \sim M$,
>
> are fulfilled, it is obvious that these conditions still hold if in them M and N are replaced by two equivalent aggregates M' and N'. Thus they express a definite relation of the cardinal numbers $\mathfrak{a}$ and $\mathfrak{b}$ to one another.
>
> Further, the equivalence of M and N, and thus the equality of $\mathfrak{a}$ and $\mathfrak{b}$, is excluded; for if we had $M \sim N$, we would have, because $N_1 \sim M$, the equivalence $N_1 \sim N$, and then, because $M \sim N$, there would exist a part M_1 of M such that $M_1 \sim M$, and therefore we should have $M_1 \sim N$; and this contradicts the condition (a).
>
> Thirdly, the relation of $\mathfrak{a}$ to $\mathfrak{b}$ is such that it makes impossible the same relation of $\mathfrak{b}$ to $\mathfrak{a}$; for if in (a) and (b) the parts played by M and N are interchanged, two conditions arise which are contradictory to the former ones.
>
> We express the relation of $\mathfrak{a}$ to $\mathfrak{b}$ characterized by (a) and (b) by saying: $\mathfrak{a}$ is "less" than $\mathfrak{b}$ or $\mathfrak{b}$ is "greater" than $\mathfrak{a}$; in signs
>
> $$\mathfrak{a} < \mathfrak{b} \text{ or } \mathfrak{b} > \mathfrak{a}.$$

Describe in modern terminology when two cardinals $\mathfrak{a} = \overline{\overline{M}}$ and $\mathfrak{b} = \overline{\overline{N}}$ are in the relation $\mathfrak{a} < \mathfrak{b}$.

11. Prove the following claim of Cantor.

> We can easily prove that,
>
> if $\mathfrak{a} < \mathfrak{b}$ and $\mathfrak{b} < \mathfrak{c}$, then we always have $\mathfrak{a} < \mathfrak{c}$.

12. Read carefully the following quote from Cantor.

> Aggregates with finite cardinal numbers are called "finite aggregates," all others we will call "transfinite aggregates" and their cardinal numbers "transfinite cardinal numbers."
>
> The first example of a transfinite aggregate is given by the totality of finite cardinal numbers v; we call its cardinal number "Aleph-zero," and denote it by $\aleph_0$;

In the modern terminology, a set whose cardinal number is $\aleph_0$ is called "countable." What symbol is used today to denote the "totality of finite cardinal numbers v"?

---

[2]The modern terminology is "subset".

13.   Prove the following claim of Cantor.

The number $\aleph_0$ is greater than any finite number μ:

$$\aleph_0 > \mu.$$

14.   Prove the following claim of Cantor.

On the other hand, $\aleph_0$ is the least transfinite cardinal number. If $\mathfrak{a}$ is any transfinite cardinal number different from $\aleph_0$, then

$$\aleph_0 < \mathfrak{a}.$$

Hint: Let $\mathfrak{a} = \overline{\overline{A}}$. Can you define a 1-1 map from $\mathbf{N}$ into A? What can you deduce from this?

15.   Let $[0, 1]$ denote the set of all real numbers between 0 and 1. Show that $\aleph_0 < \overline{\overline{[0, 1]}}$. We outline what is now known as Cantor's diagonalization method as one way to prove this. Represent real numbers in $[0, 1]$ as infinite decimals (which do not end in infinitely repeating 9's). Assume that $\mathbf{N} \sim [0, 1]$. Then to each infinite decimal one can assign a unique natural number, so the infinite decimals can be enumerated as follows:

$$.a_{11}a_{12}\ldots a_{1n}\ldots$$
$$.a_{21}a_{22}\ldots a_{2n}\ldots$$
$$\vdots$$
$$.a_{n1}a_{n2}\ldots a_{nn}\ldots$$
$$\vdots$$

Can you construct an infinite decimal $.b_1 b_2 \ldots b_n \ldots$ such that $a_{nn} \neq b_n$ for each positive n? What can you conclude from this?

16.   Let $\mathbf{R}$ denote the set of all real numbers. Is $\overline{\overline{\mathbf{R}}}$ strictly greater than $\aleph_0$? Justify your answer.

17.   For a set M, let $\mathcal{P}(M)$ denote the set of all subsets of M; that is $\mathcal{P}(M) = \{N : N \subseteq M\}$. Prove the following claim of Cantor:

$$\overline{\overline{\mathcal{P}(M)}} > \overline{\overline{M}}.$$

Hint: Employ a generalized version of Cantor's diagonalization method. Assume that $M \sim \mathcal{P}(M)$. Then there is a 1-1 and onto function $f : M \to \mathcal{P}(M)$. Consider the set $N = \{m \in M : m \notin f(m)\}$. Can you deduce that $N \subseteq M$ is not in the range of f? Does this imply a contradiction?

18.   Using the previous exercise, give an infinite increasing sequence of transfinite cardinal numbers.

Notes for the instructor

This project is designed for an undergraduate course in discrete mathematics. It could be assigned as a three-week project on naive set-theory with an emphasis on 1-1 correspondences. Since some of Cantor's writings require nontrivial interpretations, it is advisable that, in the beginning, the instructor leads the class carefully, especially in reading Cantor's "definition" of cardinal number. The instructor may also wish to lead the class in discovering that the set of rational numbers is countable, and especially in using Cantor's diagonalization method to show that the set of real numbers is not countable. There is a shadow of the axiom of choice in Cantor's claim that $\aleph_0 < \mathfrak{a}$ for any transfinite cardinal number $\mathfrak{a}$ different from $\aleph_0$ (Exercise 14). The instructor may wish to spend a little bit of class time on giving an informal explanation of the main idea behind the axiom of choice.

Bibliography

[1] Cantor, G., *Beiträge zur Begründung der transfiniten Mengenlehre. I*, Mathematische Annalen 46 (1895) 481–512.

[2] Cantor, G., *Beiträge zur Begründung der transfiniten Mengenlehre. II*, Mathematische Annalen 49 (1897) 207–246.

[3] Cantor, G., *Contributions to the Founding of the Theory of Transfinite Numbers*, Philip Jourdain (translator), Dover Publications Inc., New York, 1952.

[4] Dunham, W., *Journey Through Genius. The Great Theorems of Mathematics*, John Wiley & Sons Inc., New York, 1990.

[5] Hollingdale, S., *Makers of Mathematics*, Penguin Books, New York, 1994.

[6] Laubenbacher, R., Pengelley, D., *Mathematical Expeditions: Chronicles by the Explorers*, Springer Verlag, New York, 1999.

Early Writings on Graph Theory: Topological Connections

Janet Heine Barnett
Colorado State University - Pueblo

Introduction

The earliest origins of graph theory can be found in puzzles and game, including Euler's Königsberg Bridge Problem and Hamilton's Icosian Game. A second important branch of mathematics that grew out of these same humble beginnings was the study of position ("analysis situs"), known today as *topology*[1]. In this project, we examine some important connections between algebra, topology and graph theory that were recognized during the years from 1845–1930.

The origin of these connections lie in work done by physicist Gustav Robert Kirchhoff (1824–1887) on the flow of electricity in a network of wires. Kirchhoff showed how the current flow around a network (which may be thought of as a graph) leads to a set of linear equations, one for each circuit in the graph. Because these equations are not necessarily independent, the question of how to determine a complete set of mutually independent equations naturally arose. Following Kirchhoff's publication of his answer to this question in 1847, mathematicians slowly began to apply his mathematical techniques to problems in topology. The work done by the French mathematician Henri Poincaré (1854–1912) was especially important, and laid the foundations of a new subject now known as "algebraic topology."

This project is based on excerpts from a 1922 paper in which Oswald Veblen [1880–1960] shows how Poincaré formalized the ideas of Kirchhoff. An American mathematician born in Iowa, Veblen's father was also a mathematician who taught mathematics and physics at the State University of Iowa. At that time, graduate programs in mathematics were relatively young in the United States. A member of the first generation of American mathematicians to complete their advanced work in the United States rather than Europe, Oswald Veblen completed his Ph.D. at the University of Chicago in 1903. He remained in Chicago for two years before joining the mathematics faculty at Princeton. In 1930, he became the first faculty member of the newly founded Institute for Advanced Study at Princeton.[2] A talented fund-raiser and organizer, Veblen also served on the Institute's Board of Trustees in its early years.

During the Nazi years, Veblen was instrumental in assisting European mathematicians to find refuge in the United States. Although some American mathematicians, including George Birkhoff (1844–1944), voiced opposition to these efforts — fearing that talented young American mathematicians would lose academic positions to the immigrants — the Rockefeller Foundation and other philanthropic bodies provided financial support to these efforts as a means of

---

[1] According to Euler, the first person to discuss "analysis situs" was the mathematician and philosopher Gottfried Leibniz (1646–1716). In a 1679 letter to Christian Huygens (1629–1695), Leibniz wrote:

> I am not content with algebra, in that it yields neither the shortest proofs nor the most beautiful constructions of geometry. Consequently, in view of this, I consider that we need yet another kind of analysis, geometric or linear, which deals directly with position, as algebra deals with magnitude. [1, p. 30]

Although Leibniz himself did not appear to make contributions to the development of *analysis situs*, he did make important contributions to the development of another kind of analysis. Today, Leibniz is recognized alongside the mathematician and physicist Isaac Newton (1642–1727) as an independent co-inventor of calculus.

[2] The celebrated physicist Albert Einstein (1879–1955) was another early faculty member of the Institute, joining Veblen there in 1931.

recruiting world-class mathematicians and scientists to the United States. Veblen was also instrumental in the establishment of the American Mathematical Society's *Mathematical Reviews*, a publication aimed at providing researchers with reviews of recent mathematical papers in a timely fashion. Founded during the late 1930s when the well-known German review journal *Zentralblatt für Mathematik und ihre Grenzgebiete* was refusing to publish reviews written by Soviet and Jewish scholars, the *Mathematical Reviews* continues to play an important role in disseminating research results and promoting communication within the mathematical community.

In addition to his administrative and philanthropic work, Veblen was an active researcher who made important contributions in projective and differential geometry in addition to topology, authoring influential books in all three areas. In this project, we examine extracts from his *Analysis Situs*, the first textbook to be written on combinatorial topology. Veblen first presented this work in a series of invited Colloquium Lectures of the American Mathematical Society in 1916. Although he remained interested in topology afterwards, he published little research in this area following the 1922 publication of *Analysis Situs*. The extracts we examine are taken from [1, pp. 136–141].

Note: This project assumes the reader is familiar with basic notions of graph theory, including the definition of *isomorphism* and *isomorphism invariant*. Parts of the project (clearly marked as such) also assume familiarity with the basic linear algebra concepts of *rank*, *kernel* and *linear independence*. As needed, the reader should refer to a standard linear algebra textbook to review these concepts.

Analysis Situs

American Mathematical Society Colloquium Lectures 1916

Symbols for Sets of Cells

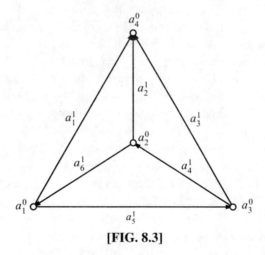

[FIG. 8.3]

14   Let us denote the 0-cells of a one-dimensional complex C_1 by $a_1^0, a_2^0, \ldots, a_{\alpha_0}^0$ and the 1-cells by $a_1^1, a_2^1, \ldots, a_{\alpha_1}^1$.

Any set of 0-cells of C_1 may be denoted by a symbol $(x_1, x_2, \ldots, x_{\alpha_0})$ in which $x_i = 1$ if a_i^0 is in the set and $x_i = 0$ if a_i^0 is not in the set. Thus, for example, the pair of points a_1^0, a_4^0 in Fig. [8.3] is denoted by $(1, 0, 0, 1)$. The total number of symbols $(x_1, x_2, \ldots, x_{\alpha_0})$ is 2^{α_0}. Hence the total number of sets of 0-cells, barring the null-set, is $2^{\alpha_0} - 1$. The symbol for a null-set, $(0, 0, \ldots, 0)$ will be referred to as zero and denoted by 0. The marks 0 and 1 which appear in the symbols just defined, may profitably be regarded as residues, modulo 2, i.e., as symbols which may be combined algebraically according to the rules

$$0 + 0 = 1 + 1 = 0, \qquad 0 + 1 = 1 + 0 = 1,$$
$$0 \times 0 = 0 \times 1 = 1 \times 0 = 0, \qquad 1 \times 1 = 1$$

Under this convention the sum (mod. 2) of two symbols, or of the two sets of points which correspond to the symbols $(x_1, x_2, \ldots, x_{\alpha_0}) = X$ and $(y_1, y_2, \ldots, y_{\alpha_0}) = Y$, may be defined as $(x_1 + y_1, x_2 + y_2, \ldots, x_{\alpha_0} + y_{\alpha_0}) = X + Y$.

Geometrically, $X + Y$ is the set of all points which are in X or in Y but not in both. For example, if $X = (1, 0, 0, 1)$ and $Y = (0, 1, 0, 1)$, $X + Y = (1, 1, 0, 0)$; i.e., X represents a_1^0 and a_4^0, Y represents a_2^0 and a_4^0, and $X + Y$ represents a_1^0 and a_2^0. Since a_4^0 appears in both X and Y, it is suppressed in forming the sum, modulo 2. This type of addition has the obvious property that if two sets contain each an even number of 0-cells, the sum (mod. 2) contains an even number of 0-cells.

15   Any set, S, of 1-cells in C_1 may be denoted by a symbol $(x_1, x_2, \ldots, x_{\alpha_1})$ in which $x_i = 1$ if a_i^1 is in the set and $x_i = 0$ if a_i^1 is not in the set. The 1-cells in the set may be thought of as labeled with 1's and those not in the set as labeled with 0's. The symbol is also regarded as representing the one-dimensional complex composed of the 1-cells of S and the 0-cells which bound them. Thus, for example, in Fig. [8.3] the boundaries of two of the faces are $(1, 0, 1, 0, 1, 0)$ and $(1, 1, 0, 0, 0, 1)$. The sum (mod. 2) of two symbols $(x_1, x_2, \ldots, x_{\alpha_1})$ is defined in the same way as for the case of symbols representing 0-cells. Correspondingly if C_1' and C_1'' are one-dimensional complexes which have a certain number (which may be zero) of 1-cells in common and have no other common points except the ends of these 1-cells, the sum $C_1' + C_1''$ (mod. 2) is defined as the one-dimensional complex obtained by suppressing all 1-cells common to C' and C'' and retaining all 1-cells which appear only in C_1' or in C_1''. For example, in Fig. [8.3], the sum of the two curves represented by $(1, 0, 1, 0, 1, 0)$ and $(1, 1, 0, 0, 0, 1)$ is $(0, 1, 1, 0, 1, 1)$ which represents the curve composed of $a_2^1, a_4^1, a_5^1, a_6^1$ and their ends.

1. The following questions are based on Section 14 of Veblen's paper, in which Veblen discusses symbols for sets of 0-cells.

   (a) In graph terminology, what is a '0-cell'? What is a '1-cell'?
   (b) How would the set of points $\{a_1^0, a_3^0, a_4^0\}$ in Fig. 8.3 be represented by Veblen?
   (c) Identify the set of points in Fig. 8.3 that are represented by the following 4-tuples.
$$W = (0, 1, 1, 0) \qquad\qquad Z = (1, 1, 1, 1)$$
   (d) Find $W + Z$ for $W = (0, 1, 1, 0)$ and $Z = (1, 1, 1, 1)$.
   Identify the set of points in Fig. 8.3 that is represented by $W + Z$.
   (e) In the last paragraph of Section 14, Veblen asserts the addition modulo 2 has a certain 'obvious property.' What property is this? How obvious is this property? Give a formal proof of the property, and interpret it in terms of sets of '0-cells'.

2. The following questions are based on Section 15 of Veblen's paper, in which Veblen discusses symbols for sets of 1-cells.

   Referring to Fig. 8.3 in Veblen's paper, let

$$M \text{ be the circuit defined by edges } a_1^1, a_3^1, a_4^1, a_6^1,$$
$$N \text{ be the circuit defined by edges } a_2^1, a_3^1, a_4^1.$$

   (a) How would Veblen represent M and N as 6-tuples?
   (b) Find $M + N$ mod 2, and identify the circuit represented by this sum in Fig. 8.3.

In sections 16 and 17 of his article, Veblen defines two important matrices associated with a graph, denoted by him as H_0 and H_1 respectively. In section 20, he provides more detail concerning the matrix H_1 for graphs that are not connected. Project questions pertaining to each of these three sections of Veblen's paper provide additional clarification of these ideas.

16   Any one-dimensional complex falls into R_0 sub-complexes each of which is connected. Let us denote these sub-complexes by $C_1^1, C_1^2, \ldots, C_1^{R_0}$ and let the notation be assigned in such a way that a_i^0 $(i = 1, 2, \ldots, m_1)$ are the 0-cells of C_1^1, a_i^0 $(i = m_1 + 1, m_1 + 2, \ldots, m_2)$ those of C_1^2, and so on.

With this choice of notation, the sets of vertices of $C_1^1, C_1^2, \ldots, C_1^{R_0}$, respectively, are represented by the symbols $(x_1, x_2, \ldots, x_{\alpha_0})$ which constitute the rows of the following matrix.

$$
H_0 = \left\|
\begin{array}{ccc}
\overbrace{11\ldots 1}^{m_1} & \overbrace{00\ldots 0}^{m_2 - m_1} & \overbrace{00\ldots 0}^{\alpha_0 - m_{R_0 - 1}} \\
00\ldots 0 & 11\ldots 1 & 00\ldots 0 \\
\vdots & & \vdots \\
\vdots & & \vdots \\
00\ldots 0 & 00\ldots 0 & 11\ldots 1
\end{array}
\right\| = \|\eta_{ij}^0\|
$$

For most purposes it is sufficient to limit attention to connected complexes. In such cases $R_0 = 1$, and H_0 consists of one row all of whose elements are 1.

3. The following questions are based on Section 16 of Veblen's paper, in which Veblen defines the matrix H_0.

   (a) In the first paragraph of this section, Veblen assumes that we may label the '0-cells of a one-dimensional complex' in a particular way. What graph isomorphism invariant allows him to make this assumption?

   (b) Find the matrix H_0 for graphs G_1 and G_2 in the appendix. Using Veblen's notation, label the vertices and edges of each graph so that the correspondence to the associated matrix H_0 is clear.

Continuing with Section 17, we consider Veblen's initial discussion of the matrix H_1:

17   By the definition ... a 0-cell is incident with a 1-cell if it is one of the ends of the 1-cell, and under the same conditions the 1-cell is incident with the 0-cell. The incidence relations between the 0-cells and the 1-cells may be represented in a table or matrix of α_0 rows and α_1 columns as follows: The 0-cells of C_1 having been denoted by a_i^0, $(i = 1, 2, \ldots, \alpha_0)$ and the 1-cells by a_j^1, $(j = 1, 2, \ldots, \alpha_1)$, let the element of the ith row and the jth column of the matrix be 1 if a_i^0 is incident with a_j^1 and let it be 0 if a_i^0 is not incident with a_j^1. For example, the table for the linear graph of Fig. [8.3] formed by the vertices and edges of a tetrahedron is as follows:

	α_1^1	α_2^1	α_3^1	α_4^1	α_5^1	α_6^1
α_1^0	1	0	0	0	1	1
α_2^0	0	1	0	1	0	1
α_3^0	0	0	1	1	1	0
α_4^0	1	1	1	0	0	0

In the case of the complex used ... to define a simple closed curve

$$
\left\|
\begin{array}{cc}
1 & 1 \\
1 & 1
\end{array}
\right\|
$$

We shall denote the element of the ith row and jth column of the matrix of incidence relations between the 0-cells and 1-cells by η_{ij}^1 and the matrix itself by

$$
\|\eta_{ij}^1\| = H_1
$$

The ith row of H_1 is the symbol for the set of all 1-cells incident with a_i^0 and the jth column is the symbol for the set of all 0-cells incident with a_j^1.

The condition which we have imposed on the graph, that both ends of every 1-cell shall be among the α_0 0-cells, implies that every column of the matrix contains exactly two 1's. Conversely, any matrix whose elements are 0's and 1's and which is such that each column contains exactly two 1's can be regarded as the incidence matrix of a linear graph. For to obtain such a graph it is only necessary to take α_0 points in a 3-space, denote them arbitrarily by $a_1^0, a_2^0, \ldots, a_{\alpha_0}^0$, and join the pairs which correspond to 1's in the same column successively by arcs not meeting the arcs previously constructed.

4. The following questions are based on Section 17 of Veblen's paper, in which Veblen defines the incidence matrix H_1.

   (a) Find the incidence matrix H_1 for graph G_3 in the appendix.

   (b) Sketch and label a graph with incidence matrix

$$H_1 = \begin{pmatrix} 1 & 1 & 1 & 1 & 0 & 0 & 0 \\ 1 & 0 & 0 & 0 & 1 & 1 & 0 \\ 0 & 0 & 0 & 0 & 0 & 1 & 1 \\ 0 & 1 & 1 & 0 & 1 & 0 & 0 \\ 0 & 0 & 0 & 1 & 0 & 0 & 1 \end{pmatrix}.$$

   (c) Give two reasons why no graph can have incidence matrix

$$H_1 = \begin{pmatrix} 1 & 1 & 1 & 1 & 0 \\ 0 & 1 & 1 & 0 & 1 \\ 1 & 1 & 0 & 0 & 1 \end{pmatrix}.$$

   (d) Explain why isomorphic graphs do not necessarily have the same incidence matrix.
   Is there some way in which we could use incidence matrices to determine if two graphs are isomorphic? Explain.

In Section 20 of his article, Veblen provides a more detailed analysis of the form and properties of H_1 for graphs that are not connected.

20 Denoting the connected sub-complexes of C_1 by $C_1^1, C_1^2, \ldots, C_1^{R_0}$ as in 16 let the notation be so assigned that $a_1^1, a_2^1, \ldots, a_{m_1}^1$ are the 1-cells in C_1, $a_{m_1+1}^1, a_{m_1+2}^1, \ldots, a_{m_2}^1$ the 1-cells in C_2; and so on. The matrix H_1 then must take the form

I	0	0	0	
0	II	0	0	
0	0	III	0	

where all the non-zero elements are to be found in the matrices I, II, III, etc., and I is the matrix of C_1, II of C_2, etc. This is evident because no element of one of the complexes C_1^i is incident with any element of any of the others.

There are two non-zero elements in each column of H_1. Hence if we add the rows corresponding to any of the blocks I, II, etc. the sum is zero (mod. 2) in every column. Hence the rows of H_1 are connected by R_0 linear relations.

Any linear combination (mod. 2) of the rows of H_1 corresponds to adding a certain number of them together. If this gave zeros in all the columns it would mean that there were two or no 1's in each column of the matrix formed by the given rows, and this would mean that any 1-cell incident with one of the 0-cells corresponding to these rows would also be incident with another such 0-cell. These 0-cells and the 1-cells incident with them would therefore form a sub-complex of C_1 which was not connected with any of the remaining 0-cells and 1-cells of C_1. Hence it would consist of one or more of the complexes C_1^i ($i = 1, 2, \ldots, R_0$) and the linear relations with which we started would be dependent on the R_0 relations already found. Hence there are exactly R_0 linearly independent linear relations among the rows of H_1, so that if ρ_1 is the rank of H_1,

$$\rho_1 = \alpha_0 - R_0.$$

5. The following questions are based on Section 20 of Veblen's paper, in which he discusses the matrix H_1 for graphs that are not connected.

   (a) In the first paragraph of this section, Veblen asserts that the matrix H_1 must take the form of a block diagonal matrix. Illustrate this by finding the matrix H_1 for graphs G_1 and G_2 in the appendix. Then explain in general why this must be true.

   (b) In the second paragraph of this section, Veblen asserts that the rows of matrix H_1 are related by R_0 linear relations. What does the value R_0 represent? Determine the R_0 linear relations for the incidence matrices H_1 of graphs G_1 and G_2 from part (a) above.

   Suggestions: Denote the ith row of H_1 by z_i. You may also wish to review Section 14 of Veblen's paper in which he explains the notation '0' as it applies to the representation of sets of 0-cells.

   (c) Verify the relationship $\rho_1 = \alpha_0 - R_0$ for the incidence matrices of graphs G_1 and G_2.

   Note.   *You will need to know how to find the rank of a matrix to complete part (c) of this question; as required, review this concept in a linear algebra textbook. Since all sums are modulo 2, you will also need to reduce the matrices to determine their rank by hand, rather than using the matrix utility on your calculator or some other computing device.*

Returning to Veblen's paper, we find the introduction of the concept of a 'one-dimensional circuit.'

One-dimensional Circuits

22   A connected linear graph each vertex of which is an end of two and only two 1-cells it called a one-dimensional circuit or a 1-circuit. Any closed curve is decomposed by any finite set of points on it into a 1-circuit. Conversely, it is easy to see that the set of all points on a 1-circuit is a simple closed curve. It is obvious, further, that any linear graph such that each vertex is an end of two and only two 1-cells is either a 1-circuit or a set of 1-circuits no two of which have a point in common.

Consider a linear graph C_1 such that each vertex is an end of an even number of edges. Let us denote by $2n_i$ the number of edges incident with each vertex at a_i^0. The edges incident with each vertex a_i^0 may be grouped arbitrarily in n_i pairs no two of which have an edge in common; let these pairs of edges be called the pairs associated with the vertex a_i^0. Let C_1' be a graph coincident with C_1 in such a way that (1) there is one and only one point of C_1' on each point of C_1 which is not a vertex and (2) there are n_i vertices of C_1' on each vertex a_i^0 of C_1 each of these vertices of C_1' being incident only with the two edges of C_1' which coincide with a pair associated with at a_i^0.

The linear graph C_1' has just two edges incident with each of its vertices and therefore consists of a number of 1-circuits. Each of these 1-circuits is coincident with a 1-circuit of C_1, and no two of the 1-circuits of C_1 thus determined have a 1-cell in common. Hence C_1 consists of a number of 1-circuits which have only a finite number of 0-cells in common.

It is obvious that a linear graph composed of a number of closed curves having only a finite number of points in common has an even number of 1-cells incident with each vertex. Hence a *necessary and sufficient condition that C_1 consist of a number of 1-circuits having only 0-cells in common is that each 0-cell of C_1 be incident with an even number of 1-cells.* A set of 1-circuits having only 0-cells in common will be referred to briefly as a set of 1-circuits.

6. The following questions are based on Section 22, in which Veblen discusses one-dimensional circuits.

   (a) First paragraph: Note that Veblen is now only interested in connected graphs. According to his definition of '1-circuit', can a 1-circuit repeat edges? vertices?

   (b) Second paragraph: Veblen outlines a method for constructing a new graph C_1' from a given linear graph C_1.

      i. What conditions on the graph C_1 are required for this construction?

      ii. The resulting graph C_1' will depend on how we pair up the edges at each vertex. Illustrate this fact using graph G_4 from the appendix. That is, apply Veblen's construction method *twice* to graph G_4, using different pairings of edges each time.

   (c) Third paragraph: Veblen claims that each 1-circuit of the graph C_1' constructed by this method will be coincident with a 1-circuit of the original graph C_1.

      Use graph G_5 from the appendix to show that this is not the case for any arbitrary pairing of edges in the original graph. That is, find a pairing of the four edges at vertex D of graph G_5 which gives us a 1-circuit of the graph C_1' that is not coincident with a 1-circuit of the graph C_1'.

   (d) In the final paragraph of Section 22, Veblen states the conclusion of this section in the form of a 'necessary and sufficient' statement. To what (familiar) theorem from graph theory is this conclusion related? Explain, and comment on Veblen's proof (especially in light of question 5c above).

In Section 24 of *Analysis Situs*, Veblen introduced an algebraic representation of one-dimensional circuits.

24 Let us now inquire under what circumstances a symbol $(x_1, x_2, \ldots, x_{\alpha_1})$ for a one-dimensional complex contained in C_1 will represent a 1-circuit or a system of 1-circuits. Consider

$$\eta_{i1}^1 x_1 + \eta_{i2}^1 x_2 + \cdots + \eta_{i\alpha_1}^1 x_{\alpha_1}$$

where the coefficients η_{ij}^1 are the elements of the ith row of H_1. Each term $\eta_{ij}^1 x_j$ of this sum is 0 if a_j^1 is not in the set of 1-cells represented by $(x_1, x_2, \ldots, x_{\alpha_1})$ because in this case $x_j = 0$; it is also zero if a_j^1 is not incident with a_i^0 because $\eta_{ij}^1 = 0$ in case. The term $\eta_{ij}^1 x_j = 1$ if a_j^1 is incident with a_i^0 and in the set represented by $(x_1, x_2, \ldots, x_{\alpha_1})$ because in this case $\eta_{ij}^1 = 1$ and $x_j = 1$. Hence there are as many non-zero terms in the sum as there are 1-cells represented by $(x_1, x_2, \ldots, x_{\alpha_1})$ which are incident with a_i^0. Hence by §22 the required condition is that the number of non-zero terms in the sum must be even. In other words if the x's and η_{ij}^1's are reduced modulo 2 as explained in §14 we must have

$$(H_1) \qquad \sum_{i=1}^{\alpha_1} \eta_{ij}^1 x_j = 0 \qquad (i = 1, 2, \ldots, \alpha_0)$$

if and only if $(x_1, x_2, \ldots, x_{\alpha_1})$ represents a 1-circuit or set of 1-circuits. The matrix of this set of equations (or congruences, mod. 2) is H_1.

7. Note that, in the algebraic representation of one-dimensional circuits discussed in Section 24, the number of non-zero terms in the sum $\eta_{i_1}^1 x_1 + \eta_{i_2}^1 x_2 + \cdots + \eta_{i\alpha_1}^1 x_{\alpha_1}$ corresponds to the degree of the vertex a_i^0.

   (a) Veblen's conclusion in the penultimate sentence of Section 24 could be re-stated in terms of solutions to the matrix-vector equation $H_1 \vec{v} = \vec{0}$, where H_1 is the incidence matrix of the graph. Complete the following example to illustrate this conclusion:

   Let H_1 be the incidence matrix for graph G_3 from the appendix. (See part a of question 4 above.) Let $\vec{X} = (1, 0, 1, 1, 0)$ and $\vec{Y} = (1, 0, 0, 1, 1)$. Determine the matrix-vector products (modulo 2) $H_1 \vec{X}$ and $H_1 \vec{Y}$. Use these results to determine whether (i) $\vec{X}$ represents a set of 1-circuits in the graph G_3; and (ii) $\vec{Y}$ represents a set of 1-circuits in the graph G_3. Explain.

   (b) What advantage might there be in representing graphs and circuits in terms of matrices and linear equations?

In the final excerpt from Veblen's paper below, a connection is made between the matrix H_1 and the problem of determining whether there exists a complete set of 1-circuits that will generate all possible 1-circuits for a given graph.

> **25**   If the rank of the matrix H_1 of the equations (H_1) be ρ_1 the theory of linear homogeneous equations (congruences, mod. 2) tells us that there is a set of $\alpha_1 - \rho_1$ linearly independent solutions of (H_1) upon which all other solutions are linearly dependent. This means geometrically that *there exists a set of $\alpha_1 - \rho_1$ 1-circuits or systems of 1-circuits from which all others can be obtained by repeated applications of the operation of adding (mod. 2) described in §14.* We shall call this a complete set of 1-circuits or systems of 1-circuits.

8. The following questions are based primarily on Section 25, in which Veblen discusses how to use the matrix H_1 to determine if there is a complete set of 1-circuits that will generate all possible 1-circuits for a given graph.

   Note.   *You will need to know about null space, basis sets and rank to complete this question; as required, review these concepts in a linear algebra textbook.*

   (a) Explain Veblen's conclusion in terms of the null space of the matrix H_1. You may find it helpful to review section 24 of Veblen's paper, and project question 7 above.

   (b) Consider the incidence matrix H_1 for graph G_6 in the appendix.

   - Find a basis for the null space of that matrix, again using modulo 2 sums.
   - Use your null-space basis to identify a complete system of 1-circuits for this graph.
   - Write the circuit $C = (0, 0, 1, 1, 1, 0, 1)$ as the sum of circuits in your complete system of 1-circuits for this graph.

Notes for the instructor

This project was developed for use in a beginning-level discrete mathematics course; it could also be used in a more advanced level discrete mathematics course. Familiarity with basic notions of graph theory, including the definition of *isomorphism* and *isomorphism invariant*, is assumed. Parts of the project (clearly marked as such) also assume familiarity with the basic linear algebra concepts of *rank*, *kernel* and *linear independence*. The instructor may either omit these questions or refer students to a standard linear algebra textbook as needed for review. In view of the more advanced nature of this project, small group work is recommended for its use within a beginning level course. Although the project could be assigned independently, assigning it following completion of one or both of the projects "Early Writings on Graph Theory: Hamiltonian Circuits and The Icosian Game" and "Early Writings on Graph Theory: Topological Connections" is also recommended for beginning-level courses; both projects appear in the current volume. Use of color pencils or markers will be helpful in answering Question 6.

Bibliography

[1] Biggs, N., Lloyd, E., Wilson, R., *Graph Theory: 1736–1936*, Clarendon Press, Oxford, 1976.

[2] James, I., *Remarkable Mathematicians: From Euler to von Neumann*, Cambridge University Press, Cambridge, 2002.

[3] Katz, V., *A History of Mathematics: An Introduction*, Second Edition, Addison-Wesley, New York, 1998.

[4] Veblen, O., "An Application of Modular Equations in Analysis Situs," *Ann. of Math.*, **14** (1912-13), 42–46.

Appendix: Graphs for Veblen Graph Theory Project

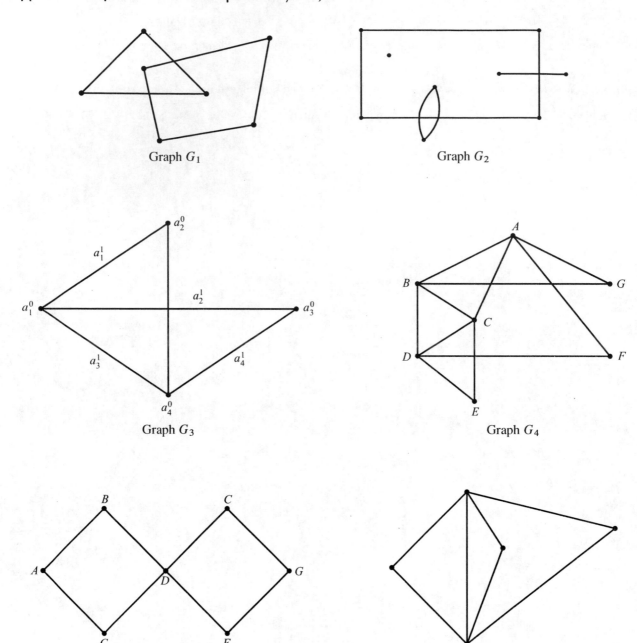

Graph G_1

Graph G_2

Graph G_3

Graph G_4

Graph G_5

Graph G_6

A Study of Logic and Programming via Turing Machines

Jerry M. Lodder
New Mexico State University

An Introduction to Turing Machines

During the International Congress of Mathematicians in Paris in 1900 David Hilbert (1862–1943), one of the leading mathematicians of the last century, proposed a list of problems for following generations to ponder [8, p. 290–329], [9]. On the list was whether the axioms of arithmetic are consistent, a question which would have profound consequences for the foundations of mathematics. Continuing in this direction, in 1928 Hilbert proposed the decision problem (das Entscheidungsproblem) [10, 11, 12], which asked whether there was a standard procedure that can be applied to decide whether a given mathematical statement is true. Both Alonzo Church (1903–1995) [2, 3] and Alan Turing (1912–1954) [13] published papers in 1936 demonstrating that the decision problem has no solution, although it is the algorithmic character of Turing's paper "On Computable Numbers, with an Application to the Entscheidungsproblem" [13] that forms the basis for the modern programmable computer. Today his construction is known as a *Turing machine*.

Let's first study a few excerpts from Turing's original paper [13, p. 231–234], and then design a few machines to perform certain tasks.

ON COMPUTABLE NUMBERS, WITH AN APPLICATION TO
THE ENTSCHEIDUNGSPROBLEM

By A. M. Turing

1. *Computing Machines.*

We have said that the computable numbers are those whose decimals are calculable by finite means. This requires more explicit definition. No real attempt will be made to justify the definitions given until we reach §9. For present I shall only say that the justification lies in the fact that the human memory is necessarily limited.

We may compare a man in the process of computing a real number to a machine which is only capable of a finite number of conditions $q_1, q_2, \ldots, q_R$, which will be called the "m-configurations". The machine is supplied with a "tape" (the analogue of paper) running through it, and divided into sections (called "squares") each capable of bearing a "symbol". At any moment there is just one square, say the r-th, bearing the symbol $S(r)$ which is "in the machine". We may call this square the "scanned square". The symbol on the scanned square may be called the "scanned symbol". The "scanned symbol" is the only one of which the machine is, so to speak, "directly aware". However, by altering its m-configuration the machine can effectively remember some of the symbols it has "seen" (scanned) previously. The possible behaviour of the machine at any moment is determined by the m-configuration q_n and the scanned symbol $S(r)$. This pair q_n, $S(r)$ will be called the "configuration"; thus the configuration determines the possible behaviour of the machine. In some of the configurations in which the scanned square is blank (i.e. bears no symbol) the machine writes down a new symbol on the scanned square; in other configurations it erases the scanned symbol. The machine may also change the square which is being scanned, but only by shifting it one place to right or left. In addition to any of these operations the m-configuration may be changed. Some of the symbols written down will form the sequence of figures which is the decimal of the real number which is being computed. The others are just rough notes to "assist the memory". It will only be these rough notes which will be liable to erasure.

It is my contention that these operations include all those which are used in the computation of a number. The defense of this contention will be easier when the theory of the machines is familiar to the reader. In the next section I therefore proceed with the development of the theory and assume that it is understood what is meant by "machine", "tape", "scanned", etc.

2. *Definitions.*

Automatic machines.

If at each stage the motion of a machine (in the sense of §1) is *completely* determined by the configuration, we shall call the machine an "automatic machine" (or a-machine).

For some purposes we might use machines (choice machines or c-machines) whose motion is only partially determined by the configuration (hence the use of the word "possible" in §1). When such a machine reaches one of these ambiguous configurations, it cannot go on until some arbitrary choice has been made by an external operator. This would be the case if we were using machines to deal with axiomatic systems. In this paper I deal only with automatic machines, and will therefore often omit the prefix a-.

Computing machines.

If an a-machine prints two kinds of symbols, of which the first kind (called figures) consists entirely of 0 and 1 (the others being called symbols of the second kind), then the machine will be called a computing machine. If the machine is supplied with a blank tape and set in motion, starting from the correct initial m-configuration the subsequence of the symbols printed by it which are of the first kind will be called the *sequence computed by the machine*. The real number whose expression as a binary decimal is obtained by prefacing this sequence by a decimal point is called the *number printed by the machine*.

At any stage of the motion of the machine, the number of the scanned square, the complete sequence of all symbols on the tape, and the m-configuration will be said to describe the *complete configuration* at that stage. The changes of the machine and tape between successive complete configurations will be called the *moves* of the machine. . . .

3. *Examples of computing machines.*

I. A machine can be constructed to compute the sequence $010101 \ldots$. The machine is to have the four m-configurations "b", "c", "f", "e" and is capable of printing "0" and "1". The behaviour of the machine is described in the following table [Example 1] in which "R" means "the machine moves so that it scans the square immediately on the right of the one it was scanning previously". Similarly for "L". "E" means "the scanned symbol is erased and "P" stands for "prints". This table (and all succeeding tables of the same kind) is to be understood to mean that for a configuration described in the first two columns the operations in the third column are carried out successively, and the machine then goes over into the m-configuration described in the last column. When the second column is blank, it is understood that the behaviour of the third and fourth columns applies for any symbol and for no symbol. The machine starts in the m-configuration b with a blank tape.

Configuration		Behaviour	
m-config.	symbol	operation	final m-config.
b	none	P(0), R	c
c	none	R	e
e	none	P(1), R	f
f	none	R	b

If (contrary to the description §1) we allow the letters L, R to appear more than once in the operations column we can simplify the table considerably.

Configuration		Behaviour	
m-config.	symbol	operation	final m-config.
b	none	P(0)	b
b	0	R, R, P(1)	b
b	1	R, R, P(0)	b

1.1. Describe the workings of a Turing machine (referred to as a "computing machine" in the original paper).

1.2. What is the precise output of the machine in Example 1? Certain squares may be left blank. Be sure to justify your answer.

1.3. Design a Turing machine which generates the following output. Be sure to justify your answer.

$$010010100101001 \ldots$$

1.4. Describe the behavior of the following machine, which begins with a blank tape, with the machine in configuration α.

Configuration		Behavior	
m-config.	symbol	operation	final m-config.
α	none	R P(1)	β
α	1	R P(0)	β
α	0	HALT	(none)
β	1	R P(1)	α
β	0	R P(0)	α

1.5. Given finite, non-empty, sets A and B, design a Turing machine which tests whether $A \subseteq B$. Suppose that the first character on the tape is a 0, simply to indicate the beginning of the tape. To the right of 0 follow the (distinct, non-blank) elements of A, listed in consecutive positions, followed by the symbol &. To the right of & follow the (distinct, non-blank) elements of B, listed in consecutive positions, followed by the symbol Z to indicate the end of the tape:

0			...		&			...			Z

The symbols 0, &, Z are neither elements of A nor B. The machine starts reading the tape in the rightmost position, at Z. If $A \subseteq B$, have the machine erase all the elements of A and return a tape with blanks for every square which originally contained an element of A. You may use the following operations for the behavior of the machine:

- R: Move one position to the right.

- L: Move one position to the left.

- S: Store the scanned character in memory. Only one character can be stored at a time.

- C: Compare the currently scanned character with the character in memory. The only operation of C is to change the final configuration depending on whether the scanned square matches what is in memory.

- E: Erase the currently scanned square.

- P( ): Print whatever is in parentheses in the current square.

You may use multiple operations for the machine in response to a given configuration. Also, for a configuration q_n, you may use the word "other" to denote all symbols $\mathcal{S}(r)$ not specifically identified for the given q_n. Be sure that your machine halts.

Turing Machines, Induction and Recursion

The logic behind the modern programmable computer owes much to Turing's "computing machines," discussed in the first section of the project. Since the state of the machine, or m-configuration as it was called by Turing, can be altered according to the symbol being scanned, the operation of the machine can be changed depending on what symbols have been written on the tape, and affords the machine a degree of programmability. The program consists of the list of configurations of the machine and its behavior for each configuration. Turing's description of his machine, however, did not include memory in its modern usage for computers, and symbols read on the tape could not be stored in any separate device. Using a brilliant design feature for the tape, Turing achieves a limited type of memory for the machine, which allows it to compute many arithmetic operations. The numbers needed for a calculation are printed on every other square of the tape, while the squares between these are used as "rough notes to 'assist the memory.' It will only be these rough notes which will be liable to erasure" [13, p. 232].

Turing continues [13, p. 235]:

> The convention of writing the figures only on alternate squares is very useful: I shall always make use of it. I shall call the one sequence of alternate squares F-squares, and the other sequence E-squares. The symbols on E-squares will be liable to erasure. The symbols on F-squares form a continuous sequence. ... There is no need to have more than one E-square between each pair of F-squares: an apparent need of more E-squares can be satisfied by having a sufficiently rich variety of symbols capable of being printed on E-squares.

Let's examine the Englishman's use of these two types of squares. Determine the output of the following Turing machine, which begins in configuration a with the tape

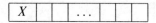

and the scanner at the far left, reading the symbol X.

Configuration		Behavior	
m-config.	symbol	operation	final m-config.
a	X	R	a
a	1	R, R	a
a	blank	P(1), R, R, P(1), R, R, P(0)	b
b	X	E, R	c
b	other	L	b
c	0	R, P(X), R	a
c	1	R, P(X), R	d
d	0	R, R	e
d	other	R, R	d
e	blank	P(1)	b
e	other	R, R	e

2.1. What is the precise output of the machine as it just finishes configuration a and enters configuration b for the first time? Justify your answer.

2.2. What is the precise output of the machine as it just finishes configuration a and enters configuration b for the second time? Justify your answer.

2.3. What is the precise output of the machine as it just finishes configuration a and enters configuration b for the third time? Justify your answer.

2.4. Guess what the output of the machine is as it just finishes configuration a and enters configuration b for the n-th time. Use induction to prove that your guess is correct. Be sure to write carefully the details of this proof by induction.

2.5. Design a Turing machine, which when given two arbitrary natural numbers, n and m, will compute the product $n \cdot m$. Suppose that the machine begins with the tape

A	1		1		$\ldots$		1		1	B	1		1		$\ldots$		1		1	C

where the number of ones between A and B is n, the number of ones between B and C is m, and the machine begins scanning the tape at the far left, reading the symbol A. The output of the machine should be:

A	1		$\ldots$		1	B	1		$\ldots$		1	C	1		1		$\ldots$		1	D

where the number of ones between C and D is $n \cdot m$. Use induction to verify that the machine produces the correct output.

Letting T denote the Turing machine which multiplies n and m together, so that the value of $T(n, m)$ is $n \cdot m$, design T so that for $n \in \mathbf{N}$,

$$T(n, 1) = n$$

and for $m \in \mathbf{N}, m \geq 2$, we have

$$T(n, m) = T(n, m - 1) + n.$$

Such an equation provides an example of a recursively defined function, an important topic in computer science. In our case, the algorithm for multiplication, T, is defined in terms of addition, a more elementary operation.

The Universal Computing Machine

The decision problem of Hilbert asked whether there is a standard procedure, an algorithm in modern terminology, which can be invoked to decide whether an arbitrary statement (within some system of logic) is valid. In answering this question, Turing introduced several fundamental concepts, certainly of importance to logic, but also pivotal to the development of the modern programmable computer. The first of these is a "computing machine," called a "Turing machine" today, which is the forerunner of a modern computer program. The next concept is the "universal computing machine," which is in fact a particular type of Turing machine that accepts the instructions of some other machine M in standard form, and outputs the same sequence as M. Turing writes [13, p. 241–242]:

> It is possible to invent a single machine which can be used to compute any computable sequence. If this machine U is supplied with a tape on the beginning of which is written the $S.D$ [standard description] of some computing machine M, then U will compute the same sequence as M.

The reader should review §3 *Examples of computing machines* reprinted above under "An Introduction to Turing Machines," and have the table from Example 1 at hand before reading the following excerpts [13, p. 239–241]:

5. *Enumeration of computable sequences.*

A computable sequence γ is determined by a description of a machine which computes γ. Thus the sequence 001011011101111... is determined by the table on p. 234, and, in fact, any computable sequence is capable of being described in terms of such a table.

It will be useful to put these tables into a kind of standard form. In the first place let us suppose that the table is given in the same form as the first table, for example, 1 on p. 233. That is to say, that the entry in the operations column is always one of the form $E: E, R: E, L: P\alpha: P\alpha, R: P\alpha, L: R: L: L:$ or no entry at all. The table can always be put into this form by introducing more m-configurations. Now let us give numbers to the m-configurations, calling them $q_1, \ldots, q_R$ as in §1. The initial m-configuration is always to be called q_1. We also give numbers to the symbols $S_1, \ldots, S_m$ and, in particular, blank $= S_0, 0 = S_1, 1 = S_2$. The lines of the table are now of form

m-config.	Symbol	Operations	Final m-config.	
q_i	S_j	PS_k, L	q_m	(N_1)
q_i	S_j	PS_k, R	q_m	(N_2)
q_i	S_j	PS_k	q_m	(N_3)

Lines such as

q_i	S_j	E, R	q_m

are to be written as

q_i	S_j	PS_0, R	q_m

and lines such as

q_i	S_j	R	q_m

to be written as

q_i	S_j	PS_j, R	q_m

In this way we reduce each line of the table to a line of one of the forms (N_1), (N_2), (N_3).

From each line of form (N_1) let us form an expression $q_i \, S_j \, S_k \, L \, q_m$; from each line of form (N_2) we form an expression $q_i \, S_j \, S_k \, R \, q_m$; and from each line of form (N_3) we form an expression $q_i \, S_j \, S_k \, N \, q_m$.

Let us write down all expressions so formed from the table for the machine and separate them by semi-colons. In this way we obtain a complete description of the machine. In this description we shall replace q_i by the letter "D" followed by the letter "A" repeated i times, and S_j by "D" followed by "C" repeated j times. This new description of the machine may be called the *standard description* (S.D). It is made up entirely from the letters "A", "C", "D", "L", "R", "N", and from "$;$". ...

Let us find a description number for the machine I of §3. When we rename the m-configurations its table becomes:

m-config.	Symbol	Operations	Final m-config.
q_1	S_0	PS_1, R	q_2
q_2	S_0	PS_0, R	q_3
q_3	S_0	PS_2, R	q_4
q_4	S_0	PS_0, R	q_1

Other tables could be obtained by adding irrelevant lines such as

$$q_1 \quad S_1 \quad PS_1, R \quad q_2$$

Our first standard form would be

$$q_1 \, S_0 \, S_1 \, R \, q_2; \; q_2 \, S_0 \, S_0 \, R \, q_3; \; q_3 \, S_0 \, S_2 \, R \, q_4; \; q_4 \, S_0 \, S_0 \, R \, q_1; \; .$$

The standard description is

$$DADDCRDAA; DAADDRDAAA; DAAADDCCRDAAAA; DAAAADDRDA;$$

Continuing from [13, p. 243], we read:

Each instruction consists of five consecutive parts

(i) "D" followed by a sequence of letters "A". This describes the relevant m-configuration.

(ii) "D" followed by a sequence of letters "C". This describes the scanned symbol.

(iii) "D" followed by another sequence of letters "C". This describes the symbol into which the scanned symbol is to be changed.

(iv) "L", "R", or "N", describing whether the machine is to move to left, right, or not at all.

(v) "D" followed by a sequence of letters "A". This describes the final m-configuration.

3.1. What is the output of the following machine, T, if T begins in configuration a with a blank tape, scanning the blank at the far left?

Configuration		Behavior	
m-config.	symbol	operation	final m-config.
a	1	R	c
a	blank	P(1), R	b
b	0	R	a
b	blank	P(1)	a
c	blank	P(0)	b

3.2. Rewrite the output of machine T using the "standard description" for the output symbols, i.e., $S_0 =$ blank, $S_1 = 0$, $S_2 = 1$, and then replace each S_j with D followed by C repeated j times.

3.3. What is the standard description ($S.D$) of machine T? Be sure that every instruction, including the last one, is followed by a semi-colon. Make sure that you have the correct answer to this before continuing.

3.4. Suppose that the number of configurations for a given machine M is limited to nine, while the number of symbols which M can recognize or write is limited to four. The machine M begins in its first listed configuration (a) with a blank tape, reading the blank at the far left. Now suppose that the standard description of M is written on every other square (in fact the F-squares) of a second tape, with each instruction followed by a semi-colon (on an F-square as well), with the last semi-colon followed by the symbol "::" on an F-square. Initially all other squares on this second tape are blank. If the standard description for machine I of §3 in Turing's paper is entered on tape in this way, what is the output of the following machine U, which begins in configuration one reading the tape at the far left? Note that 20R is shorthand for "move 20 squares to the right," and similarly for 10R. How does this compare with the actual output of machine I, §3?

3.5. If the standard description of machine T from question 3.1 is written on tape as described in 3.4, what is the output of U when applied to this tape? How does this compare to the actual output of T?

3.6. Describe in words the operation of configuration one from machine U. Describe separately the operations of configurations two, three, and four.

3.7. Note that only the first 16 configurations of U are listed. From this point, if U was originally supplied with the standard description of a machine M (as specified in 3.4), then U should output the same sequence as M, except coded according to the standard description of the output symbols. Moreover, U keeps track of the current configuration of M in the first 18 squares immediately following "::". The position of the scanner for M is recorded via the symbol "T" on the tape which U processes. Beginning with configuration 17, outline the remaining operations of U. You may use phrases such as "match m-configuration," "match scanned symbol," or " move scanner for M," etc. Be sure to include a written explanation of these and any other operations you decide to use in your outline. For this part, you do not need to design an actual Turing machine to perform these tasks. Does recursion occur in your outline? How?

Configuration		Behavior	
m-config.	symbol	operation	final m-config.
1	D	R	1
1	A	R	2
1	::	none	none
1	other	R	1
2	blank	R	2
2	A	R	4
2	D	R, R	3
3	D	R, P(X)	5
3	C	R	4
4	;	R	1
4	::	none	none
4	other	R	4
5	::	20R, P(Y), R, R, P(D), 10R, P(Z)	6
5	other	R	5
6	X	E, R	7
6	other	L	6
7	C	R, P(X)	8
7	R	R, P(X)	10
7	L	none	none
7	N	R, P(X)	11
8	Y	4R	9
8	other	R	8
9	blank	P(C)	6
9	C	R, R	9
10	Z	E, P(T)	12
10	other	R	10
11	Y	R, P(T)	12
11	other	R	11
12	X	E, 4R, P(X)	13
12	other	L	12
13	::	R, R	14
13	other	R	13
14	blank	P(A)	15
14	Y	none	none
14	other	R, R	14
15	X	E, R	16
15	other	L	15
16	;	none	17
16	A	R, P(X)	13

3.8.   Beginning with configuration 17, find the actual machine instructions of U so that U finds a match between an arbitrary configuration stored on the 18 squares to the right of "::" and a configuration at the beginning of a coded instruction to the left of "::". Suppose that the standard description entered on tape is that of a machine M for which there is always a well-defined configuration to follow every move of M. Use only the operations "R," "L," "E," "P( )," "none," and be sure to explain the new steps of U. What is the present-day terminology used to describe U?

Extra Credit: Write a computer program for a universal computing machine (in the language of your choice). Demonstrate with several examples that your universal machine functions properly.

The Decision Problem, *Das Entscheidungsproblem*

Turing's paper "On Computable Numbers with an Application to the Entscheidungsproblem" proved most influential not only for mathematical logic, but also for the development of the programmable computer, and together with work of Alonzo Church (1903–1995) [2, 3] and others [7], inaugurated a new field of study known today as computability. Recall that Turing's original motivation for writing the paper was to answer the decision problem of David Hilbert, posed in 1928 along with a list of other problems [10, 11] dealing with the consistency, completeness and independence of the axioms of a logical system in general. The solutions to these problems, in particular Kurt Gödel's (1906–1978) demonstration of the incompleteness of arithmetic with the existence of statements that are not provable (as true or false) [6], had profound consequences for mathematics, and brought mathematical logic to the fore as a separate field of study [7]. In this section, however, we deal primarily with the problem of deciding whether a given statement is valid within a logical system, the decision problem, which Hilbert expressed as [12, p. 112–113]:

> ... [T]here emerges the fundamental importance of determining whether or not a given formula of the predicate calculus is universally valid. ... A formula ... is called *satisfiable* if the sentential variables can be replaced with the values truth and falsehood ... in such a way that the formula [becomes] a true sentence. ... It is customary to refer to the equivalent problems of *universal validity* and *satisfiability* by the common name of the *decision problem*.

Following Gödel's results, the decision problem remained, although it must be reinterpreted as meaning whether there is a procedure by which a given proposition can be determined to be either "provable" or "unprovable". In the text *Introduction to Mathematical Logic* [4, p. 99], Church formulated this problem as:

> The *decision problem* of a logistic system is the problem to find an effective procedure or algorithm, a *decision procedure*, by which for an arbitrary well-formed formula of the system, it is possible to determine whether or not it is a theorem ....

To be sure, Church proves that the decision problem has no solution [2, 3], although it is the algorithmic character of Turing's solution that is pivotal to the logical underpinnings of the programmable computer. Moreover, the simplicity of a Turing machine provides a degree of accessibility to logic and computability ideal for readers new to this material.

The concept of a universal computing machine, studied above, has evolved into what now is known as a compiler or interpreter in computer science, and is indispensable for the processing of any programming language. The question then arises, does the universal computing machine provide a solution to the decision problem? The universal machine is the standard procedure for answering all questions that can in turn be phrased in terms of a computer program.

First, study the following excerpts from Turing's paper [13, p. 232–233]:

> *Automatic machines.*
>
> If at each stage the motion of a machine is *completely* determined by the configuration, we shall call the machine an "automatic machine" (or *a*-machine). ...
>
> *Computing machines.*
>
> If an *a*-machine prints two kinds of symbols, of which the first kind (called figures) consists entirely of 0 and 1 (the others being called symbols of the second kind), then the machine will be called a computing machine. If the machine is supplied with a blank tape and set in motion, starting from the correct initial *m*-configuration, the subsequence of symbols printed by it which are of the first kind will be called the *sequence computed by the machine*. ...
>
> *Circular and circle-free machines.*
>
> If a computing machine never writes down more than a finite number of symbols of the first kind, it will be called *circular*. Otherwise it is said to be *circle-free*. ...
>
> A machine will be circular if it reaches a configuration from which there is no possible move, or if it goes on moving and possibly printing symbols of the second kind, but cannot print any more symbols of the first kind.

Computable sequences and numbers.

A sequence is said to be computable if it can be computed by a circle-free machine. A number is computable if it differs by an integer from the number computed by a circle-free machine. ...

4.1. Consider the following machine, T_1, which begins in m-configuration a with a blank tape, reading the blank at the far left. Is T_1 circle-free? Justify your answer.

T_1 :

Configuration		Behavior	
m-config.	symbol	operation	final m-config.
a	blank	R, P(1)	b
a	0	R	b
b	1	R, R, P(0)	a
b	blank	(none)	a

4.2. Consider the following machine, T_2, which begins in m-configuration a with a blank tape, reading the blank at the far left. Is T_2 circle-free? Justify your answer.

T_2 :

Configuration		Behavior	
m-config.	symbol	operation	final m-config.
a	blank	R, P(1)	b
a	0	R	b
b	1	R, R, P(0)	a
b	0	R	a

4.3. Describe in your own words the key feature which distinguishes a circle-free machine from a circular machine.

4.4. Is the sequence 101001000100001 ... computable? If so, find a circle-free machine (with a finite number of m-configurations) that computes this sequence on every other square (the F-squares) of a tape which is originally blank. If not, prove that there is no circle-free machine that computes the above sequence.

Turing's insight into the decision problem begins by listing all computable sequences in some order:

$$\phi_1, \phi_2, \phi_3, \ldots, \phi_n, \ldots,$$

where ϕ_n is the n-th computable sequence. Moreover, let $\phi_n(k)$ denote the k-th figure (0 or 1) of ϕ_n. For example, if

$$\phi_2 = 101010 \ldots,$$

then $\phi_2(1) = 1, \phi_2(2) = 0, \phi_2(3) = 1$, etc. Turing then considers the sequence β' defined by $\beta'(n) = \phi_n(n)$. If the decision problem has a solution, then [13, p. 247]:

> We can invent a machine D which, when supplied with the $S.D$ [standard description] of any computing machine M will test this $S.D$ and if M is circular will mark the $S.D$ with the symbol "u" [unsatisfactory] and if it is circle-free will mark it with "s" [satisfactory]. By combining the machines D and U [the universal computing machine] we could construct a machine H to compute the sequence β'.

4.5. Is the number of computable sequences finite or infinite? If finite, list the computable sequences. If infinite, find a one-to-one correspondence between the natural numbers, **N**, and a subset of the computable sequences. Use the result of this question to carefully explain why H must be circle-free.

4.6. Since H is circle-free, the sequence computed by H must be listed among the ϕ_n's. Suppose this occurs for $n = N_0$. In a written paragraph, explain how $\beta'(N_0)$ should be computed. Is it possible to construct a machine H that computes β'? If so, find the configuration table for H. If not, what part of H, i.e., D or U, cannot be constructed? Justify your answer.

4.7. Does the universal computing machine solve the decision problem? Explain.

4.8. By what name is the decision problem known today in computer science? Support your answer with excerpts from outside sources.

Notes for the instructor

The project contains four sub-projects "An Introduction to Turing Machines," "Turing Machines, Induction and Recursion," "The Universal Computing Machine," and "The Decision Problem," with the first two of these ideal for an introductory undergraduate course in discrete mathematics or computer programming. The first could be assigned as a separate two-week project while covering naïve set theory, while the second could be assigned while the class studies inductive or recursive constructions. A variant of this second sub-project used recently in a beginning programming course requires the construction of a Turing machine to compute the sum of two positive binary integers. This version of the project, "Turing Machines and Binary Addition," is available from the web resource [1].

The latter two sub-projects, which sketch the main results of Turing's paper, are well suited for an advanced or intermediate course in discrete mathematics, logic, or programming. The universal machine is a Turing machine that accepts as its input any other machine, T, and computes the same output as T. This foreshadows the development of a compiler or interpreter in computer science. The decision problem asks whether there is a decision procedure that can be applied to any well-formulated mathematical statement and determine whether the statement is true or false. Turing's negative solution to this problem forms the historical legacy of his paper, not to mention the notion of "Turing computable" to refer to that which can be computed via a Turing machine.

Bibliography

[1] Bezhanishvili, G., Leung, H., Lodder, J., Pengelley, D., Ranjan, D., "Teaching Discrete Mathematics via Primary Historical Sources," www.math.nmsu.edu/hist_projects/

[2] Church, A., "An Unsolvable Problem of Elementary Number Theory," *Amer. Journal Math.*, **58** (1936), 345–363. This paper with a short foreword by Davis was reprinted on pages 88–107 of [5].

[3] Church, A., "A Note on the Entscheidungsproblem," *Journal of Symbolic Logic*, **1** (1936), 40–41.

[4] Church, A., *Introduction to Mathematical Logic*, Princeton University Press, Princeton, New Jersey, 1996.

[5] Davis, M., *The Undecidable. Basic Papers on Undecidable Propositions, Unsolvable Problems and Computable Functions*, Martin Davis (editor), Raven Press, Hewlett, N.Y., 1965.

[6] Gödel, K., "Über Formal Unentscheidbare Sätze der Principia Mathematica und Verwandter Systeme I," *Monatsh. Math. Phys.*, **38** (1931), 173–198. The English translation of this paper by Mendelson with foreword by Davis was reprinted on pages 4–38 of [5].

[7] Grattan-Guiness, I., *The Search for Mathematical Roots, 1870–1940: Logics, Set Theories and the Foundations of Mathematics from Cantor through Russell to Gödel*, Princeton University Press, Princeton, New Jersey, 2000.

[8] Hilbert, D., *Gesammelte Abhandlungen*, Vol. III, Chelsea Publishing Co., New York, 1965.

[9] Hilbert, D., "Mathematical Problems," Newson M., (translator) *Bulletin of the American Mathematical Society*, **8** (1902), 437–439.

[10] Hilbert, D., "Probleme der Grundlegung der Mathematik," *Mathematische Annalen*, **102**, (1930), 1–9.

[11] Hilbert, D., Ackermann, W., *Grundzüge der Theoretischen Logik*, Dover Publications, New York, 1946.

[12] Hilbert, D., Ackermann, W., *Principles of Mathematical Logic*, L. Hammond, G. Leckie, F. Steinhardt, translators, Chelsea Publishing Co., New York, 1950.

[13] Turing, A. M., "On Computable Numbers with an Application to the Entscheidungsproblem," *Proceedings of the London Mathematical Society* **42** (1936), 230–265. A correction, **43** (1937), 544–546. This paper with a short foreword by Davis was reprinted on pages 115–154 of [5].

Church's Thesis[1]

Guram Bezhanishvili
New Mexico State University

Introduction

In this project we will learn about both primitive recursive and general recursive functions. We will also learn about Turing computable functions, and will discuss why the class of general recursive functions coincides with the class of Turing computable functions. We will introduce the effectively calculable functions, and the ideas behind Alonzo Church's (1903–1995) proposal to identify the class of effectively calculable functions with the class of general recursive functions, known as "Church's thesis." We will analyze Kurt Gödel's (1906–1978) initial rejection of Church's thesis, together with the work of Alan Turing (1912–1954) that finally convinced Gödel of the validity of Church's thesis. We will learn much of this by studying and working with primary historical sources by Gödel, Stephen Cole Kleene (1909–1994), and Turing.

We begin by asking the following question: What does it mean for a function f to be *effectively calculable*? Obviously if we can find an algorithm to calculate f, then f is effectively calculable. For example, the famous Euclidean algorithm tells us that the binary function producing the greatest common divisor of two integers is effectively calculable. But what if we can not find an algorithm that calculates f? The reason could be that there is no algorithm calculating f; or it could be that f is effectively calculable but we were not successful in finding an algorithm. Thus, it is evident that we need better means to identify effectively calculable functions.

The problem of identifying the effectively calculable functions (of natural numbers) was at the center stage of mathematical research in the twenties and thirties of the twentieth century. In the early thirties at Princeton, Church and his two gifted students Kleene and John Barkley Rosser (1907–1989) were developing the theory of λ-*definable* functions. Church proposed to identify the effectively calculable functions with the λ-definable functions. Here is Kleene's description of these events, taken from page 59 of [12]:

> The concept of λ-definability existed full-fledged by the fall of 1933 and was circulating among the logicians at Princeton. Church had been speculating, and finally definitely proposed, that the λ-definable functions are all the effectively calculable functions—what he published in [2], and which I in [11] Chapter XII (or almost in [10]) called "Church's thesis".

> When Church proposed this thesis, I sat down to disprove it by diagonalizing out of the class of the λ-definable functions. But, quickly realizing that the diagonalization cannot be done effectively, I became overnight a supporter of the thesis.

Though Kleene became an "overnight" supporter of the thesis, it was a different story with Gödel. Gödel arrived at Princeton in the fall of 1933. Church proposed his thesis to Gödel early in 1934, but, according to a November 29, 1935 letter from Church to Kleene, Gödel regarded it "as thoroughly unsatisfactory." Instead, in his lectures during the spring of 1934 at Princeton [6], Gödel generalized the notion of *primitive recursive* functions, which was introduced

---

[1] Thanks are due to Joel Lucero-Bryan for team-teaching this project in an advanced undergraduate course in mathematical logic at New Mexico State University.

by him in his epoch-making paper on undecidable propositions [5].[2] He did this by modifying a suggestion made by Jacques Herbrand (1908–1931) in a 1931 letter, to obtain the notion of general recursive functions (also known as the Herbrand-Gödel general recursive functions). Below we give an excerpt from the abovementioned letter from Church to Kleene (taken from [4]) that gives an account of his discussion of effective calculability with Gödel:

> In regard to Gödel and the notions of recursiveness and effective calculability, the history is the following. In discussion [sic] with him the notion of lambda-definability, it developed that there was no good definition of effective calculability. My proposal that lambda-definability be taken as a definition of it he regarded as thoroughly unsatisfactory. I replied that if he would propose any definition of effective calculability which seemed even partially satisfactory I would undertake to prove that it was included in lambda-definability. His only idea at the time was that it might be possible, in terms of effective calculability as an undefined notion, to state a set of axioms which would embody the generally accepted properties of this notion, and to do something on that basis. Evidently it occurred to him later that Herbrand's definition of recursiveness, which has no regard to effective calculability, could be modified in the direction of effective calculability, and he made this proposal in his lectures. At that time he did specifically raise the question of the connection between recursiveness in this new sense and effective calculability, but said he did not think that the two ideas could be satisfactorily identified "except heuristically."

It appears that Gödel's rejection of λ-definability as a possible "definition" of effective calculability was the main reason behind Church's announcement of his thesis in terms of general recursive functions [1]. Church made his announcement at a meeting of the American Mathematical Society in New York City on April 19, 1935. Below is an excerpt from his abstract:

> Following a suggestion of Herbrand, but modifying it in an important respect, Gödel has proposed (in a set of lectures at Princeton, N. J., 1934) a definition of the term *recursive function*, in a very general sense. In this paper a definition of *recursive function of positive integers* which is essentially Gödel's is adopted. And it is maintained that the notion of an effectively calculable function of positive integers should be identified with that of a recursive function, since other plausible definitions of effective calculability turn out to yield notions which are either equivalent to or weaker than recursiveness.

Note that in the abstract Church relegated λ-definability to "other plausible definitions of effective calculability" that were "either equivalent to or weaker than recursiveness," which indicates that, at the time, Church was not yet certain whether λ-definability was equivalent to general recursiveness. Kleene filled in this gap in [8] by showing that these two notions were indeed equivalent. Thus, in the full version of his paper [2], Church was already fully aware that the two notions of general recursiveness and λ-definability coincide.

Kleene's theorem that identified general recursive and λ-definable functions, together with Kleene's famous Normal Form Theorem[3] were beginning to convince Gödel of the validity of Church's thesis. However, it wasn't until the work of Turing that he finally accepted Church's thesis.

Turing's famous paper [15] appeared in 1936 (a correction to it was published in 1937). Turing introduced what we now call *Turing machines*, and defined a function to be *computable* if it can be computed on a Turing machine. His work was entirely independent of the related research being done in Princeton. According to [12], page 61:

> Turing learned of the work at Princeton on λ-definability and general recursiveness just as he was ready to send off his manuscript, to which he then added an appendix outlining a proof of the equivalence of his computability to λ-definability. In [16] he gave a proof of the equivalence in detail.

Thus, Turing introduced his notion of computability in 1936–1937 and, using some of the results of Kleene, showed that the three notions of Turing computable, general recursive, and λ-definable functions coincide.

---

[2]It has to be noted that what we now call "primitive recursive" functions Gödel simply called "recursive." The term "primitive recursive" was introduced by Kleene in [7].

[3]The Normal Form Theorem appeared in [7] and considerably simplified the notion of general recursive functions.

On page 72 of Gödel's "postscriptum" to his 1934 lecture notes which he prepared in 1964 for [3], Gödel states:

> Turing's work gives an analysis of the concept of "mechanical procedure" (alias "algorithm" or "computation procedure" or "finite combinatorial procedure"). This concept is shown to be equivalent with that of a "Turing machine".

Thus, Gödel made it clear that, in his view, Turing's work was of fundamental importance in establishing the validity of Church's thesis. In particular, it influenced Gödel to accept it.

Our account of effective calculability would be incomplete if we did not mention that around the same time and independently of Turing, but not of the work in Princeton, Emil Leon Post (1897–1954) formulated yet another equivalent version of computability [13]. However, his work was less detailed than Turing's.

Lastly we mention that *partial* recursive functions were introduced by Kleene in [9]. In [11] he also generalized the notion of Turing computable functions to partial functions and showed that a partial function is Turing computable if, and only if, it is partial recursive. The importance of his work is underlined in footnote 20 of [4] quoted below:

> It is difficult for those who have learned about recursive functions via a treatment that emphasized partial functions from the outset to realize just how important Kleene's contribution was. Thus Rogers' excellent and influential treatise [14], p. 12, contains an historical account which gives the impression that the subject had been formulated in terms of partial functions from the beginning.

To summarize, in the 1930s there were several versions proposed to formalize the intuitive concept of effective calculability. These were λ-definability (Church), general recursiveness (Gödel), Turing computability (Turing), and Post computability (Post). All these concepts were seen to be equivalent to each other, thus producing evidence for general acceptance of Church's thesis.

This project will only scratch the surface of the subject. Instead of working with partial functions, we will restrict our attention to total functions. We will learn about primitive recursive and general recursive functions from the work of Gödel and Kleene. We will also learn about Turing machines, Turing computable sequences (of natural numbers), and Turing computable real numbers (in the interval $[0, 1]$) from the work of Turing and Kleene. We will examine Turing's and Kleene's definitions of computable functions (of natural numbers), and show that every general recursive function is computable on a Turing machine. The fact that every Turing computable function is general recursive requires the relatively advanced technique of Gödel numbers and will not be addressed in this project. Instead we refer the interested reader to [5], [6] or [11] for the definition of Gödel numbers, and [16] or [11] for the proof that every Turing computable function is general recursive.

The four sources by Gödel, Turing, and Kleene that will be used in the project are [6], [15], [10], and [11]. Edited versions of the first three with forewords have been reprinted in [3].

Part I. Primitive recursive functions

Note: For the reading in part I we will use excerpts of Gödel's 1934 lecture notes [6] reprinted in Davis [3] and edited by Gödel himself. We decided to choose the reprinted version over the original since its definition of rule (1) is more convenient for our purposes.

1.(a)   Read carefully the following excerpt of Gödel's 1934 lecture notes [6] reprinted in Davis [3].

> The function $\phi(x_1, \ldots, x_n)$ shall be *compound* with respect to $\psi(x_1, \ldots, x_m)$ and $\chi_i(x_1, \ldots, x_n)$
> $(i = 1, \ldots, m)$ if, for all natural numbers $x_1, \ldots, x_n$,
>
> (1) $\phi(x_1, \ldots, x_n) = \psi(\chi_1(x_1, \ldots, x_n), \ldots, \chi_m(x_1, \ldots, x_n))$.
>
> $\phi(x_1, \ldots, x_n)$ shall be said to be *recursive* with respect to $\psi(x_1, \ldots, x_{n-1})$ and $\chi(x_1, \ldots, x_{n+1})$ if,
> for all natural numbers $k, x_2, \ldots, x_n$,
>
> (2) $\begin{aligned} &\phi(0, x_2, \ldots, x_n) = \psi(x_2, \ldots, x_n) \\ &\phi(k + 1, x_2, \ldots, x_n) = \chi(k, \phi(k, x_2, \ldots, x_n), x_2, \ldots, x_n). \end{aligned}$

In both (1) and (2), we allow the omission of each of the variables in any (or all) of its occurrences on the right side (e.g., $\phi(x, y) = \psi(\chi_1(x), \chi_2(x, y))$ is permitted under (1))[4]. We define the class of *recursive* functions to be the totality of functions which can be generated by substitution, according to the scheme (1), and recursion, according to the scheme (2), from the successor function $x + 1$, constant functions $f(x_1, \ldots, x_n) = c$, and identity functions $U_j^n(x_1, \ldots, x_n) = x_j$ $(1 \leq j \leq n)$. In other words, a function ϕ shall be recursive if there exists a finite sequence of functions $\phi_1, \ldots, \phi_n$ which terminates with ϕ such that each function of the sequence is either the successor function $x + 1$ or a constant function $f(x_1, \ldots, x_n) = c$, or an identity function $U_j^n(x_1, \ldots, x_n) = x_j$, or is compound with respect to preceding functions, or is recursive with respect to preceding functions.

1.(b)　Rewrite rule (1) for $n = 1$ and $m = 2$. Also rewrite rule (2) for $n = 1$. Explain in your own words what the rules express.

1.(c)　Give a definition of primitive recursive functions in your own words.

1.(d)　On page 44 of [3] Gödel states:

The functions $x + y$, xy, x^y and $x!$ are clearly [primitive] recursive.

Give an argument for why this is so. Hint: Review your answer to 1.(c) and make sure that you have a good grasp of Gödel's definition of primitive recursive functions.

1.(e)　Is every function of natural numbers primitive recursive? Hint: Use a cardinality argument.

1.(f)　Read carefully the excerpts from Church's 1935 letter to Kleene and Church's 1935 abstract appearing above. Also read carefully the following excerpt of Gödel's 1934 lecture notes [6] reprinted in Davis [3].

Recursive functions have the important property that, for each given set of values of the arguments, the value of the function can be computed by a finite procedure[5].

Do you see any connection between Gödel's writing and Church's thesis? Explain your answer.

Part II.　General recursive functions

2.(a)　Read carefully the following excerpt of section 1 of Kleene [10].

We consider the following schemata as operations for the definition of a function ϕ from given functions appearing in the right members of the equations (c is any constant natural number):

(I)　　$\phi(x) = x'$,

(II)　　$\phi(x_1, \ldots, x_n) = c$,

(III)　　$\phi(x_1, \ldots, x_n) = x_i$,

(IV)　　$\phi(x_1, \ldots, x_n) = \theta(\chi(x_1, \ldots, x_n), \ldots, \chi_m(x_1, \ldots, x_n))$,

(Va)　　$\begin{cases} \phi(0) = c \\ \phi(y') = \chi(y, \phi(y)), \end{cases}$

(Vb)　　$\begin{cases} \phi(0, x_1, \ldots, x_n) = \psi(x_1, \ldots, x_n) \\ \phi(y', x_1, \ldots, x_n) = \chi(y, \phi(y, x_1, \ldots, x_n), x_1, \ldots, x_n), \end{cases}$

---

[4][This sentence could have been] omitted, since the removal of any of the occurrences of variables on the right may be effected by means of the function U_j^n. [This footnote occurs in the original source.]

[5]The converse seems to be true, if, besides recursions according to the scheme (2), recursions of other forms (e.g., with respect to two variables simultaneously) are admitted. This cannot be proved, since the notion of finite computation is not defined, but it serves as a heuristic principle. [This footnote occurs in the original source.]

Schema (I) introduces the successor function, Schema (II) the constant functions, and Schema (III) the identity functions. Schema (IV) is the schema of definition by substitution, and Schema (V) the schema of primitive recursion. Together we may call them (and more generally, schemata reducible to a series of applications of them) the *primitive recursive* schemata.

A function ϕ that can be defined from given functions $\psi_1, \ldots, \psi_k$ by a series of applications of these schemata we call *primitive recursive* in the given functions; and in particular, a function ϕ definable ab initio[6] by these means, *primitive recursive*.

Is your answer to 1.(c) equivalent to Kleene's definition of primitive recursive functions? Explain why.

2.(b)   Recall (or look up) the definitions of *total* and *partial* functions. Discuss briefly the difference between the two. Read carefully the following definition of the μ-operator taken from the beginning of section 3 of Kleene [10].

Consider the operator: μy (the least y such that). If this operator is applied to a predicate $R(x_1, \ldots, x_n, y)$ of the $n + 1$ variables $x_1, \ldots, x_n, y$, and if this predicate satisfies the condition

$$(2) \qquad (\forall x_1) \cdots (\forall x_n)(\exists y) R(x_1, \ldots, x_n, y),$$

we obtain a function $\mu y R(x_1, \ldots, x_n, y)$ of the remaining n free variables $x_1, \ldots, x_n$.

Thence we have a new schema,

$$(\text{VI}_1) \qquad \phi(x_1, \ldots, x_n) = \mu y [\rho(x_1, \ldots, x_n, y) = 0],$$

for the definition of a function ϕ from a given function ρ which satisfies the condition

$$(3) \qquad (\forall x_1) \cdots (\forall x_n)(\exists y)[\rho(x_1, \ldots, x_n, y) = 0].$$

Give an example of a function obtainable by the application of the μ-operator that is not total. What does condition (3) guarantee about the function obtained by using schema (VI_1)?

2.(c)   Read carefully excerpts of theorem III and the corollary to it taken from page 51 of Kleene [10].

THEOREM III. The class of general recursive functions is closed under applications of Schemata (I)–(VI) with (3) holding for applications of (VI).

COROLLARY. Every function obtainable by applications of Schemata (I)–(VI) with (3) holding for applications of (VI) is general recursive.

Based on the corollary, give a definition of general recursive functions. Explain whether or not every primitive recursive function is general recursive.

2.(d)   Is every function of natural numbers general recursive? Explain your answer.

2.(e)   (Extra Credit) Give a reasonable argument for why the class of general recursive functions is *strictly* larger than the class of primitive recursive functions.

---

[6]From the beginning.

Part III. Turing machines

3.(a)   Read carefully the following excerpt of section 1 of Turing [15].

We may compare a man in the process of computing a real number to a machine which is only capable of a finite number of conditions $q_1, q_2, \ldots, q_R$ which will be called "m-configurations". The machine is supplied with a "tape" (the analogue of paper) running through it, and divided into sections (called "squares") each capable of bearing a "symbol". At any moment there is just one square, say the r-th, bearing the symbol $\mathfrak{S}(r)$ which is "in the machine". We may call this square the "scanned square". The symbol on the scanned square may be called the "scanned symbol". The "scanned symbol" is the only one of which the machine is, so to speak, "directly aware". However, by altering its m-configuration the machine can effectively remember some of the symbols which it has "seen" (scanned) previously. The possible behaviour of the machine at any moment is determined by the m-configuration q_n and the scanned symbol $\mathfrak{S}(r)$. This pair $q_n, \mathfrak{S}(r)$ will be called the "configuration": thus the configuration determines the possible behaviour of the machine. In some of the configurations in which the scanned square is blank (i.e. bears no symbol) the machine writes down a new symbol on the scanned square: in other configurations it erases the scanned symbol. The machine may also change the square which is being scanned, but only by shifting it one place to right or left. In addition to any of these operations the m-configuration may be changed. Some of the symbols written down will form the sequence of figures which is the decimal of the real number which is being computed. The others are just rough notes to "assist the memory". It will only be these rough notes which will be liable to erasure.

Also read carefully the following excerpt of section 2 of Turing [15].

Automatic machines.

If at each stage the motion of a machine (in the sense of §1) is *completely* determined by the configuration, we shall call the machine an "automatic machine" (or a-machine).

For some purposes we might use machines (choice machines or c-machines) whose motion is only partially determined by the configuration (hence the use of the word "possible" in §1). When such a machine reaches one of these ambiguous configurations, it cannot go on until some arbitrary choice has been made by an external operator. This would be the case if we were using machines to deal with axiomatic systems. In this paper I deal only with automatic machines, and will therefore often omit the prefix a-.

Computing machines.

If an a-machine prints two kinds of symbols, of which the first kind (called figures) consists entirely of 0 and 1 (the others being called symbols of the second kind), then the machine will be called a computing machine. If the machine is supplied with a blank tape and set in motion, starting from the correct initial m-configuration, the subsequence of the symbols printed by it which are of the first kind will be called the *sequence computed by the machine*. The real number whose expression as a binary decimal is obtained by prefacing this sequence by a decimal point is called the *number computed by the machine*.

At any stage of the motion of the machine, the number of the scanned square, the complete sequence of all symbols on the tape, and the m-configuration will be said to describe the *complete configuration* at that stage. The changes of the machine and tape between successive complete configurations will be called the *moves* of the machine.

Circular and circle-free machines.

If a computing machine never writes down more than a finite number of symbols of the first kind, it will be called *circular*. Otherwise it is said to be *circle-free*.

A machine will be circular if it reaches a configuration from which there is no possible move, or if it goes on moving, and possibly printing symbols of the second kind, but cannot print any more symbols of the first kind...

Computable sequences and numbers.

A sequence is said to be computable if it can be computed by a circle-free machine. A number is computable if it differs by an integer from the number computed by a circle-free machine.

Note: The machines that Turing designed will be called Turing machines.

3.(b) What are the primary components of a Turing machine? Describe m-configurations, configurations, and complete configurations of a Turing machine, and their differences.

3.(c) Formulate in your own words Turing's definition of a computing machine, a computable sequence (of natural numbers), and a computable real number (in the interval $[0, 1]$).

3.(d) Read carefully the following excerpt of section 3 of Turing [15] where it is shown that the sequence $010101 \ldots$ is computable.

A machine can be constructed to compute the sequence $010101 \ldots$. The machine is to have the four configurations "$\mathfrak{b}$", "$\mathfrak{c}$", "$\mathfrak{f}$", "$\mathfrak{e}$" and is capable of printing "0" and "1". The behaviour of the machine is described in the following table in which "R" means "the machine moves so that it scans the square immediately on the right of the one it was scanning previously". Similarly for "L". "E" means "the scanned symbol is erased" and "P" stands for "prints". This table (and all succeeding tables of the same kind) is to be understood to mean that for a configuration described in the first two columns the operations in the third column are carried out successively, and the machine then goes over into the m-configuration described in the last column. When the second column is left blank, it is understood that the behaviour of the third and fourth columns applies for any symbol and for no symbol. The machine starts in the m-configuration $\mathfrak{b}$ with a blank tape.

Configuration		Behaviour	
m-config.	*symbol*	*operations*	*final m-config.*
$\mathfrak{b}$	None	$P0, R$	$\mathfrak{c}$
$\mathfrak{c}$	None	R	$\mathfrak{e}$
$\mathfrak{e}$	None	$P1, R$	$\mathfrak{f}$
$\mathfrak{f}$	None	R	$\mathfrak{b}$

If (contrary to the description in §1) we allow the letters L, R to appear more than once in the operations column we can simplify the table considerably.

m-config.	*symbol*	*operations*	*final m-config.*
	None	$P0$	$\mathfrak{b}$
$\mathfrak{b}$	0	$R, R, P1$	$\mathfrak{b}$
	1	$R, R, P0$	$\mathfrak{b}$

What can you conclude about the real number $\frac{1}{3}$? Design a Turing machine that computes the real number $\frac{1}{7}$. Give an argument for why the machine you just designed does what it is supposed to do.

3.(e) (Extra Credit) Read carefully the following excerpt of section 3 of Turing [15] where it is shown that the sequence $001011011101111011111 \ldots$ is computable.

As a slightly more difficult example we can construct a machine to compute the sequence 001011011101111011111 .... The machine is to be capable of five m-configurations, viz. "o", "q", "p", "f", "b" and of printing "∂", "x", "0", "1". The first three symbols on the tape will be "$\partial\partial0$"; the other figures follow on alternate squares. On the intermediate squares we never print anything but "x". These letters serve to "keep the place" for us and are erased when we have finished with them. We also arrange that in the sequence of figures on alternate squares there shall be no blanks.

Configuration		Behaviour	
m-config.	*symbol*	*operations*	*final m-config.*
b		$P\partial, R, P\partial, R, P0, R, R, P0, L, L$	o
o	1	R, Px, L, L, L	o
	0		q
q	Any (0 or 1)	R, R	q
	None	$P1, L$	p
p	x	E, R	q
	∂	R	f
	None	L, L	p
f	Any	R, R	f
	None	$P0, L, L$	o

To illustrate the working of this machine a table is given below of the first few complete configurations. These complete configurations are described by writing down the sequence of symbols which are on the tape, with the m-configuration written below the scanned symbol. The successive complete configurations are separated by colons.

```
: ∂∂0  0 : ∂∂0  0 : ∂∂0  0 : ∂∂0  0     : ∂∂0  0   1
  b    o        q          q            q          p
∂∂0  0   1 : ∂∂0  0  1 : ∂∂0  0  1 : ∂∂0  0  1 : ∂∂0  0  1 :
       p         p              f                          f
∂∂0  0  1 : ∂∂0  0  1     : ∂∂0  0  1  0 :
          f             f              o
∂∂0  0  1x0 : ....
           o
```

In this example Turing uses the symbols "∂" and "x", which are the symbols of the second kind. Explain the need for these symbols, and their use in this particular Turing machine. Give an argument for why the machine described above does what it is supposed to do.

Part IV. Turing computable functions

4.(a)   Read carefully Turing's definition of computable functions of natural numbers taken from page 254 of [15].

If γ is a computable sequence in which 0 appears infinitely[7] often, and n is an integer, then let us define $\xi(\gamma, n)$ to be the number of figures 1 between the n-th and the $(n+1)$-th figure 0 in γ. Then $\phi(n)$ is computable if, for all n and some γ, $\phi(n) = \xi(\gamma, n)$.

---

[7]If $\mathcal{M}$ computes γ, then the problem whether $\mathcal{M}$ prints 0 infinitely often is of the same character as the problem whether $\mathcal{M}$ is circle-free. [This footnote occurs in the original source.]

Note: We will call these functions Turing computable.

4.(b) In your own words explain what it means for a function of natural numbers of one variable to be Turing computable. Extra Credit. Generalize the concept of Turing computable functions to functions of two variables. Hint: Can you code every function of two variables by a function of one variable? Generalize the concept of Turing computable functions to functions of multiple variables.

4.(c) In your own words explain Kleene's definition of a Turing machine by reading carefully the following excerpt from section 67 of Kleene [11].

> The machine is supplied with a linear *tape*, (potentially) infinite in both directions (say to the *left* and *right*). The tape is divided into *squares*. Each square is capable of being *blank*, or of having *printed* upon it any one of a finite list $s_1, \ldots, s_j$ ($j \geq 1$) of *symbols*, fixed for a particular machine. If we write "s_0" to stand for "blank", a given square can thus have any one of $j + 1$ conditions $s_0, \ldots, s_j$. The tape will be so employed that in any "situation" only a finite number (≥ 0) of squares will be printed.
>
> The tape will pass through the machine so that in a given "situation" the machine *scans* just one square (the *scanned square*). The symbol on this square, or s_0 if it is blank, we call the *scanned symbol* (even though s_0 is not properly a symbol).
>
> The machine is capable of being in any one of a finite list $q_0, \ldots, q_k$ ($k \geq 1$) of (*machine*) *states* (called by Turing "machine configurations" or "m-configurations"). We call q_0 the *passive* (or *terminal*) *state*; and $q_1, \ldots, q_k$ we call *active states*. The list $q_0, \ldots, q_k$ is fixed for a particular machine.
>
> A (*tape vs. machine*) *situation* (called by Turing "complete configuration") consists in a particular printing on the tape (i.e. which squares are printed, and each with which of the j symbols), a particular position of the tape in the machine (i.e. which square is scanned), and a particular state (i.e. which of the $k + 1$ states the machine is in). If the state is active, we call the situation *active*; otherwise, *passive*.
>
> Given an active situation, the machine performs an (*atomic*) *act* (called a "move" by Turing). The act performed is determined by the scanned symbol s_a and the machine state q_c in the given situation. This pair (s_a, q_c) we call the *configuration*. (It is *active* in the present case that q_c is active; otherwise *passive*.) The act alters the three parts of the situation to produce a resulting situation, thus. First, the scanned symbol s_a is changed to s_b. (But $a = b$ is permitted, in which case the "change" is identical.) Second, the tape is shifted in the machine (or the machine shifts along the tape) so that the square scanned in the resulting situation is either one square to the left of, or the same square as, or one square to the right of, the square scanned in the given situation. Third, the machine state q_c is changed to q_d. (But $c = d$ is permitted.)
>
> No act is performed, if the given situation is passive.
>
> The machine is used in the following way. We choose some active situation in which to start the machine. We call this the *initial situation* or *input*. Our notation will be chosen so that the state in this situation (the *initial state*) is q_1. The machine then performs an atomic act. If the situation resulting from this act is active, the machine acts again. The machine continues in this manner, clicking off successive acts, as long and only as long as active situations result. If eventually a passive situation is reached, the machine is said then to *stop*. The situation in which it stops we call the *terminal situation* or *output*.
>
> The change from the initial situation to the terminal situation (when there is one) may be called the *operation* performed by the machine.
>
> To describe an atomic act, we use an expression of one of the three following forms:
>
> $$s_b L q_d, \qquad s_b C q_d, \qquad s_b R q_d.$$

The "L", "C", "R", indicate that the resulting scanned square is to the left of, the same as ("center"), or to the right of, respectively, the given scanned square.

The first part of the act (i.e. the change of s_a to s_b) falls into four cases: when $a = 0$ and $b > 0$, it is "prints s_b"; when $a > 0$ and $b = 0$, "erases s_a"; when $a, b > 0$ and $a \neq b$, "erases s_a and prints s_b" or briefly "overprints s_b"; when $a = b$, "no change". We often describe this part of the act as "prints s_b" without regard to the case.

To define a particular machine, we must list the symbols $s_1, \ldots, s_j$ and the active states $q_1, \ldots, q_k$, and for each active configuration (s_a, q_c) we must specify the atomic act to be performed. These specifications may be given by displaying the descriptions of the required acts in the form of a (*machine*) *table* with k rows for the active states and $j + 1$ columns for the square conditions.

EXAMPLE I. The following table defines a machine ("Machine $\mathfrak{A}$") having only one symbol s_1 and only one active state q_1.

Name of machine	Machine state	Scanned symbol s_0	s_1
$\mathfrak{A}$	q_1	$s_1 C q_0$	$s_1 R q_1$

Suppose the symbol s_1 is actually a tally mark "|". Let us see what the machine does, if a tape of the following appearance is placed initially in the machine so that the square which we identify by writing the machine state q_1 over it is the scanned square. The conditions of all squares not shown will be immaterial, and will not be changed during the action.

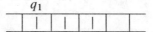

The machine is in the state q_1, and is scanning a square on which the symbol s_1 is printed. In this configuration, the atomic act ordered by the table is $s_1 R q_1$; i.e. no change is made in the condition of the scanned square, the machine shifts right, and again assumes state q_1. The resulting situation appears as follows.

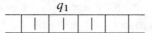

The next three acts lead successively to the following situations, in the last of which the machine stops.

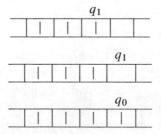

Machine $\mathfrak{A}$ performs the following operation: It seeks the first blank square at or to the right of the scanned square, prints a | there, and stops scanning that square.

4.(d) Explain the similarities and differences of Turing's and Kleene's definitions. Which definition of Turing machines do you prefer? Explain why.

4.(e) The following is Kleene's definition of Turing computable functions of multiple variables (see [11], p. 359).

Now we define how a machine shall 'compute' a partial number-theoretic function ϕ of n variables (cf. §63). The definition for an ordinary (i.e. completely defined) number-theoretic function is obtained by omitting the reference to the possibility that $\phi(x_1, \ldots, x_n)$ may be undefined.

We begin by agreeing to represent the natural numbers $0, 1, 2, \ldots$ by the sequence of tallies $|, ||, |||, \ldots$, respectively, the tally "$|$" being the symbol s_1. There are $y + 1$ tallies in the representation of the natural number y.

Then to represent an m-tuple $y_1, \ldots, y_m$ ($m \geq 1$) of natural numbers on the tape, we print the corresponding numbers of tallies, leaving a single blank between each two groups of tallies and before the first and after the last.

EXAMPLE 2. The triple $3, 0, 2$ is represented thus:

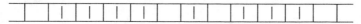

We say that (the representation of) a number y (or of any m-tuple $y_1, \ldots, y_m$) on the tape is (*scanned*) *in the standard position*, when the scanned square is the one bearing the last tally in the representation of y (or of y_m).

Now we say that a given machine $\mathfrak{A}$ *computes* a given partial function ϕ of n variables ($n \geq 1$), if the following holds for each n-tuple $x_1, \ldots, x_n$ of natural numbers. (For the case $n = 0$, cf. Remark 1 below.) Let $x_1, \ldots, x_n$ be represented on the tape, with the tape blank elsewhere, i.e. outside of the $x_1 + \cdots + x_n + 2n + 1$ squares required for the representation. Let $\mathfrak{A}$ be started scanning the representation of $x_1, \ldots, x_n$ in standard position. Then $\mathfrak{A}$ will eventually stop with the $n + 1$-tuple $x_1, \ldots, x_n, x$ represented on the tape and scanned in standard position, if and only if $\phi(x_1, \ldots, x_n)$ is defined and $\phi(x_1, \ldots, x_n) = x$. (If $\phi(x_1, \ldots, x_n)$ is undefined, $\mathfrak{A}$ may fail to stop. It may stop but without an $n + 1$-tuple $x_1, \ldots, x_n, x$ scanned in standard position.)

Example 2 (concluded). If $\phi(3, 0, 2) = 1$ and $\mathfrak{A}$ computes ϕ, then when $\mathfrak{A}$ is started in the situation

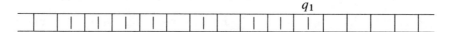

with all squares other than those shown blank, it must eventually stop in the situation

where the condition of the squares other than those shown is immaterial.

Although only one symbol s_1 or "$|$" is used in stating the arguments and in receiving the function value, others may be used in the progress of the computation. For each $n \geq 1$, each machine (with its first symbol s_1 serving as the tally) computes a certain partial function of n variables.

A partial function ϕ is *computable*, if there is a machine $\mathfrak{A}$ which computes it.

Remark 1 that Kleene refers to is on page 363 of Kleene [11]. Below we give an excerpt from it.

REMARK 1. In this chapter, outside the present remark and passages referring to it, we shall understand that we are dealing with functions of $n \geq 1$ variables. Since we have not provided for representing n-tuples of natural numbers on the tape for $n = 0$, we say a machine computes a function ϕ of 0 variables, if it computes the function $\phi(x)$ of 1 variable such that $\phi(x) \simeq \phi$.

State in your own words Kleene's definition of Turing computable functions of multiple variables. Kleene mentions that the representation of the tuple $x_1, \ldots, x_n$ on a tape requires $x_1 + \cdots + x_n + 2n + 1$ squares. Explain why.

4.(f) Are Turing's and Kleene's definitions of computable functions of one variable equivalent? Explain why.

4.(g) (Extra Credit) Is your generalization of Turing computability to functions of multiple variables equivalent to Kleene's definition of computable functions of multiple variables? Explain why.

Part V. Turing computable and general recursive functions

5.(a)   Explain why the successor function $s(x) = x + 1$ is Turing computable.

5.(b)   Explain why the constant functions $f(x_1, \ldots, x_n) = c$ are Turing computable.

5.(c)   Explain why the identity functions $U_j^n(x_1, \ldots, x_n) = x_j$ are Turing computable.

5.(d)   Explain why, whenever $\psi(x_1, \ldots, x_m)$ and $\chi_1(x_1, \ldots, x_n), \ldots, \chi_m(x_1, \ldots, x_n)$ are Turing computable, then $\phi(x_1, \ldots, x_n)$ defined by

$$\phi(x_1, \ldots, x_n) = \psi(\chi_1(x_1, \ldots, x_n), \ldots, \chi_m(x_1, \ldots, x_n))$$

is also Turing computable. In other words, explain why the schema of substitution preserves Turing computability.

5.(e)   Explain why, whenever $\psi(x_1, \ldots, x_{n-1})$ and $\chi(x_1, \ldots, x_{n+1})$ are Turing computable, then $\phi(x_1, \ldots, x_n)$ defined by

$$\phi(0, x_2, \ldots, x_n) = \psi(x_2, \ldots, x_n)$$

and

$$\phi(k + 1, x_2, \ldots, x_n) = \chi(k, \phi(k, x_2, \ldots, x_n), x_2, \ldots, x_n)$$

is also Turing computable. In other words, explain why the schema of primitive recursion preserves Turing computability.

5. (f)   Explain why, whenever $\rho(x_1, \ldots, x_n, y)$ is Turing computable and

$$(\forall x_1) \ldots (\forall x_n)(\exists y)[\rho(x_1, \ldots, x_n, y) = 0],$$

then $\phi(x_1, \ldots, x_n)$ defined by

$$\phi(x_1, \ldots, x_n) = \mu y[\rho(x_1, \ldots, x_n, y) = 0]$$

is also Turing computable. In other words, explain why Kleene's μ-operator preserves Turing computability.

5.(g)   Using exercises 5.(a)–5.(f) make a deduction concerning primitive recursive and general recursive functions. How is your conclusion related to Church's thesis?

Notes for the instructor

This project is designed for an advanced undergraduate or graduate course in mathematical logic. It is also well suited for an advanced undergraduate or graduate course in computability theory. It could be assigned as a three to four week project on primitive recursive, general recursive, and Turing computable functions. If there is any extra time available, the instructor may wish to elaborate on why every Turing computable function is general recursive.

Bibliography

[1] Church, A., "An Unsolvable Problem of Elementary Number Theory, Preliminary Report (abstract)," *Bull. Amer. Math. Soc.*, **41** (1935), 332–333.

[2] Church, A., "An Unsolvable Problem of Elementary Number Theory," *Amer. Journal Math.*, **58** (1936), 345–363. This paper with a short foreword by Davis was reprinted on pages 88–107 of [3].

[3] Davis, M., *The Undecidable. Basic Papers on Undecidable Propositions, Unsolvable Problems and Computable Functions*, Martin Davis (editor), Raven Press, Hewlett, NY, 1965.

[4] Davis, M., "Why Gödel Didn't Have Church's Thesis," *Inform. and Control*, **54** (1982), no. 1-2, 3–24.

[5] Gödel, K., "Über Formal Unentscheidbare Sätze der Principia Mathematica und Verwandter Systeme I," *Monatsh. Math. Phys.*, **38** (1931), 173–198. The English translation of this paper by Mendelson with foreword by Davis was reprinted on pages 4–38 of [3].

[6] Gödel, K., *On Undecidable Propositions of Formal Mathematical Systems*, Mimeographed lecture notes by S. C. Kleene and J. B. Rosser, Institute for Advanced Study, Princeton, N.J., 1934. These lecture notes with foreword by Davis and postscriptum by Gödel were reprinted in [3], pages 39–74.

[7] Kleene, S. C., "General Recursive Functions of Natural Numbers," *Math. Ann.*, **112** (1936), 727–742. This paper with a short foreword by Davis was reprinted on pages 236–253 of [3].

[8] Kleene, S. C., "λ-Definability and Recursiveness," *Duke Math. Journal*, **2** (1936), 340–353.

[9] Kleene, S. C., "On Notation for Ordinal Numbers," *Journal of Symbolic Logic*, **3** (1938), 150–155.

[10] Kleene, S. C., "Recursive Predicates and Quantifiers," *Trans. Amer. Math. Soc.*, **53** (1943), 41–73. This paper with foreword by Davis was reprinted on pages 254–287 of [3].

[11] Kleene, S. C., *Introduction to Metamathematics*, D. Van Nostrand Co., Inc., New York, 1952.

[12] Kleene, S. C., "Origins of Recursive Function Theory," *Ann. Hist. Comput.*, **3** (1981), no. 1, 52–67.

[13] Post, E. L., "Finite Combinatory Processes, Formulation I," *Journal of Symbolic Logic*, **1** (1936), 103–105. This paper with a short foreword by Davis was reprinted on pages 288–291 of [3].

[14] Rogers, H., Jr., *Theory of Recursive Functions and Effective Computability*, McGraw-Hill Book Co., New York, 1967.

[15] Turing, A. M., "On Computable Numbers with an Application to the Entscheidungsproblem," *Proceedings of the London Mathematical Society* **42** (1936), 230–265. A correction, **43** (1937), 544–546. This paper with a short foreword by Davis was reprinted on pages 115–154 of [3].

[16] Turing, A. M., "Computability and λ-Definability," *Journal of Symbolic Logic*, **2** (1937), 153–163.

Two-Way Deterministic Finite Automata

Hing Leung
New Mexico State University

Introduction

In 1943, McCulloch and Pitts [4] published a pioneering work on a model for studying the behavior of the nervous systems. Following up on the ideas of McCulloch and Pitts, Kleene [2] wrote the first paper on finite automata, which proved a theorem that we now call Kleene's theorem. A finite automaton can be considered as the simplest machine model in that the machine has a finite memory; that is, the memory size is independent of the input length. In a 1959 paper [5], Michael Rabin and Dana Scott presented a comprehensive study on the theory of finite automata, for which they received the Turing award in 1976, the highest award in computer science. The citation for the Turing Award states that the award was granted:

> For their joint paper "Finite Automata and Their Decision Problem," which introduced the idea of nondeterministic machines, which has proved to be an enormously valuable concept. Their (Scott & Rabin) classic paper has been a continuous source of inspiration for subsequent work in this field.

In this project, we will not discuss nondeterministic machines. We consider two-way finite automata which is another concept that was introduced in the seminal paper by Rabin and Scott [5].

In an early stage, the theory of finite automata was developed as a mathematical theory for sequential circuits. A sequential circuit maintains a current state from a finite number of possible states. The circuit logic (which is a finite state control) decides the new state based on the current state of the circuit and the given input symbol. Once an input symbol is processed, the circuit will not be able to read it again.

A main application of finite automata is text processing. In compiler design, finite automata are used to capture the logic of lexical analysis. Other applications include string matching, natural language processing, text compression, etc.

A one-way deterministic finite automata (DFA) is defined as the mathematical model of a machine with a finite amount of memory where the input is processed once from left to right. After an input has been read, the DFA decides whether the input is accepted or rejected.

Two-Way Finite Automata

A Turing machine is an abstract mathematical model of a computer. Recall that a Turing machine can move back and forth in the working tape while reading and/or writing.

There are several variants of Turing machines in the literature. Our variant of a Turing machine consists of a finite-state control (also called the program), a read-only input tape where the input is given and a working tape (also called memory) where computations are performed.

In computing an answer, intermediate results are computed and kept in the working tape so that it can be referenced later for further computations.

Unlike a real computer, a Turing machine has no limit to the amount of memory that it can use. As the input becomes more complicated (say longer or of larger value), a Turing machine may use more memory to compute.

Consider the problem of primality testing. That is, given a positive integer, we want to test if it is a prime. As the given integer becomes larger, we need to use more memory to test for primality. We say that the primality testing problem requires an unbounded amount of memory to solve.

However, there are other problems that may require only a finite amount of memory to solve. Consider the problem of computing the parity of a binary sequence. The parity is the number of occurrences of 1's in the binary sequence modulo 2. (Note: The concept defined is not the same as whether the binary sequence, when considered as a binary number, is odd or even.) One can keep reading the binary sequence bit by bit, and maintain the intermediate result of the current parity in the working tape. That is, the memory usage is one bit (thus, finite) in the working tape no matter how long the binary sequence is. Note that the length of the input is not counted in the memory use. We assume that the input tape allows only read operations. The input tape cannot be over-written.

One may wonder what computational problems can be solved using finite memory Turing machines. Another interesting question is whether we can simplify the model of Turing machines using finite memory.

Since the memory usage is finite (independent of the size of the input), we can incorporate[1] the finite memory into the finite state control so that the Turing machine no longer uses a working tape. The resulting model is called the two-way deterministic finite automaton[2] (2DFA) in that it consists of a finite state control and a read-only input tape that allows an input to be read back and forth. As in the case of DFA, the 2DFA decides whether a given input is accepted or rejected.

One can see that a 2DFA can be equivalently defined as a read only Turing machine (without the ability to write symbols on the tape).

DFA and 2DFA

In comparison, DFA and 2DFA differ in that an input can be read only once from left to right by a DFA, whereas a 2DFA can read the input back and forth with no limit on how many times an input symbol can be read.

In computer science, DFA has been studied heavily, since many problems are found to be solvable by DFA. Many textbooks in discrete mathematics discuss the DFA model. However, no textbooks in discrete mathematics discuss the model of 2DFA.

Is 2DFA unrealistic? As we have discussed before, 2DFA is a very interesting model in that it captures/solves problems that are solvable by a computer using finite memory. In fact, in most real life computing tasks performed by a computer, the input has been saved into the computer's hard disk, or is kept in a CD-ROM or DVD-ROM. Thus, there is no reason why a program cannot read the input more than once using two-way processing. So, 2DFA is indeed a very meaningful and arguably more realistic model than DFA.

How does DFA compare to 2DFA? Clearly, DFA is a restricted version of 2DFA. Therefore, 2DFA can solve any problems that are solvable by DFA. Next, are there problems that can be solved by 2DFA but cannot be solved by DFA?

This question is one of many fundamental questions answered in Rabin and Scott's paper [5] on the theory of finite automata. It is proved that 2DFA can be simulated by DFA. That is, whatever problems can be solved by 2DFA can also be solved by DFA.

One may wonder why we should study 2DFA given that DFA, being a simpler model, can do the same job. Are there advantages of 2DFA over DFA? It turns out that 2DFA can be significantly[3] simpler in design for solving the same problem than DFA. In this project, we are going to illustrate the advantages of 2DFA using a number of examples.

It is difficult to explain why most textbooks[4] in automata theory are not covering 2DFA. One possible reason is that the equivalence proof (that 2DFA can be simulated by the simpler model DFA) given by Rabin and Scott is too difficult.

---

[1] The technique involved, which we omit, is not really direct or immediate. But it is not difficult either.

[2] There is another machine model called two-way nondeterministic finite automata. But we are not going to discuss nondeterministic automata in this project.

[3] Technically (and, more accurately), we say that 2DFA can be *exponentially* more succinct in descriptional size than DFA for solving the same problems.

[4] Two textbooks ([1], [3]) cover 2DFA. One [1] follows the approach by Rabin and Scott, and the other [3] follows Shepherdson's ideas.

John C. Shepherdson [6] was able to offer another proof of this important result. It is a very clean proof that we want to present in this project. Instead of going through the proof steps, we emphasize the technique for constructing[5] a DFA from a 2DFA.

Shepherdson is a retired professor in mathematics at the University of Bristol, Great Britain. He published many papers in symbolic logic, computability, logic programming and fuzzy logic.

In fact, Rabin and Scott referred the readers to Shepherdson's proof in their pioneering paper, and decided to give only a sketch of their proof of the equivalence result. Following is an excerpt from Rabin and Scott's paper about Shepherdson's proof:

> The result, with its original proof, was presented to the Summer Institute of Symbolic Logic in 1957 at Cornell University. Subsequently J. C. Shepherdson communicated to us a very elegant proof which also appears in this Journal. In view of this we confine ourselves here to sketching the main ideas of our proof.

We hope that the students using this project will not find Shepherdson's construction tricky, but instead will find that it makes a lot of sense and is the logical way to go for proving the equivalence result. Students will be required to read from the verbal explanations given by Shepherdson, and derive from it computer programs for solving problems in this project.

Project

We assume that students are familiar with the concept and formal definition of DFA.

Following are the definitions of one-way and two-way deterministic finite automata (adapted) from the paper by Shepherdson [6].

> Definition 1. A one-way finite automaton (DFA) over a finite alphabet[6] Σ is a system $A = (Q, \delta, q_0, F)$, where Q is a finite non-empty set (the internal states of A), δ is a function from $Q \times \Sigma$ into Q (the table of moves of A), q_0 is an element of Q (the initial state of A), and F is a subset of Q (the designated final states of A). The class $T(A)$ of tapes accepted[7] by A is the class of all finite sequences $\sigma_1, \ldots, \sigma_n$ of symbols of Σ for which the sequence q_0 (initial state), $q_1, \ldots, q_n$ defined by $q_{i+1} = \delta(q_i, \sigma_{i+1})$ $(i = 0, \ldots, n-1)$ satisfies $q_n \in F$. A set of tapes is said to be definable by a one-way automaton if it is equal to $T(A)$ for some A.

> Definition 2. A two-way finite automaton (2DFA) over Σ is a system $A = (Q, \delta, q_0, F)$ as in Definition 1 with the difference that now δ is a function from $Q \times \Sigma$ into $Q \times D$ where $D = \{L, S, R\}$. A operates as follows: It starts on the leftmost square of the given tape in state q_0. When its internal state is q and it scans the symbol σ, then if $\delta(q, \sigma) = (q', d)$ it goes into the new state q' and moves one square to the left, stays where it is, or moves one square to the right according as $d = L, S,$ or R. The class $T(A)$ of tapes accepted by A is the class of those tapes t such that A eventually moves off the right-hand edge of t in a state belonging to F.

We want to design a 2DFA over $\Sigma = \{0, 1\}$ that accepts tapes containing two 1's separated by four symbols in between them. That is, the 2DFA accepts finite sequences $\sigma_1 \ldots \sigma_n$ of symbols of Σ such that $\sigma_i = \sigma_{i+5} = 1$ for some $i \in \{1, \ldots, n-5\}$. For example, the sequence 001010101100 should be accepted as the 5th and 10th symbols are both 1.

The following 10-state 2DFA A_1, where q_0 is the starting state and q_9 is the only final state, systematically checks every possible 6-symbol subsequence to see if it begins and ends with 1's.

---

[5] In contrast, it is difficult to construct mechanically a DFA from a 2DFA based on the proof of Rabin and Scott.

[6] Σ is the set of input symbols. In the examples considered in this project, the input is always a binary sequence with $\Sigma = \{0, 1\}$.

[7] The word 'accepted' is used as the DFA is considered a machine. One can replace the word 'accepted' by the word 'denoted' or 'defined' in the definition of $T(A)$.

current state	symbol	new state	go to
q_0	0	q_0	R
q_0	1	q_1	R
q_1	0 or 1	q_2	R
q_2	0 or 1	q_3	R
q_3	0 or 1	q_4	R
q_4	0 or 1	q_5	R
q_5	0	q_6	L
q_5	1	q_9	R
q_6	0 or 1	q_7	L
q_7	0 or 1	q_8	L
q_8	0 or 1	q_0	L
q_9	0 or 1	q_9	R

[1]   Demonstrate the steps performed by A_1 in accepting the tape 11001010.

[2]   Demonstrate the steps performed by A_1 in processing the tape 11001001. Conclude that the tape is not accepted by A_1.

Another example 2DFA A_2 (taken from Example 2.14 of the textbook by Hopcroft and Ullman [1]) is given as follows, where q_0 is the starting state and $F = \{q_0, q_1, q_2\}$:

current state	symbol	new state	go to
q_0	0	q_0	R
q_0	1	q_1	R
q_1	0	q_1	R
q_1	1	q_2	L
q_2	0	q_0	R
q_2	1	q_2	L

[3]   Demonstrate that the input 101001 is accepted by A_2.

[4]   Demonstrate that the input 10111 is not accepted by A_2. Indeed, A_2 will run into an infinite loop.

Given the description of a 2DFA, Shepherdson explained how to derive the description of an equivalent DFA that accepts the same set of inputs. The following is an excerpt (modified slightly) from Shepherdson [6] describing the construction:

> The only way an initial portion t of the input tape can influence the future behaviour of the two-way machine A when A is not actually scanning this portion of the tape is via the state transitions of A which it causes. The external effect of t is thus completely determined by the transition function, or "table", τ_t which gives (in addition to the state in which the machine originally exits from t), for each state q of A in which A might re-enter t, the corresponding state q' which A would be in when it left t again. This is all the information the machine can ever get about t however many times it comes back to refer to t; so it is all the machine needs to remember about t. But there are only a finite number of different such transition tables (since the number of states of A is finite), so the machine has no need to use the input tape to supplement its own internal memory; a one-way machine $\bar{A}$ with a sufficiently large number of internal states could store the whole transition table τ_t of t as it moved forward, and would then have no need to reverse and refer back to t later. If we think of the different states which A could be in when it re-entered t as the different questions A could ask about t, and the corresponding states A would be in when it subsequently left A again, as the answers, then we can state the result more crudely and succinctly thus: A machine can spare itself the necessity of coming back to refer to a piece of tape t again, if, before it leaves t, it thinks of all the possible questions it might later

come back and ask about t, answers these questions now and carries the table of question-answer combinations forward along the tape with it, altering the answers where necessary as it goes along.

To summarize Shepherdson's idea, we need to maintain for each prefix t two pieces of information: (1) the state in which the machine exits from t (when starting at q_0 in the leftmost square of the input), and (2) the external effect τ_t.

Let us refer to the processing of the input tape 101001 by A_2. Consider the first symbol of the input which is a 1. That is, let $t = 1$.

To answer the first question, we observe that when A_2 begins with the initial state q_0 at the symbol 1, it will exit t to its right at state q_1.

Next, we summarize the external effect τ_t where $t = 1$ by answering the following questions:

1. Suppose later in the processing of the input, the first symbol 1 is revisited by a left move from the right. What will be the effect if it is revisited with a state q_0?

2. Suppose later in the processing of the input, the first symbol 1 is revisited by a left move from the right. What will be the effect if it is visited with a state q_1?

3. Suppose later in the processing of the input, the first symbol 1 is revisited by a left move from the right. What will be the effect if it is visited with a state q_2?

Verify that the answers to the three questions are: (1) move right at state q_1, (2) move left at state q_2 and (3) move left at state q_2. In short the external effect can be succinctly summarized as a 3-tuple $[(q_1, R), (q_2, L), (q_2, L)]$.

We combine the two pieces of information computed in a 4-tuple $[(q_1, R), (q_1, R), (q_2, L), (q_2, L)]$. Let us call this 4-tuple the effect of t where $t = 1$. Note that the first entry is the answer to the first question, whereas the next three entries are the external effect of t. That is, the effect of t is the total effect, which is more than just the external effect.

Next, given the effect computed for the first symbol 1, we want to compute the effect of the first two symbols 10 of the input 101001.

From the effect $[(q_1, R), (q_1, R), (q_2, L), (q_2, L)]$ of the symbol 1, we know that the machine exits 1 to its right with the state q_1. Thus, the machine is at state q_1 when visiting the second symbol 0. According to the transition table, the machine will again move to the right of the second symbol at state q_1.

To compute the external effect of $t = 10$, we have to answer the following questions:

1. Suppose later in the processing of the input, the second symbol 0 is revisited by a left move from the right. What will be the effect if it is revisited with a state q_0?

2. Suppose later in the processing of the input, the second symbol 0 is revisited by a left move from the right. What will be the effect if it is visited with a state q_1?

3. Suppose later in the processing of the input, the second symbol 0 is revisited by a left move from the right. What will be the effect if it is visited with a state q_2?

The answers are as follows:

1. When the second symbol 0 is revisited with a state q_0, the machine according to the transition table will move to the right with the state q_0.

2. When the second symbol 0 is revisited with a state q_1, the machine according to the transition table will move to the right with the state q_1.

3. When the second symbol 0 is revisited with a state q_2, the machine according to the transition table will move to the right with the state q_0.

Therefore, the effect of $t = 10$ is summarized in the 4-tuple $[(q_1, R), (q_0, R), (q_1, R), (q_0, R)]$.

Note that in the above computations, we do not make use of the external effect computed for the first prefix 1 to compute the effect for the prefix 10. This is not usually the case. In general, the external effect for t is needed in computing the new effect when t is extended by one more symbol.

[5] Next, with the answers $[(q_1, R), (q_0, R), (q_1, R), (q_0, R)]$ for the effect for 10, compute the effect when the input is extended by the third symbol 1.

From the effect for 10, we know that the machine arrives at the third symbol 1 at state q_1. According to the transition table, the machine will move to the left to the 2nd symbol at state q_2. Next, consulting the last entry of the effect for 10, we know the machine will leave 10 to its right at state q_0. Thus, the machine revisits the third symbol 1 at state q_0. Again, from the transition table, the machine moves to the right at state q_1. Verify that the external effect of 101 is $[(q_1, R), (q_1, R), (q_1, R)]$. That is, the effect for 101 is $[(q_1, R), (q_1, R), (q_1, R), (q_1, R)]$.

In answering question [5], you should provide the steps involved in computing the external effect for 101.

Note that given the effect for t and a new symbol 1, we can compute the effect for $t' = t1$ by referring to the transition table for the machine. There is no need to know what t is. Only the effect for t is needed in computing the new effect for t'.

[6] Repeatedly, compute the effects by extending the current input 101 with symbols 0, 0, 1. From the effect computed for 101001, conclude that the input is accepted. Recall that 101001 was shown in [3] to be accepted by A_2.

[7] Repeat the whole exercise with the input 10111 as in [4]. Conclude that the input 10111 is not accepted.

To summarize, in simulating A_2 by a DFA, we need only remember the effect of the sequence of input symbols that has been processed so far.

[8] How many different effects can there be in simulating A_2 by a DFA? How many states are needed for a DFA to accept the same set of inputs as A_2? Note that there are only finitely many different effects even though we have an infinite number of inputs of finite lengths.

[9] Applying Shepherdson's method to A_1, how many states are there in the DFA constructed?

The work involved is very tedious. It is impossible to do it by hand. You should write a program to perform the computation.

Note that the smallest DFA[8] accepting the same set of inputs as A_1 has 33 states.

In computer programming, we can detect if the end-of-input has been reached. For example, in C programming, we write

```
while ((input=getChar()) != EOF)
```

Thus, it is very reasonable to extend the 2DFA model so that the machine can detect the two ends of an input tape.

Read the following excerpt (modified slightly) from Shepherdson's paper [6] regarding extending the 2DFA model by providing each input tape with special marker symbols b, e at the beginning and end of the input. Let us call this new model 2DFA-with-endmarkers.

> At first sight it would appear that with a two-way automaton more general sets of tapes could be defined if all tapes were provided with special marker symbols b, e (not in Σ) at the beginning and end, respectively. For then it would be possible, e.g., to design a two-way machine which, under certain conditions, would go back to the beginning of a tape, checking for a certain property, and then return again. This would appear to be impossible for an unmarked tape because of the danger of inadvertently going off the left-hand edge of the tape in the middle of the computation. In fact, the machine has no way of telling when it has returned to the first symbol of the tape. However, the previous construction result implies that this is not so; that the addition of markers does not make any further classes of tapes definable by two-way automata. For if the set $\{b\}U\{e\}$ (of all tapes of the form bte for $t \in U$) is definable by a two-way automaton then it is definable by a one-way automaton; and it is easy to prove that $\{b\}U\{e\}$ is definable by a one-way automaton if and only if U is.

---

[8]There is a mechanical method for computing the number of states of a smallest DFA.

We assume that a 2DFA-with-endmarkers starts on the left endmarker b at the initial state.

Suppose we want to design a 2DFA-with-endmarkers over $\Sigma = \{0, 1\}$ to accept input tapes that have a symbol 1 in the sixth position from the right end. For example, the input 10101011 has eight symbols with the third symbol being a 1, which is the sixth position from the last (eighth) position. The following 2DFA-with-endmarkers A_3, where q_0 is the starting state and q_9 is the accepting state, accepts the input tapes described. Observe that an accepted input must begin with b and end with e. Furthermore, b and e do not appear in other positions of the string accepted.

current state	symbol	new state	go to
q_0	b	q_1	R
q_1	0 or 1	q_1	R
q_1	e	q_2	L
q_2	0 or 1	q_3	L
q_3	0 or 1	q_4	L
q_4	0 or 1	q_5	L
q_5	0 or 1	q_6	L
q_6	0 or 1	q_7	L
q_7	1	q_8	R
q_8	0 or 1	q_8	R
q_8	e	q_9	R

[10] Following the discussion by Shepherdson and based on the definition of A_3, explain the construction of a DFA equivalent to A_3 accepting the same set of input tapes where the inputs are delimited by b and e. How many states does the DFA have? Next, modify the DFA constructed to accept inputs with the endmarkers b and e removed. What is the change in the number of states of the DFA?

As in [9], you should write a program to perform the computation.

Note that it can be shown that the smallest DFA accepting the same set of inputs (without the endmarkers b and e) as A_3 has 64 states.

Consider another 2DFA-with-endmarkers A_4, where q_0 is the starting state and q_9 is the accepting state, as defined below.

current state	symbol	new state	go to	
q_0	b	q_1	R	
q_1	0 or 1	q_1	R	
q_1	e	q_2	L	
q_i	0	q_{i+1}	L	$i = 2, 3, 4$
q_i	1	q_i	L	$i = 2, 3, 4$
q_5	0	q_7	L	
q_5	1	q_6	L	
q_6	0	q_6	L	
q_6	1	q_7	L	
q_7	0	q_7	L	
q_7	1	q_5	L	
q_7	b	q_8	R	
q_8	0 or 1	q_8	R	
q_8	e	q_9	R	

[11] Re-do part [10] for A_4.

Note that it can be shown that the smallest DFA accepting the same set of inputs (without the endmarkers b and e) as A_4 has 64 states.

Notes for the instructor

The project is suitable to be used in a senior undergraduate theory of computation course in computer science. While the programming skills needed to solve some of the project questions are not particularly challenging, they are not trivial either. It is suggested that a student needs two years of programming training before attempting the programming tasks given in this project. As extra credit problem, one can ask the students to deduce (using the Myhill-Nerode theorem) the sizes of smallest DFAs that are equivalent to A_1, A_2, A_3 and A_4, respectively.

Bibliography

[1] Hopcroft, J. E., and Ullman, J. D., *Introduction to Automata Theory, Languages, and Computation*, Addison-Wesley, Reading, MA, 1979.

[2] Kleene, S. C., "Representation of Events in Nerve Nets and Finite Automata," in *Automata Studies*, Shannon, S. C., McCarthy, J. (editors) Princeton University Press, NJ, 1956, 3–41.

[3] Kozen, D. C., *Automata and Computability*, Springer-Verlag, New York, 1997.

[4] McCulloch, W. S., and Pitts, W., "A Logical Calculus of Ideas Immanent in Nervous Activity." *Bull. Math. BioPhys.*, **5** (1943), 115–133.

[5] Rabin, M. O., Scott, D., "Finite Automata and Their Decision Problems," *IBM Journal of Research and Development*, **3** (1959), 114–125.

[6] Shepherdson, J., "The Reduction of Two-Way Automata to One-Way Automata," *IBM Journal of Research and Development*, **3** (1959), 198–200.

Part III

Articles Extending Discrete Mathematics Content

A Rabbi, Three Sums, and Three Problems

Shai Simonson
Stonehill College

1 Introduction

We present a slice of discrete math history and connect it to three neat problems and their solutions. The solutions emphasize the value of exploring and tabulating data in order to help discover theorems. Often, the exploration not only helps discover theorems, but suggests a direction and idea for a proof.

The methodology is a model of how to use mathematical history and data patterns in teaching discrete mathematics. Proofs in discrete math tend to be constructive. Discovering patterns can suggest an algorithm, which in turn can suggest a proof. Furthermore, many theorems in discrete math make use of basic sums and combinatorial identities that have been known for hundreds of years. Familiarity with the history of an identity can also suggest relevant proof methods.

Underlying the content in this article is the implicit theme of how mathematics and computers support each other. Computers can be used to generate data for pattern searching, or to help with calculations that are otherwise too cumbersome. Symbiotically, these same theorems, whose statements and proofs are motivated by the data gathered with the use of a computer, are themselves applicable to computer science in the analysis of algorithms as well as other areas.

2 A Rabbi's Identities

Rabbi Levi ben Gershon (1288–1344, also known as Gersonides) lived in southern France and was a well-known scientist, mathematician, and philosopher among both Jewish and Christian scholars of his day [Si1], [Si2]. Levi discovered formulas for certain kinds of sums. These include the well-known formulas for sums of consecutive integers, sums of consecutive squares, and sums of consecutive cubes.

$$1 + 2 + \cdots + n = \frac{n(n+1)}{2} \tag{A}$$

$$1^2 + 2^2 + 3^2 + \cdots + n^2 = \frac{n(n+1)(2n+1)}{6} \tag{B}$$

$$1^3 + 2^3 + 3^3 + \cdots + n^3 = \left(\frac{n(n+1)}{2}\right)^2 \tag{C}$$

Levi was certainly not the first to discover these formulas, but he was the first to use mathematical induction style arguments in his proofs [Ka]. Levi used induction to prove (C), but different methods for (A) and (B). His methods serve as a good review of techniques for generating and proving these closed formulas, as well as a source of clever exercises for students.

Levi's work also highlights the pros and cons of symbolic algebra versus the Euclid-style "prose"-algebra of the middle ages. In symbolic algebra we write $x^2 + y^2$, but for Euclid, Levi, and all mathematicians before the 16th century, this formula would have been written out as "the sum of the square of x with the square of y." Did the limitations of "prose"-algebra prevent Levi from extending his induction proofs to other sums? Perhaps Levi simply modeled each proof on whatever constructive ideas led him to the discovery of the sum. No one knows the answers to these questions, but they make for good debate in the classroom. Levi's work is not widely known, and we hope the reader enjoys the details of this discussion.

Each of Levi's three formulas (A), (B), and (C) plays a role in the solution of three different problems discussed in Section 3.

2.1   Triangle Numbers

The numbers generated by formula (A) are called *triangle numbers*, because they look like triangles if you draw them with dots. Triangle numbers have a natural combinatorial interpretation. The nth triangle number is equal to the number of ways to choose two items from $n + 1$. This is usually written as $\binom{n+1}{2}$. If you try to count these pairs and number the items 1 through $n + 1$, then the first item can be paired with n others, the second item with $n - 1$ others, and so on, until the $(n - 1)$st item can be paired with only the nth item. This sum is $1 + 2 + \cdots + n$. The first five triangle numbers are 1, 3, 6, 10, and 15.

Although Levi understood the distributive property, and he could essentially perform algebraic manipulations like "FOIL"-ing, he had no symbolic algebra as we know it today. All his work was written out in Euclid-style prose just like the work of his Jewish, Christian, and Islamic contemporaries. Without symbolic algebra, we lose the compact form of formula (A) but we gain three new ways to read that formula. It is a lesson worth remembering.

Indeed Levi has three versions of formula (A) each with its own proof. These are theorems (26), (27), and (28) in his book *Maaseh Hoshev*, literally "the way of calculation." The title of his book is also a pun on a biblical phrase that describes labor on the tabernacle needing intensive thought, analysis, and calculation.

> (26) The sum of consecutive numbers from one up to an even number is equal to half the number of terms times the number of terms plus one.

Levi proves this identity by adding the numbers up in pairs, each successive pair coming from the front and back of the series. He notes that each pair sums to $n + 1$ and there are $n/2$ such pairs. He had two other proofs to take care of the case when n is odd.

> (27) The sum of consecutive terms from one up to an odd number is equal to the middle term times the number of terms.

The proof pairs up numbers from the inside working outward, showing that each pair sums to twice the middle term.

> (28) The sum of consecutive numbers from one up to an odd number is equal to half the last term times the number following the last term.

The proof uses (27) with some additional algebra arguments.

2.2   Square Pyramid Numbers

A number generated by formula (B) is called a *square pyramid number*, because it can be constructed by stacking squares of diminishing size on top of one another, forming a square-base pyramid. The first five square pyramid numbers are 1, 5, 14, 30, and 55. Levi's proof of this formula is longwinded and complex, building the formula logically from many simpler algebraic identities. A sketch of his proof with many details omitted is shown below. Filling in the missing details for each step is a good source of exercises for students.

The numbering of the lemmas is faithful to his book *Maaseh Hoshev*. Going forward, all the lemmas are translated from Levi's prose into modern symbolic notation.

First he proves the two identities, (29) and (30), by "pairing" tricks. In the first identity (29), he pairs numbers from back and front as he did for triangle numbers. In the second identity (30), he pairs 1 from the left group with n from the right, 2 from the left group with $n - 1$ from the right, etc.

$$(1 + 3 + \cdots + (2n - 1)) = n^2, \tag{29}$$

$$(1 + 2 + \cdots + n) + (1 + 2 + \cdots + n + (n + 1)) = (n + 1)^2. \tag{30}$$

Using (30) Levi proves

$$1 + (1 + 2) + (1 + 2 + 3) + \cdots + (1 + 2 + \cdots + n) = \begin{cases} 2^2 + 4^2 + 6^2 + \cdots + n^2 & \text{if } n \text{ is even,} \\ 1^2 + 3^2 + 5^2 + \cdots + n^2 & \text{if } n \text{ is odd.} \end{cases} \tag{32}$$

He puts lemma (32) aside for later. Then he proves three more simple lemmas.
He proves

$$(1 + 2 + 3 + \cdots + n) + (2 + 3 + 4 + \cdots + n) + \cdots + n = 1^2 + 2^2 + 3^2 + \cdots + n^2 \tag{33}$$

by rearranging the numbers.

He proves

$$\begin{aligned} &(1 + 2 + 3 + \cdots + n) + (2 + 3 + 4 + \cdots + n) + \cdots + n \\ &+ 1 + (1 + 2) + (1 + 2 + 3) + \cdots + (1 + 2 + \cdots + (n - 1)) \\ &= n(1 + 2 + 3 + \cdots + n) \end{aligned} \tag{34}$$

not by using (33) but by another simple rearranging argument.

And, he proves

$$(n + 1)^2 + n^2 - (n + 1 + n) = 2n^2 \tag{35}$$

by essentially multiplying it out.

Levi then uses (33) and (35) to prove

$$(1 + 2 + 3 + \cdots + n) + (2 + 3 + 4 + \cdots + n) + \cdots + n - (1 + 2 + 3 + \cdots + n) \tag{36}$$
$$= \begin{cases} 2(2^2 + 4^2 + 6^2 + \ldots + (n - 1)^2) & \text{when } n - 1 \text{ is even,} \\ 2(1^2 + 3^2 + 5^2 + \cdots + (n - 1)^2) & \text{when } n - 1 \text{ is odd.} \end{cases}$$

And, he combines (32), (34), and (36) to prove

$$n(1 + 2 + 3 + \cdots + (n + 1)) = \begin{cases} 3(1^2 + 3^2 + 5^2 + \cdots + n^2) & \text{when } n \text{ is odd,} \\ 3(2^2 + 4^2 + 6^2 + \cdots + n^2) & \text{when } n \text{ is even.} \end{cases} \tag{37}$$

Finally, Levi uses (32), (33), (34), and (37) to prove

$$\left(n - \frac{n - 1}{3}\right)(1 + 2 + 3 + \cdots + n) = 1^2 + 2^2 + 3^2 + \cdots + n^2. \tag{38}$$

This last identity is formula (B) in disguise. Notice that formula (A) can be used to replace $(1 + 2 + 3 + \cdots + n)$ with $\frac{n(n+1)}{2}$.

2.3 Sums of Cubes

Levi proves formula (C) by mathematical induction. The proof by induction for formula (C) comes directly from the inductive lemma

$$(1 + 2 + 3 + \cdots + n)^2 = n^3 + (1 + 2 + 3 + \cdots + (n - 1))^2 \tag{41}$$

which he proves essentially by "FOIL"-ing out the left side. The proof of (C) is remarkably short and clear in comparison to the proof of (B).

Levi was one of the first to use the idea of mathematical induction, 300 years before Pascal [Ka]. It is interesting that he chose induction to prove formula (C), but not for formulas (A) and (B). Why do you think this was? His proof for (A) is fairly simple, but considering the complexity of his proof for (B), one naturally wonders why he did not use induction there as well. These are the kinds of questions that interest historians of mathematics.

3   Three Problems

Building on Levi's work, we tackle three counting problems. The first two are fairly well-known and have elegant closed-form solutions. The third problem is not well-known and the solution requires extra effort. Before we start on the three problems it is helpful to solve a little warm-up problem, and add one more formula, (D), to Levi's (A), (B), and (C).

3.1   Triangular Pyramid Numbers — A Warm-Up Problem

The sum of the first n triangle numbers is called a *triangular pyramid number* because it is formed by stacking triangles of diminishing size on top of each other into a triangular pyramid. Let's warm up by deriving a closed form for the nth triangular pyramid number. The easy way to do this is to use Levi's formulas (A) and (B) to help:

$$\sum_{i=1}^{n} \frac{i(i+1)}{2} = \frac{1}{2}\left(\sum_{i=1}^{n} i^2 + \sum_{i=1}^{n} i\right) = \frac{1}{2}\left(\frac{n(n+1)(2n+1)}{6} + \frac{n(n+1)}{2}\right).$$

Simplifying the right side proves that the nth triangular pyramid number is equal to

$$1 + 3 + 6 + \cdots + \frac{n(n+1)}{2} = \frac{n(n+1)(n+2)}{6}. \tag{D}$$

With formulas (A), (B), (C), and (D), we are armed to attack our three counting problems. However, before we move ahead to the three problems promised, there is a short interesting tangent showing a connection between the relatively simple discrete math formulas (B) and (D), and state of the art number theory.

Is there a number (besides 1) that is both a triangular pyramid number and a square pyramid number? That is, do formulas (B) and (D) generate any common integers besides 1? It would make a good children's Sherlock Holmes mystery, where somebody knocks down a pyramid of oranges in the store, and when the oranges are rebuilt into a pyramid, it looks like none are missing, but the grocer insists that some oranges are missing. A young Holmes interviews the grocer and confirms that the original pyramid was square while the rebuilt pyramid is triangular. Holmes declares, "The grocer is telling the truth; there are oranges missing. Because as everyone knows, there is no number besides 1 that is both a triangular pyramid number and a square pyramid number."

Well, in fact, nobody knew this for sure until 1988 when Beukers and Top proved it using some high powered methods [BT], [HL]. Their work is a special case of finding rational points on elliptic curves.

Now let's begin on the three problems.

3.2   Problem 1 — Counting Squares in a Grid

How many squares are there in an n by n grid? See Figure 1.

Figure 1. A 4 by 4 grid.

Before one tackles any counting problem, it never hurts to get some data. Let's make a chart for the first four values of n, and count the squares explicitly. You can generate the first few cases by hand pretty quickly, but if you are a clever programmer, you might enjoy writing a program to count the squares. The advantage of a program is that once it's finished, it generates lots more data very quickly. Moreover, sometimes the process itself of writing a program reveals an organized way to think about the counting. We won't need a program in these examples, but the use of programs in similar problems can be exciting and helpful for students and researchers alike.

n	Number of Squares in an n by n Grid
1	1
2	5
3	14
4	30

Compare the numbers in the table above with those generated from identity (B), and observe that they are identical. Indeed, it looks as though the number of squares in an n by n grid is the same as the sum of the first n squares. And, we can prove it. There are n^2 squares of size 1, and $(n-1)^2$ squares of size 2, ..., and one square of size n.

3.3 Problem 2 — Counting Rectangles in a Grid

Counting the number of rectangles explicitly is a lot more work than counting squares. Doing it for grids of size 1 by 1, 2 by 2, 3 by 3, and 4 by 4, we get the chart below:

n	Number of Rectangles in an n by n Grid
1	1
2	9
3	36
4	100

The pattern is interesting. The numbers are squares, and moreover they are squares of triangle numbers. Levi's identity (C) says that the square of the nth triangle number is the sum of the first n cubes. It seems, therefore, that the number of rectangles in an n by n grid is the sum of the first n cubes.

We can construct a combinatorial proof of this discovery. Indeed, attempting to write a program to count the rectangles might lead to this proof. Each rectangle in an n by n grid is uniquely defined by selecting two of the $n+1$ vertical lines and two of the $n+1$ horizontal lines. The area outlined by these four lines is the rectangle that they define. Hence the number of rectangles is

$$\binom{n+1}{2}^2 = (1 + 2 + \cdots + n)^2 = 1^3 + 2^3 + \cdots + n^3.$$

3.4 Problem 3 — Counting Triangles

How many triangles in the triangular "grid" with four edges per side? See Figure 2.

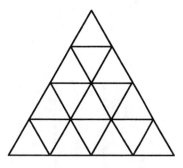

Figure 2. A triangular grid.

This last puzzle does not yield its secret so easily. Our exploration builds conjectures, looks for patterns, tests the conjectures and revises them.

First we gather some data by counting the number of triangles explicitly for different size triangular figures with the number of edges on a side varying from 1 to 5. To make sure we don't miss any triangles, it is a good idea to organize the counting by the size of the triangles, or by whether or not the triangles point up or down. The total results are summarized below:

n	Number of Triangles in a Triangular Figure with Sides of Length n
1	1
2	5
3	13
4	27
5	48

These numbers are not so familiar. We aren't as lucky as we were last time, but if we look harder we may still see a pattern that suggests a closed form and a proof. Let's count the triangles by size. That idea worked for squares and rectangles.

	Number of Triangles in a Triangular Figure with Sides of Length n				
n	Size 1	Size 2	Size 3	Size 4	Size 5
1	1	0	0	0	0
2	4	1	0	0	0
3	9	3	1	0	0
4	16	7	3	1	0
5	25	13	6	3	1

This table has one nice column — the square numbers of column 1 — but column 2 is not familiar, and the other columns are too short to reveal any patterns. Let's explore further. You may have noticed that some of the triangles point upward and some point downward. Let's separately count triangles that point up, and triangles that point down.

	Number of Triangles in a Triangular Figure with Sides of Length n		
n	Triangles that Point Up	Triangles that Point Down	Total
1	1	0	1
2	4	1	5
3	10	3	13
4	20	7	27
5	35	13	48

The first column should look familiar. The numbers in it are the triangular pyramid numbers. However, the second column is still a mystery. Let's break it down a little more, this time by type of triangle (up or down) and size.

	Number of Triangles in a Triangular Figure with Sides of Length n										
	Triangles that Point Up					Triangles that Point Down					
n	Size 1	Size 2	Size 3	Size 4	Size 5	Size 1	Size 2	Size 3	Size 4	Size 5	Total
1	1	0	0	0	0	0	0	0	0	0	1
2	3	1	0	0	0	1	0	0	0	0	5
3	6	3	1	0	0	3	0	0	0	0	13
4	10	6	3	1	0	6	1	0	0	0	27
5	15	10	6	3	1	10	3	0	0	0	48

Now we are getting somewhere. It's a table of triangle numbers! Let T_n be the nth triangle number. The sum of the nth row of the triangles that point up is T_n larger than the sum of the $(n-1)$st row of triangles that point up. Less obvious is that the sum of the nth row of the triangles that point down is T_{n-1} larger than the sum of the $(n-2)$nd row of triangles that point down. This means we would expect the next two rows in the table to be

6   21   15   10   6   3   1   0   15   6   1   0   0   0   0
7   28   21   15   10   6   3   1   21   10   3   0   0   0   0

Once we discover patterns like these, we are directed by our own exploration and serendipitous discovery to try to explain why they might be so. Let U_n and D_n, respectively, be the number of up-pointing and down-pointing triangles in a triangular "grid" with n edges per side. Figure 2 shows the grid when $n = 4$.

Lemma 1. $U_n = U_{n-1} + T_n$.

Proof. A triangle figure with n edges per side contains all the up-pointing triangles of a similar figure with $n - 1$ edges per side, plus whatever extra up-pointing triangles we get by including one or more of the n edges from the bottom of the triangle. There is one up-pointing triangle that uses all these edges (the whole figure), two that use $n - 1$ of these edges, three that use $n - 2$ of these edges, ..., and n triangles that use exactly one of these edges. This sum is T_n. See Figure 3 for an example when $n = 5$.   □

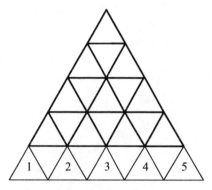

Figure 3. A triangular grid with a new row containing five edges.

The next proof is similar but slightly more difficult than the previous one. The difficulty is surmountable because our patterns tell us exactly what we must look for in the proof. And, when you know what to look for, then a proof is easier to find.

Lemma 2. $D_n = D_{n-2} + T_{n-1}$.

Proof. A triangle figure with n edges per side contains all the down-pointing triangles of a similar figure with $n - 2$ edges per side, plus whatever extra down-pointing triangles we get by including a point on any one of the two new rows. The number of down-pointing triangles with i edges per side that include a point on the bottom (nth) row equals $n - (2i - 1)$. This is because there are $n - 1$ down-pointing triangles with one edge per side that use points on the bottom row, and every time you add an edge to the side of the triangle, you lose the triangles that used the right and leftmost points on the nth row. See Figure 4 for an example when $n = 6$. For a similar reason, the number of down-pointing

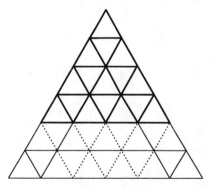

Figure 4. A triangular grid with two new rows. There are 5 one-sided, 3 two-sided, and 1 three-sided triangles pointing downward with the point on the bottom (sixth) row. There are 4 on-sided and 2 two-sided triangles pointing downward, with the point on the fifth row.

triangles with i edges per side that include a point on the next to bottom ($(n-1)$st) row equals $n - 2i$. The sum of these two as i ranges from 1 to $\frac{n}{2}$ is T_{n-1} (when i is greater than $\frac{n}{2}$, there are no triangles). $\square$

Let $M_n = U_n + D_n$, i.e., the total number of triangles in a triangular grid with n edges per side (M is the transliterated first letter of the Hebrew word for triangle). Lemmas 1 and 2 imply that $M_n = M_{n-2} + T_n + 2T_{n-1}$. Also, $M_0 = 0$ and $M_1 = 1$. Recall that $T_n = \frac{n(n+1)}{2}$. Then we have

Theorem 1.

$$M_n = M_{n-2} + \frac{n(n+1)}{2} + 2\frac{(n-1)n}{2} = M_{n-2} + \frac{3n^2}{2} - \frac{n}{2}.$$

This seems to solve our problem but it is not as satisfying as the solutions for squares and rectangles. It would be nice to get a *closed* formula for the number of triangles; that is, a formula without any recurrence or loops, like we had for squares and rectangles.

Recurrence equations are easier to discover than closed formulas, but closed formulas are often easier to use for calculations. Fortunately, many recurrence equations can be solved and turned into closed formulas by simply "unraveling" the formulas. This means repeatedly applying the recurrence over and over again, starting with the base case, constructing a summation that we then turn into a closed form. In this case, since the recurrence equation skips two indices with each iteration, we get two different summations, one for the even indices and one for the odd.

Starting with even indices, we have

$$M_0 = 0,$$

$$M_2 = M_0 + \frac{3}{2}\left(2^2\right) - \frac{1}{2}(2),$$

$$M_4 = M_2 + \frac{3}{2}\left(4^2\right) - \frac{1}{2}(4).$$

This generalizes to

$$
\begin{aligned}
M_{2n} &= \frac{3}{2}\left(2^2 + 4^2 + 6^2 + \cdots + (2n)^2\right) - \frac{1}{2}(2 + 4 + 6 + \cdots + 2n) \\
&= \frac{3}{2} \times 4\left(1^2 + 2^2 + 3^2 + \cdots + n^2\right) - (1 + 2 + 3 + \cdots + n) \\
&= 6\frac{(n(n+1)(2n+1))}{6} - \frac{n(n+1)}{2} \\
&= n(n+1)(2n+1) - \frac{n(n+1)}{2} \\
&= \frac{(4n^3 + 5n^2 + n)}{2}.
\end{aligned}
$$

The first line above comes from unraveling the recurrence of Theorem 1; the third line is derived from the second line using Levi's identities (A) and (B); the fourth line is due to identity (A); and the rest is straightforward algebra.

As we did for the even indices, let's find the pattern and closed form for the odd values, M_{2n+1}.

$$
\begin{aligned}
M_{2n+1} &= 1 + \frac{3}{2}\left(3^2 + 5^2 + \cdots + (2n+1)^2\right) - \frac{1}{2}(3 + 5 + \cdots + (2n+1)) \\
&= \frac{3}{2}\left(1^2 + 3^2 + 5^2 + \cdots + (2n+1)^2\right) - \frac{1}{2}(1 + 3 + 5 + \cdots + (2n+1)) \\
&= \frac{(2n+1)(1 + 2 + 3 + \cdots + (2n+2))}{2} - \frac{(n+1)^2}{2} \\
&= \frac{(2n+1)(n+1)(2n+3)}{2} - \frac{(n+1)^2}{2} \\
&= \frac{4n^3 + 11n^2 + 9n + 2}{2}.
\end{aligned}
$$

As before, the first line above comes from unraveling Theorem 1. The third line is derived from the second line using Levi's identities (29) and (37). And, as before, the fourth line is due to identity (A), and the rest is straightforward algebra.

The two formulas for the number of triangles in a grid are not as pretty as the closed forms for the number of squares and rectangles. But they are the truth, and you can't choose the truth! An equivalent but much prettier way to write these formulas is

$$M_n = \begin{cases} \dfrac{n(n+2)(2n+1)}{8} & \text{when } n \text{ is even,} \\ \dfrac{n(n+2)(2n+1)-1}{8} & \text{when } n \text{ is odd.} \end{cases}$$

or simply

$$M_n = \frac{n(n+2)(2n+1)}{8} + \frac{1}{16}\left((-1)^n - 1\right).$$

The stuff at the end of the formula just takes care of the minor difference between the odd and even cases. Notice the similarity between the main part of this formula $\frac{n(n+2)(2n+1)}{8}$ and the closed form for the number of squares in a grid, $\frac{n(n+1)(2n+1)}{6}$.

4   Conclusion

We presented a trio of problems that focus on the benefits of discovery and experiment in discrete mathematics. The solutions to the problems use recurrence equations, summations, and counting concepts, and involve some famous identities.

There are many ways this material can be used in the classroom. Several identities of Levi ben Gershon (1288–1344) can be used with discussion centered on contrasting different style proofs and analyzing why someone might use one technique over another. The three problems (squares, rectangles and triangles in a grid) can be used in group exploration and discovery, with the instructor as the guide in a Pólya-style lecture.

5   Acknowledgements

Thanks to Victor Katz for introducing me to the mathematics of Levi ben Gershon, and Hendrik Lenstra for directing me to the paper of Beukers and Top. Many thanks to Ralph Bravaco, Brian Hopkins, and an anonymous referee for their careful review and excellent suggestions.

Bibliography

[BT]   Frits Beukers and Jaap Top, "On oranges and integral points on certain plane cubic curves," Nieuw Arch. Wisk. (4) 6, No. 3 (1988) 203–210.

[HL]   Hendrick Lenstra, personal communication.

[Ka]   Victor J. Katz, *A History of Mathematics: An Introduction*, New York: HarperCollins, 1993, 278–280.

[Si1]   S. Simonson, "The Missing Problems of Gersonides – A Critical Edition, Part I," Historia Mathematica, Vol. 27, No. 3 (2000) 243–302.

[Si2]   S. Simonson, "Gems of Levi ben Gershon," Mathematics Teacher, 93, 8 (2000) 659–663.

Storing Graphs in Computer Memory

Larry E. Thomas
Saint Peter's College

1 Introduction

Many of the structures we use in discrete mathematics are nonlinear when they are drawn in the usual way on paper. For example, a graph will often have many edges leaving or entering a given vertex, with the number of edges changing from vertex to vertex. Drawing such a graph on paper could produce a rather messy diagram which takes two dimensions to represent. However, most graph algorithms used in problems of any size are executed on a computer which has only a linear (one-dimensional) memory. The question naturally arises: How do we represent an inherently two-dimensional object, such as a graph or a matrix, in one dimension? This note explores some of the ways this question can be answered.

2 A computer's memory

We will begin with a simplified explanation of the way the memory of a regular desktop computer might be organized. The memory consists of consecutive *locations* where data, represented in a binary code, can be stored. In a memory with n consecutive locations, the locations are numbered with *addresses* beginning with address 0: $0, 1, 2, \ldots, n$. If, for example, a single location can hold one byte (eight bits) of binary code, a one-kilobyte memory will have locations numbered $0, 1, 2, \ldots, 1024$, since 1Kb of memory contains $2^{10} = 1024$ bytes.[1] Pictorially, we can think of the memory as a long collection of numbered boxes as seen in Figure 1.

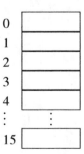

Figure 1. A 16-location memory.

This long collection of numbered boxes is essentially a linear construct. Our problem is to find a way to store information about nonlinear discrete structures, such as matrices and graphs, in this linear construct.

---

[1] It is important to note that a single piece of data stored in memory may occupy more than one location. Typically, it takes one location to store a character, such as the letter R, in memory, but it might take four locations to store an integer, and six or more to store a decimal number.

3   Representing matrices

Most modern computer languages have built-in facilities representing vectors and matrices. In this section we will have a behind-the-scenes look at how this representation is accomplished in standard procedural languages such as C++.

3.1   Locator functions

Locator functions are at the heart of *formula-based* representations. A locator function maps an element of a discrete structure to a specific location in a computer's memory. Locator functions are commonly used to represent vectors and matrices, both of which are usually called *arrays* in programming languages. A vector is a one-dimensional array and a matrix is a two-dimensional array.

3.2   Vectors

Finding a locator function for a vector is straightforward. Suppose we want to compute the addresses in which to store the elements of the vector $\mathbf{v} = (v_1, v_2, \ldots, v_n)$. The computer will decide where to place v_1 in memory; let α be the address of the location holding v_1. The computer also knows s, the number of locations it takes to store each element of $\mathbf{v}$. Then the locator function

$$\text{addr}(i) = \alpha + (i - 1)s, \qquad 1 \le i \le n$$

gives the address of v_i.

3.3   Matrices

The locator function for a matrix is a bit more complicated. We can view the problem as storing the first row of the matrix in sequential locations, just as we stored a single vector above, then storing the second row, then the third, and so on.[2] Consider finding the locator function for the elements of an $m \times n$ matrix A. As before, let α be the address of A_{11} and let s be the number of locations it takes to store one element of A. Then $r = ns$ is the number of locations it takes to store one row of A. It follows that the locator function

$$\text{addr}(i, j) = \alpha + (i - 1)r + (j - 1)s, \qquad 1 \le i \le m, \quad 1 \le j \le n$$

gives the address of a_{ij}.

To see that this formula does indeed give sequential, contiguous locations in which to store the elements of A, let us consider an example. Suppose that A is a 3×4 matrix. The following matrix shows the *addresses* in which the elements of A will be stored.

$$\begin{bmatrix} \alpha & \alpha + s & \alpha + 2s & \alpha + 3s \\ \alpha + 4s & \alpha + 5s & \alpha + 6s & \alpha + 7s \\ \alpha + 8s & \alpha + 9s & \alpha + 10s & \alpha + 11s \\ \alpha + 12s & \alpha + 13s & \alpha + 14s & \alpha + 15s \end{bmatrix}$$

We can see that the addresses are indeed sequential and contiguous, beginning at α and ending at $\alpha + 15s$.

4   The treasure hunt game

The rigid structure of vectors and matrices makes it fairly easy to find locator functions for them. While it is possible to find the equivalent of locator functions for unruly structures such as trees and graphs, it would be useful to have an intuitive way to represent vertices and edges. To gain flexibility and intuitive appeal, we have to give up the idea of storing data sequentially, but we still need to use the one-dimensional computer memory as shown in Figure 1. We need a way to jump around in this memory to follow the edges of a graph wherever they lead us.

The method computer programmers use to jump around in memory can be introduced using the Treasure Hunt Game. In this game someone hides boxes in various places. Each box contains a letter of the alphabet and a clue about

---

[2]This is the *row-major* order and is commonly used now. The order which goes column-by-column is *column-major* order.

where to find the next box. The players are given a clue about where to find the first box; then they set out to find all of the other boxes. When they find a box they record the letter in the box and then they use the new clue (the pointer) to find the next box. When they have found the last box, the letters they recorded will spell out the Magic Word.

Since we will eventually be using a computer's memory to store the boxes, we require that all of the boxes be hidden someplace along a straight road, representing the linear nature of the memory. Figure 2 shows an example.

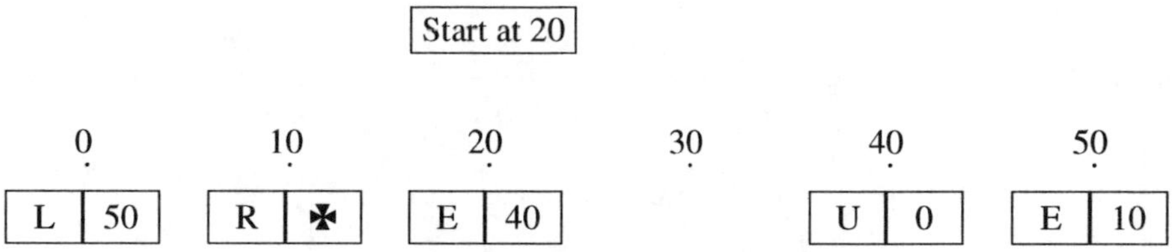

Figure 2. The Treasure Hunt game.

The numbers along the top of the "road" give the distance in feet, say, from the beginning. The boxes are shown below the road; each box contains a letter and the position of the next box. When the game begins, the players are told to go 20 miles from the beginning to find the first box. They find a box containing the letter E and the number 40. The number 40 points them to the next box, which is located 40 miles from the beginning. When they arrive at the position 40 miles from the beginning, they find another box containing the letter U and the *pointer* 0, telling them to look at position 0 for the next box. They do so and find the letter L and a pointer to position 50. At 50 they find the letter E and the pointer 10. When they go to position 10, they find the letter R, but instead of a numerical pointer, they find the symbol ✠. This symbol indicates that there are no more boxes. If the players recorded the letters as they found them, they found the Magic Word to be EULER.

There is a complete analogy between the road in the Treasure Hunt game and a computer's memory. In terms of a computer program retrieving data that has been stored in memory, the above paragraph could be rewritten as follows: The program starts by looking in location 20 for the first letter. It finds the letter E and a *pointer* to location 40. So it looks in location 40 where it finds the letter U and a pointer to location 0. In location 0 it finds the letter L and a pointer to location 50, where it finds an E and a pointer to location 10. At location 10 it finds the letter R and the special symbol ✠, which is used here to indicated that there is no more data.

The structure that results from storing a piece of data along with a pointer to the next piece of data is called a *linked list*. In practice it is the programmer's responsibility to indicate which pieces of data are to be linked, but the programmer *does not* decide on the actual memory locations that will be used to stored the linked data. These locations are chosen by the computer's systems programs. In fact, if the same program is run twice the locations chosen on each run may be different.

Fortunately, it is rare that a programmer needs to know the actual memory addresses in the linked list. All the programmer needs to know is which nodes are linked. In the example, it is important for the programmer to know that the first E is linked to the U, but the programmer does not need to know which addresses are involved in establishing the link. That means that it is not necessary, or even possible, to show the numerical values of the addresses in drawing diagrams to represent linked lists. We will still use boxes as in Figure 2, but we will leave the "clue" or "pointer" section blank and we will arrange the boxes so that linked boxes follow each other. Figure 3 shows the standard way of representing a linked list in computer science texts.

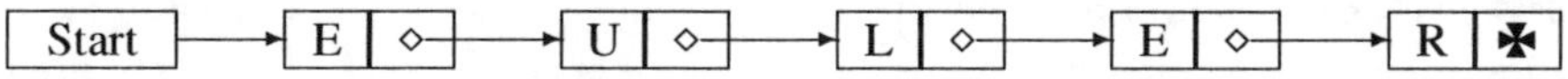

Figure 3. The treasure hunt's linked list.

5   Graphs

Now that we know something about pointers we can look at some of the standard ways of storing a graph in a computer's memory. We will start with the *adjacency matrix* and then consider the *vector-based adjacency list* and the *linked adjacency list*. The last two methods rely on the *adjacency list* of a vertex. If v is a vertex in a graph, its adjacency list, adj(v), is just a set of the vertices that are adjacent to v. In Figure 4(a), for example, adj(2) = $\{1, 3, 4, 5\}$.

5.1   Adjacency matrices

In texts on discrete mathematics the most common way to represent a graph seems to be the *adjacency matrix* of the graph. For a graph with n vertices numbered $1, 2, 3, \ldots, n$, the adjacency matrix $A = [a_{ij}]$ is defined by

$$a_{ij} = \begin{cases} 1 & \text{if vertex } i \text{ is adjacent to vertex } j, \\ 0 & \text{otherwise.} \end{cases}$$

The adjacency matrix for the graph in Figure 4(a) is

$$A = \begin{bmatrix} 0 & 1 & 0 & 0 & 0 \\ 1 & 0 & 1 & 1 & 1 \\ 0 & 1 & 0 & 0 & 1 \\ 0 & 1 & 0 & 0 & 0 \\ 0 & 1 & 1 & 0 & 0 \end{bmatrix}.$$

Standard matrix operations such as multiplication and transposition can be used to analyze a graph using its adjacency matrix.

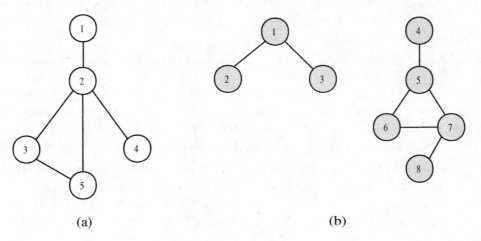

(a)   (b)

Figure 4. Two graphs.

5.2   Vector-based adjacency lists

One way to represent a graph is to use two vectors to keep track of the edges between vertices. First the vertices have to be numbered as shown, for example, in the graphs of Figure 4. Then a vector $\mathbf{e}$ is created to record adj(v) for each vertex v: first all of the elements of adj(1) are listed in $\mathbf{e}$, then all elements of adj(2), and so on. The order in which the vertices are listed is immaterial. Another vector, $\mathbf{p}$ holds the indices in $\mathbf{e}$ where the individual adjacency lists begin. $\mathbf{p}$ is constructed so that the vertices in adj(i) lie in positions $\mathbf{p}[i]$ through $\mathbf{p}[i + 1] - 1$. We can think of the elements of $\mathbf{p}$ as pointers to positions in $\mathbf{e}$. So the graph in part (a) of the Figure 4 is represented as shown in Figure 5.

Note that, for example, the vertices that can be reached from the vertex numbered 2 are in positions $\mathbf{p}[2] = 2$ through $\mathbf{p}[3] - 1 = 5$ in $\mathbf{e}$. The definition of $\mathbf{p}$ leads to an "extra" last element that references one index position past the end of $\mathbf{e}$. The list for Figure 4(b) can be constructed similarly to produce Figure 6.

$$\mathbf{p} = [1, 2, 6, 8, 9, 11].$$

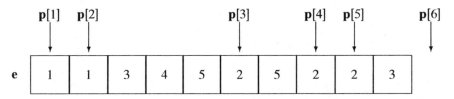

Figure 5. Vector-based adjacency list for Figure 4(a).

$$\mathbf{p} = [1, 3, 4, 5, 6, 9, 11, 14, 15].$$

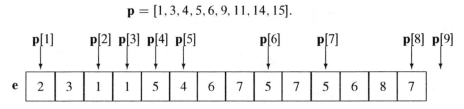

Figure 6. Vector-based adjacency list for Figure 4(b).

The size of vector **e** is twice the number of edges in the graph and the size of **p** is one more than the number of vertices. The definition of **p** makes it easy to calculate the degree of a vertex since the degree of vertex i is just $\mathbf{p}[i+1] - \mathbf{p}[i]$. The number of edges in the graph is $(\mathbf{p}[n+1] - 1)2$, where n is the number of vertices. If the number of edges in the graph is less than n^2, this method of storage takes less room than using a traditional adjacency matrix. This method of representation is not much used since the method based on pointers, which we present next, is generally easier to deal with. But it is useful in languages such as COBOL and FORTRAN, which do not have pointers.[3]

5.3 Linked adjacency lists

The most common way to represent graphs in standard graph algorithms is the *linked adjacency list* based on the link lists we saw in Section 4. For each vertex v in the graph, the elements in adj(v) are placed in a linked list. It is easy to traverse these lists to find all of the vertices adjacent to v. Since it is easy to add and delete elements in these lists, it is easy to add and delete edges in the graph. A vector of n pointers is used to gain access to the individual linked lists. Figure 7 shows the representation of the graph in Figure 4(a).

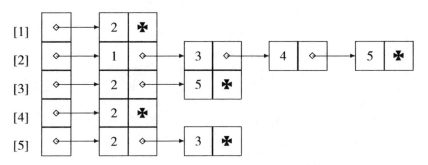

Figure 7. The linked representation of Figure 4(a).

---

[3]The vectors **p** and **e** can be combined into one longer vector by adding $n+1$ to each element of **p** and prepending it to **e**.

6   Additional Reading

Vector-based adjacency lists and linked adjacency lists are common topics in texts on data structures and algorithms. Two excellent choices are:

Cormen, Thomas H., Leiserson, Charles E., Rivest, Ronald L., Stein, Clifford, *Introduction to Algorithms*, McGraw-Hill, New York, 2001.

Sahni, Sartaj, *Data Structures, Algorithms, and Applications in C++*, McGraw-Hill, New York, 1998.

Inclusion-Exclusion and the
Topology of Partially Ordered Sets

Eric Gottlieb
Rhodes College

1 Introduction

It is surprising and satisfying that the principle of inclusion-exclusion, a well-known combinatorial tool, is a special case of a more general phenomenon that also gives rise to the classical Möbius function of number theory. Specifically, the principle of inclusion-exclusion is equivalent to Möbius inversion on the Boolean lattice, while the classical Möbius function arises from Möbius inversion on the lattice of divisors. Generalized Möbius inversion was developed by Gian-Carlo Rota [8] in 1964.

To use Möbius inversion, one must be able to compute the value of what is known as the *generalized Möbius function*. This function provides a link between combinatorics and topology and motivates a broader exploration of the topology of partially ordered sets.

We will begin by using the principle of inclusion-exclusion to derive a formula for the Euler totient function. This will lead us to define the classical Möbius function, which turns out to be the generalized Möbius function of $\mathcal{D}$, the lattice of divisors. We will use $\mathcal{D}$ as a point of departure to explain generalized Möbius inversion. Then, we will compute the Möbius function for the Boolean lattice $\mathcal{B}$ and show that Möbius inversion on $\mathcal{B}$ gives rise to the principle of inclusion-exclusion. We will conclude with a discussion of poset topology.

In this purely expository article, our intent is to convey some connections between number theory, combinatorics, and topology. For this reason, we do not work in full generality and leave unproved some of the results that we use. The reader who seeks a fuller treatment of the topics presented here is encouraged to consult Stanley [9].

2 The Principle of Inclusion-Exclusion

The principle of inclusion-exclusion (PIE) is a way to count the number of elements in the union of a collection of sets $A_1, \ldots, A_n$. A good first approximation to this number would be $|A_1| + \cdots + |A_n|$. The problem is that some elements get counted more than once. This is true, for example, of elements in $A_i \cap A_j$ where $i \neq j$. To improve our approximation, we need to subtract all of the $|A_i \cap A_j|$'s. But then the elements in $A_i \cap A_j \cap A_l$ are undercounted when i, j, and l are distinct, so the number of elements in sets of this type needs to be added back in. This process continues until we finally add or subtract (depending on whether k is odd or even) $|A_1 \cap \cdots \cap A_n|$ to the alternating sum.

In Figure 1, it is easily verified that $|A \cup B \cup C| = 34$. Our first approximation to this number is $|A| + |B| + |C| = 15 + 15 + 21 = 51$, which is too big. We must subtract $|A \cap B| + |A \cap C| + |B \cap C| = 3 + 5 + 10 = 18$, obtaining 33. The final correction comes from adding $|A \cap B \cap C| = 1$ to give 34.

293

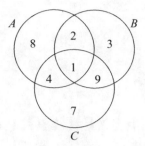

Figure 1. $34 = 15 + 15 + 21 - 3 - 5 - 10 + 1$.

Theorem 2.1 (PIE)**.** Let $A_1, \ldots, A_n$ be finite sets. Then

$$|A_1 \cup \cdots \cup A_n| = \sum_{\substack{1 \le i_1 < \cdots < i_k \le n \\ k \ge 1}} (-1)^{k+1} |A_{i_1} \cap \cdots \cap A_{i_k}|.$$

The proof can be found in any elementary combinatorics book (see [5], for example), so we do not include it here. Later in this article, we will show that it follows from generalized Möbius inversion.

We will now use the PIE to compute the value of the Euler totient function, which is useful in number theory and algebra. An integer $n > 1$ is *prime* if the only divisors of n are 1 and n. For example, 2, 3, 23, and 137 are prime, while 1, 4, 16, and 133 are not. A *totative* of a positive integer n is a positive integer not exceeding n that has no prime factor in common with n. The *Euler totient function* ϕ gives the number of totatives of n. For example, $\phi(12) = 4$ since 12 has four totatives, namely 1, 5, 7, and 11. See [1] or any other elementary number theory text for further details.

Theorem 2.2. Let n be a positive integer with prime factorization $n = p_1^{\alpha_1} \ldots p_k^{\alpha_k}$. Then $\phi(n) = n \prod_{i=1}^{k} \left(1 - \frac{1}{p_i}\right)$.

Proof. The formula works if $n = 1$. If $n > 1$, let A_i denote the set of numbers in $[n] = \{1, \ldots, n\}$ that are divisible by p_i. Then $A_1 \cup \cdots \cup A_k$ is the set of positive integers not exceeding n having a prime factor in common with n, so $\phi(n) = n - |A_1 \cup \cdots \cup A_k|$.

It is easy to see that $|A_{i_1} \cap \cdots \cap A_{i_r}| = n/(p_{i_1} \cdots p_{i_r})$. By the PIE we have

$$|A_1 \cup \cdots \cup A_k| = \sum_{\substack{1 \le i_1 < \cdots < i_r \le k \\ r \ge 1}} (-1)^{r+1} |A_{i_1} \cap \cdots \cap A_{i_r}|$$

$$= \sum_{\substack{1 \le i_1 < \cdots < i_r \le k \\ r \ge 1}} (-1)^{r+1} \frac{n}{p_{i_1} \cdots p_{i_r}}$$

so that

$$\phi(n) = n - |A_1 \cup \cdots \cup A_k|$$

$$= n - \sum_{\substack{1 \le i_1 < \cdots < i_r \le k \\ r \ge 1}} (-1)^{r+1} \frac{n}{p_{i_1} \cdots \cdots p_{i_r}} \tag{8}$$

$$= n \left(1 + \sum_{\substack{1 \le i_1 < \cdots < i_r \le k \\ r \ge 1}} (-1)^r \frac{1}{p_{i_1} \cdots p_{i_r}}\right)$$

$$= n \sum_{\substack{1 \le i_1 < \cdots < i_r \le k \\ r \ge 0}} (-1)^r \frac{1}{p_{i_1} \cdots p_{i_r}} \tag{9}$$

$$= n \prod_{i=1}^{k} \left(1 - \frac{1}{p_i}\right). \quad \square$$

There are other appealing applications of the PIE, including an elegant demonstration that the proportion of permutations on n letters with no fixed point approaches $1/e$ as n grows. This application is included in many combinatorics texts (for example, it is outlined in the exercises of [5]), so we do not give it here.

3   The Classical Möbius Function

Examining expression (9) for $\phi(n)$ we see a natural motivation to define the function $\mu(d)$ for a positive integer d by

$$\mu(d) = \begin{cases} (-1)^r & \text{if } d \text{ is the product of } r \text{ distinct primes,} \\ 0 & \text{otherwise.} \end{cases}$$

for then we get the cleaner summation formula

$$\phi(n) = \sum_{d \mid n} \mu(d)\frac{n}{d}. \tag{10}$$

Here $d \mid n$ means that d is a factor of n. The function μ is called the *classical Möbius function*. Some of its many interesting properties are mentioned at the website [6]. Most important for us is the following theorem.

Theorem 3.1 (Classical Möbius inversion). Let f and g be two integer-valued functions on the positive integers. Then the following two conditions are equivalent:

- $f(n) = \sum_{d \mid n} g(d)$ for all positive integers n.

- $g(n) = \sum_{d \mid n} \mu(d) f(\frac{n}{d})$ for all positive integers n.

We will show in the next section that this result follows from generalized Möbius inversion.

Applying classical Möbius inversion to equation (10) with $f(n) = n$ and $g(n) = \phi(n)$ gives the unexpected and attractive result

$$n = \sum_{d \mid n} \phi(d).$$

For example, $12 = 1 + 1 + 2 + 2 + 2 + 4 = \phi(1) + \phi(2) + \phi(3) + \phi(4) + \phi(6) + \phi(12)$.

4   Partially ordered sets

We now develop some basic definitions and conventions concerning partially ordered sets. We will use this terminology in the next section, when we show how classical Möbius inversion is a special case of a more general phenomenon occurring in the context of locally finite partially ordered sets.

A *partially ordered set*, or *poset*, is a set P together with a relation $\leq_P$ satisfying the following conditions.

- $x \leq_P x$ for all $x \in P$. This property is known as *reflexivity*.

- If $x \leq_P y$ and $y \leq_P x$ then $x = y$ for all $x, y \in P$. This property is known as *antisymmetry*.

- If $x \leq_P y$ and $y \leq_P z$ then $x \leq_P z$ for all $x, y, z \in P$. This property is known as *transitivity*.

For those not familiar with partially ordered sets, perhaps their most counterintuitive aspect is the fact that two elements may fail to be comparable. That is, there may be distinct $x, y \in P$ so that $x \not\leq_P y$ and $y \not\leq_P x$.

For example, the *Boolean lattice*[1] *on n elements* is denoted by $\mathcal{B}_n$ and is defined to be the set of subsets of $[n] = \{1, \ldots, n\}$ with $A \leq_{\mathcal{B}_n} B$ whenever $A \subseteq B$. Thus $\emptyset \leq_{\mathcal{B}_4} \{1, 3\}$ and $\{1, 3\} \leq_{\mathcal{B}_4} \{1, 3, 4\}$. However, there is no order relation between $\{1, 2\}$ and $\{4\}$ in $\mathcal{B}_4$ since neither set is contained in the other.

Let $\mathcal{D}_n$ denote the set of divisors of the positive integer n. Order the elements of $\mathcal{D}_n$ by $l \leq_{\mathcal{D}_n} m$ if and only if $l \mid m$. The poset $\mathcal{D}_n$ is called the *lattice of divisors of n*. In $\mathcal{D}_{60}$, we have $3 \leq_{\mathcal{D}_{60}} 6$ but there is no order relation between 10 and 12 since neither divides the other. Observe that $\mathcal{D}_n$ generalizes $\mathcal{B}_n$ since $\mathcal{D}_{p_1 \cdots p_r}$ is the same as $\mathcal{B}_r$ when $p_1, \ldots, p_r$ are distinct primes.

If $x \leq_P y$ and there is no element $z \neq x, y$ so that $x \leq_P z \leq_P y$, then we say that y *covers* x and write $x \lessdot_P y$. The *Hasse diagram* is a useful way to represent posets visually. To construct the Hasse diagram of a poset P, the elements of P are written on the page in such a way that if $x \lessdot_P y$, then x appears lower than y on the page and there is a line joining x and y. Figure 2 shows the Hasse diagrams of $\mathcal{B}_4$ and $\mathcal{D}_{60}$.[2]

---

[1] A lattice is a special kind of poset.

[2] Find the skeleton of a four-dimensional cube in the Hasse diagram of $\mathcal{B}_4$ and of two 3-dimensional cubes that share a face in the Hasse diagram of $\mathcal{D}_{60}$. Hint for the first one: look at the set of subsets of $\{1, 2, 3\}$ and at the set of supersets of $\{4\}$.

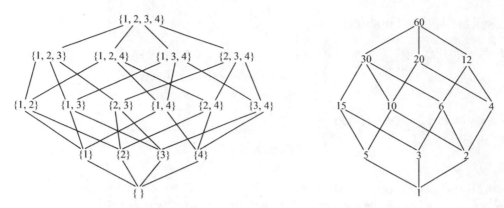

Figure 2. The Hasse diagrams of $\mathcal{B}_4$ and $\mathcal{D}_{60}$.

The sets underlying $\mathcal{B}_4$ and $\mathcal{D}_{60}$ are finite. Any such poset is called a *finite poset*. Sometimes it is more convenient to work with posets that are infinite but for which every subset of the form $[x, y] = \{z \in P \mid x \leq_P z \text{ and } z \leq_P y\}$ is finite. Such a subset is called an *interval* of P. A poset in which every interval is finite is called *locally finite*. For example, let $\mathcal{B}$ denote the set of finite subsets of $\mathbb{N}$, the natural numbers, ordered by inclusion. Let $\mathcal{D}$ denote the set of positive integers ordered by divisibility. Then both $\mathcal{B}$ and $\mathcal{D}$ are locally finite.

The Boolean lattices $\mathcal{B}$ and $\mathcal{D}$ share another useful property. A *minimum element* of a poset P is denoted by $\hat{0}_P$ and is defined by the property that $\hat{0} \leq_P x$ for all $x \in P$. A poset is *lower bounded* if it has a minimum element. Both $\mathcal{B}$ and $\mathcal{D}$ are lower bounded with $\hat{0}_{\mathcal{B}} = \emptyset$ and $\hat{0}_{\mathcal{D}} = 1$. The reader is encouraged to consult [5] or [9] for more on posets, lattices, and related terms.

5   Möbius Inversion on the Lattice of Divisors

We can rewrite Theorem 3.1 as follows.

Theorem 5.1. Let f and g be any two integer-valued functions on the set of positive integers. Then the following two conditions are equivalent:

- $f(n) = \sum_{d \leq_{\mathcal{D}} n} g(d)$,

- $g(n) = \sum_{d \leq_{\mathcal{D}} n} \mu(d) f(n/d)$.

Note that the sums are finite since $\mathcal{D}$ is lower bounded. This version of the theorem might lead the reader to wonder whether analogous theorems might hold for other locally finite lower bounded posets. The affirmative conclusion, which appears here as Theorems 5.3 and 5.4, was given in Rota's 1964 paper [8].

Incidence algebras are the most natural way to understand the generalized Möbius function and generalized Möbius inversion so we will use them here. This discussion is not difficult, but it is outside the scope of most undergraduate combinatorics classes. Readers wishing to forego it could skip ahead to equation (11), which can be treated as a recursive definition of the generalized Möbius function, and to Theorems 5.3 and 5.4.

Let P be a locally finite poset. A complex-valued function f on the closed nonempty intervals of P is called an *incidence function* of P. We write $f(a, b) = f([a, b])$. We denote the set of incidence functions on P by $\mathfrak{I}(P)$. The set $\mathfrak{I}(P)$ is closed under sums and scalar multiples. Since P is locally finite, we can define a product[3] on $\mathfrak{I}(P)$ by

$$(fg)(a, b) = \sum_{c \in [a, b]} f(a, c) g(c, b),$$

making $\mathfrak{I}(P)$ into an associative algebra over $\mathbb{C}$. This algebra, which is called the *incidence algebra* of P, has identity element

$$\delta_P(a, b) = \begin{cases} 1 & a = b, \\ 0 & a <_P b. \end{cases}$$

---

[3] Such a product is known as a *convolution*.

We can express the conditions from Theorem 5.1 in the incidence algebra using the *zeta function*, which is defined on the intervals of an arbitrary poset P by $\zeta_P(a, b) = 1$ for all $a \le b$. Notice that if f and g are in $\mathfrak{I}(P)$ and P is locally finite, then $f(a, b) = (g\zeta_P)(a, b)$ if and only if $f(a, b) = \sum_{c \in [a,b]} g(a, c)$. In the lattice of divisors, this is the same as the first condition of Theorem 5.1; just take $f(n) = f(1, n)$ and $g(d) = g(1, d)$.

It can be shown (see Prop. 3.6.2 of [9]) that for any locally finite poset P, the function ζ_P is invertible. The inverse function is called the *Möbius function* of P and is denoted μ_P. Thus $f = g\zeta_P$ is equivalent to $g = f\mu_P$, and in particular $f = g\zeta_D$ if and only if $g = f\mu_D$. By reasoning similar to that above, this is equivalent to the second condition of Theorem 5.1 provided that $\mu_D(m, n) = \mu(n/m)$. To prove this identity we use a recursive formula, which we now derive. If $a \le_P b$ then

$$\sum_{c \in [a,b]} \mu_P(a, c) = \sum_{c \in [a,b]} \mu_P(a, c)\zeta_P(c, b) = (\mu_P \zeta_P)(a, b) = \delta_P(a, b) = \begin{cases} 1 & a = b, \\ 0 & a < b, \end{cases}$$

so that

$$\mu_P(a, b) = \begin{cases} 1 & a = b, \\ -\sum_{a \le c < b} \mu_P(a, c) & a <_P b. \end{cases} \tag{11}$$

Now it is a simple matter to calculate the Möbius function of any interval.

Theorem 5.2. Suppose $m|n$. Then $\mu_D(m, n) = \mu(n/m)$.

Proof. We proceed by induction on the number of divisors of n/m. The base step is established by the observation that $\mu_D(n, n) = 1 = \mu(1)$.

For the induction step, suppose n/m has more than one divisor and write

$$\mu_D(m, n) = -\sum_{m \le_D x <_D n} \mu(m, x).$$

There are fewer divisors of x/m than of n/m so, by induction hypothesis, we can write

$$\mu_D(m, n) = -\sum_{m \le_D x <_D n} \mu(x/m).$$

Let r be the number of primes dividing m/n. If m/n is squarefree, then each x/m is also squarefree.[4] In constructing x/m we are free to choose any subset of the primes dividing n/m except all r of them. Using the fact that the alternating sum of the binomial coefficients is 0, we obtain

$$\mu_D(m, n) = -\sum_{i=0}^{r-1} (-1)^i \binom{r}{i} = (-1)^r - \sum_{i=0}^{r} (-1)^i \binom{r}{i} = (-1)^r = \mu(n/m).$$

If m/n contains a square, then the only nonzero terms in the sum are those for which x/m is squarefree. In constructing a squarefree x/m we are free to use any subset of the primes, including all r of them. Thus we get

$$\mu_D(m, n) = -\sum_{i=0}^{r} (-1)^i \binom{r}{i} = 0 = \mu(n/m). \quad \square$$

We have seen that the equivalence of the relations $f = g\zeta_P$ and $g = f\mu_P$ gives rise to classical Möbius inversion by taking $P = D$. Generalized Möbius inversion is precisely this equivalence with P unspecified. However, it is more elementary and more natural to express generalized Möbius inversion without reference to the incidence algebra and in terms of complex-valued functions on P instead of using incidence functions.

Let f and g be in $\mathfrak{I}(P)$. Then $f = g\zeta_P$ if and only if $g = f\mu_P$. In particular, $f(\hat{0}, x) = (g\zeta_P)(\hat{0}, x) = \sum_{y \le_P x} g(\hat{0}, y)$ if and only if $g(\hat{0}, x) = (f\mu_P)(\hat{0}, x) = \sum_{y \le x} f(\hat{0}, y)\mu_P(y, x)$. Setting $f(x) = f(\hat{0}, x)$ and $g(x) = g(\hat{0}, x)$ gives the following theorem.

---

[4] An integer is *squarefree* if it is not divisible by the square of any prime.

Theorem 5.3 (Lower bounded Möbius inversion). Let P be a locally finite lower bounded poset and let f and g be complex-valued functions on P. The following two conditions are equivalent.

- $f(x) = \sum_{y \leq x} g(y)$ for all $x \in P$.

- $g(x) = \sum_{y \leq x} f(y)\mu_P(y, x)$ for all $x \in P$.

A poset is *upper bounded* if it has a maximum element, usually denoted by $\hat{1}$. The following upper bounded Möbius inversion theorem can be proved similarly to the lower bounded version.

Theorem 5.4 (Upper bounded Möbius inversion). Let P be a locally finite upper bounded poset and let f and g be complex-valued functions on P. The following two conditions are equivalent.

- $f(x) = \sum_{y \geq x} g(y)$ for all $x \in P$.

- $g(x) = \sum_{y \geq x} f(y)\mu_P(x, y)$ for all $x \in P$.

6   Möbius Inversion on the Boolean Lattice

We will now see that Möbius inversion on $\mathcal{B}$ gives rise to the PIE. To do this, we must compute $\mu_\mathcal{B}$. Once again, we use formula (11).

Theorem 6.1. $\mu_\mathcal{B}(A, B) = (-1)^{|B \setminus A|}$ for all finite subsets $A \subseteq B$ of $\mathbb{N}$.

Proof. The result obviously holds if $A = B$. Suppose $A \subsetneq B$. Then

$$
\begin{aligned}
\mu_\mathcal{B}(A, B) &= -\sum_{A \subseteq C \subsetneq B} \mu_\mathcal{B}(A, C) \\
&= -\sum_{A \subseteq C \subsetneq B} (-1)^{|C \setminus A|} \\
&= -\sum_{i=0}^{|B \setminus A|-1} (-1)^i \binom{|B \setminus A|}{i} \\
&= (-1)^{|B \setminus A|}. \quad \square
\end{aligned}
$$

Intervals in $\mathcal{B}$ and $\mathcal{D}$ are products of simpler posets, and Möbius functions of poset products are well behaved. These facts can be used to give cleaner calculations of the Möbius functions of these posets. See Proposition 3.8.2 of [9] for details.

Theorems 5.4 and 6.1 say that $f(B) = \sum_{A \supseteq B} g(A)$ if and only if $g(B) = \sum_{A \supseteq B} (-1)^{|A \setminus B|} f(A)$. We will now show that the PIE follows from this statement. Let $C_1, \dots, C_n$ be finite sets and let $C = C_1 \cup \cdots \cup C_n$. For $B \subseteq [n]$ define $f(B) = |\cap_{i \in B} C_i|$. Let $g(B)$ denote the number of elements $x \in C$ such that $x \in C_i$ if and only if $i \in B$. The sets A between B and $[n]$ partition the elements x counted by f; the part containing x consists of those elements of C having the same index set $A_x = \{i \mid x \in C_i\} \supseteq B$ as x. These elements are counted by $g(A_x)$. Thus $f(B) = \sum_{A \supseteq B} g(A)$. Applying Theorem 5.4 yields $g(B) = \sum_{A \supseteq B} (-1)^{|A \setminus B|} f(A)$. Taking $B = \emptyset$ gives

$$
0 = g(\emptyset) = \sum_{A \subseteq [n]} (-1)^{|A|} f(A).
$$

Note that $f(\emptyset) = |C|$. Solving for $|C|$ we obtain

$$
|C| = \sum_{\emptyset \neq A \subseteq [n]} (-1)^{|A|+1} f(A)
$$

which is equivalent to the statement of Theorem 2.1.

7 Poset topology

We now introduce some notions from topology that will allow us to understand the Möbius function in a new way. There are several places in this section where we use terms (e.g., homotopic and barycentric subdivision) without defining them. The reader seeking further explanation is encouraged to consult [4] or any other text introducing algebraic topology. We begin with the notion of simplicial complexes, which are used to compute the homology and cohomology of triangulated topological objects.

A 0-dimensional geometric simplex is a point in Euclidean space. An $(n + 1)$-dimensional geometric simplex Σ_{n+1} is obtained from an n-dimensional geometric simplex Σ_n by placing a new point P outside of a hyperplane containing Σ_n and joining P by line segments to each point in Σ_n. Thus, a 1-dimensional geometric simplex is a line segment, a 2-dimensional geometric simplex is a triangle, and a 3-dimensional geometric simplex is a tetrahedron. See Figure 3. For technical reasons, we must also define a (-1)-dimensional simplex to be the empty set.

Figure 3. Simplices of dimensions 0, 1, 2, and 3.

An *n-dimensional face* of a simplex Σ is an n-dimensional simplex that is the intersection of Σ with a hyperplane that does not intersect the interior of Σ. If Σ is n-dimensional, then it is convenient to think of Σ as having Σ as an n-dimensional face.

Thus, if Σ is 3-dimensional, then its unique (-1)-dimensional face is the empty set; its four 0-dimensional faces are its corners; its six 1-dimensional faces are the segments joining its corners; its four 2-dimensional faces are the triangles on its boundary determined by the corners; and its unique 3-dimensional face is all of Σ.

The binomial coefficients in this sequence provide motivation to move the notion of simplex out of the realm of geometry and into the realm of combinatorics. An *abstract simplex* Δ is the power set of some finite set V. The $(n + 1)$-sets in Δ are likened to the n-dimensional faces of its geometric counterpart.

A geometric simplicial complex is a collection of simplices that are "glued" together in a "nice" way. Specifically, the intersection of any two simplices in the set should be a face of both simplices in the intersection. See Figure 4 for an example of a connected simplicial complex.

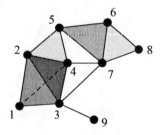

Figure 4. A connected simplicial complex.

Moving again to the combinatorial, an *abstract simplicial complex* is a collection Δ of subsets of some finite set V which contains $\{v\}$ whenever $v \in V$ and which is closed under containment. That is, if $A \subseteq B \in \Delta$ then $A \in \Delta$. For example, the geometric simplex shown in Figure 4 corresponds to the abstract simplex

$$\mathfrak{P}(\{1, 2, 3, 4\}) \cup \mathfrak{P}(\{2, 4, 5\}) \cup \mathfrak{P}(\{5, 6, 7\}) \cup \mathfrak{P}(\{6, 7, 8\}) \cup \{\{4, 7\}, \{3, 7\}, \{3, 9\}, \{9\}\}.$$

Here $\mathfrak{P}(A)$ denotes the set of all subsets of A. We don't need $\mathfrak{P}(\{4, 7\})$ here since its elements $\emptyset$, $\{4\}$, and $\{7\}$ are in other sets in the union. The same applies to $\{3, 7\}$ and $\{3, 9\}$, though we do need to include $\{9\}$ since it is in none of the other sets.

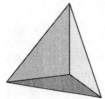
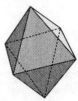

Figure 5. The surfaces of the tetrahedron and octahedron as triangulations of the 2-sphere.

It is often the case that natural geometric objects, like the 2-sphere or the torus, can be viewed as simplicial complexes. This is called a *triangulation* of the object. The sphere, for example, is homeomorphic to (i.e., topologically the same as) the simplicial complex that is the boundary of a 3-simplex. It is also homeomorphic to the boundary of the octahedron. See Figure 5. One can similarly triangulate the torus by viewing it as three triangular antiprisms (i.e., octahedra) stuck together at opposite faces. See Figure 6.

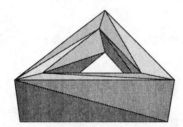

Figure 6. The torus as a simplicial complex.

A *topological invariant* is some algebraic entity, such as a number or a group, that is associated to a geometric object. Often, invariants are computed from a triangulation of the object. By definition, invariants depend only on the geometric object and not on which triangulation of the object we choose.

To show that the sphere and the torus are topologically distinct, one can show that some topological invariant assumes different values on the two objects. One such invariant is the *Euler characteristic*, which is obtained by taking the sum $\chi = f_0 - f_1 + f_2$. Here f_0, f_1, and f_2 are the number of 0-, 1-, and 2-dimensional faces of a simplicial complex homeomorphic to the object in question.

The definition of the Euler characteristic generalizes to higher dimensions in the obvious way. For example, the simplicial complex shown in Figure 4 has Euler characteristic $f_0 - f_1 + f_2 - f_3 = 9 - 16 + 7 - 1 = -1$.

The Euler characteristic of the sphere can be computed as $4 - 6 + 4 = 2$ or as $6 - 12 + 8 = 2$ depending on whether the triangulation in question is the boundary of the tetrahedron or the octahedron. The fact that the Euler characteristic of the sphere has value 2 is known as Euler's formula.

The Euler characteristic of the torus is $9 - 27 + 18 = 0$ as can be seen by examining the triangulation in Figure 6. Thus, the sphere and the torus are not homeomorphic. Had the numbers turned out to be the same, we would be unable to draw any conclusions about whether or not the sphere and the torus are homeomorphic since it is possible for nonhomeomorphic objects to have the same Euler characteristic.

The *reduced Euler characteristic* $\tilde{\chi}$ is defined to be $\chi - 1$. The curious practice of creating a new topological invariant that is obviously equivalent to the Euler characteristic in terms of its ability to distinguish between objects can be justified for technical reasons. We mention $\tilde{\chi}$ because of its connection to the Möbius function.

A *chain* in a poset P is a subset of P for which every two elements are comparable. A chain is *maximal* if there is no other chain that strictly contains it. For example, the elements $\emptyset, \{1, 3\}, \{1, 2, 3, 7\}$ form a chain in $\mathcal{B}$. If P is finite, then the set of chains of P form an abstract simplicial complex since every singleton of P is a chain and every subset of a chain is a chain. This simplicial complex is called the *order complex* of P and is denoted by $\Delta(P)$.

P is called *bounded* if it has elements $\hat{0}$ and $\hat{1}$ that satisfy $\hat{0} \leq x \leq \hat{1}$ for all $x \in P$. If P is bounded let $\hat{P} = P \setminus \{\hat{0}, \hat{1}\}$. Figures 7 and 8 show the order complexes of $\hat{\mathcal{B}}_4$ and $\hat{\mathcal{D}}_{60}$, respectively. The reader is encouraged to verify that the triangles of these order complexes correspond to maximal chains in $\hat{\mathcal{B}}_4$ and $\hat{\mathcal{D}}_{60}$.

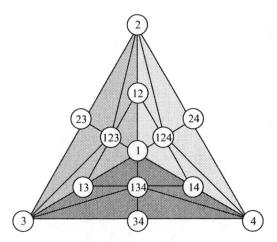

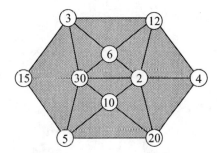

Figure 7. The order complex of $\mathcal{B}_4$. Six triangles on the bottom and the vertex labeled 234 are obscured from view.

Figure 8. The order complex of $\mathcal{D}_{60}$.

It is an amazing observation that if P is bounded, then $\mu_P(\hat{0}, \hat{1})$ is the reduced Euler characteristic of the order complex of $\Delta(\hat{P})$. This realization has spurred researchers to investigate finer topological aspects of order complexes.

Björner and Wachs (see [2] and [3]) adapted and generalized a tool used in polytope theory called shellability for use on order complexes. The resulting technique, known as *lexicographical shellability*, allows one to compute more refined invariants of $\Delta(\hat{P})$ by finding a certain kind of labeling of the edges of the Hasse diagram of P. It is not always easy to find a suitable labeling, but working directly with the poset instead of its order complex makes the problem more accessible.

We conclude by pointing out that abstract simplicial complexes are ubiquitous in combinatorics. Examples include antichains in posets; cliques, matchings, and spanning forests in graphs; monotonic subsequences of sequences of real numbers, in the spirit of Erdös-Szekeres; and non-attacking rook placements on chessboards. Topological aspects of some of these have been investigated; others, to the best of our knowledge, have not.

Bibliography

[1] A. Baker, *A Concise Introduction to the Theory of Numbers*, Cambridge University Press, 1984.

[2] A. Björner and M. L. Wachs, *Nonpure shellable complexes and posets I*, Trans. AMS **348** (1996), 1299–1327.

[3] A. Björner and M. L. Wachs, *Nonpure shellable complexes and posets II*, Trans. AMS **349** (1997), 3945–3975.

[4] M. Henle, *A Combinatorial Introduction to Topology*, Dover Publications, 1979.

[5] K. Bogart, *Introductory Combinatorics*, 3rd edition, Harcourt Academic Press, 2000.

[6] http://mathworld.wolfram.com/MoebiusFunction.html.

[7] http://planetmath.org/encyclopedia/MobiusInversionFormula.html.

[8] G.-C. Rota, *On the Foundations of Combinatorial Theory I. Theory of Möbius Functions*, Z. Wahrscheinlichkeit-
      stheorie und Verw. Gebiete **2** (1964), 340–368.

[9] R. P. Stanley, *Enumerative Combinatorics vol. 1*, vol. 1, Wadsworth & Brooks/Cole, 1986.

Part IV

Articles on Discrete Mathematics Pedagogy

Guided Group Discovery in a Discrete Mathematics Course for Mathematics Majors[1]

Mary E. Flahive
Oregon State University

1 Introduction

In this article we discuss the use of guided group discovery in Oregon State University's discrete mathematics course for math majors. Since Fall 2003 this course has been taught at Oregon State using an ongoing modification of Kenneth P. Bogart's successful group discovery method and notes [2], "Teaching Introductory Combinatorics by Guided Group Discovery." Section 2 summarizes Ken's notes and method, and Sections 3 and 4 respectively contain the adaptation of his notes and the implementation of his method at Oregon State.

Ken's prototype was a small elective course in which the average entering student was very motivated to learn the material. In our department the course is required, and it is financially unrealistic for us to expect either very small classes or in-class assistance from a senior student. We think our modification is sufficiently general that it can be successfully used by other mathematics departments with similar student demographics. Since the adaptation is an ongoing project, the interested reader is referred to [6] for current information on its status.

Ken generously served as a consultant for our adaptation of his method, both informally in Fall 2003 and during the first year of the grant. We are grateful for his help.

2 A Short Overview of the Bogart Course

The goal of Ken's project was to design notes and a method for teaching enumerative combinatorics in which "a large majority of the students would learn a large majority of the material of beginning combinatorics." The method was informed by earlier work in mathematics education, including Davidson's work on small group discovery [3]; Schoenfeld's research on problem solving [8]; and the work of Dubinsky and his collaborators on the genetic decomposition of mathematical knowledge [1, 4, 5].

The set of class notes is the only written resource for the course, and is comprised of inter-connected problem sequences in combinatorial mathematics and graph theory. Through working the problems students learn combinatorial processes and are guided to the discovery of general principles. Students are expected to complete 90% of the problems, with the proviso that they should not systematically avoid the more challenging problems. Most of the problems are preceded by symbols, whose meaning is summarized in the following table:

·	essential for this or the next section
•	essential
○	motivational material
+	summary
→	especially interesting
*	difficult

---
[1]This material is based upon work supported by the National Science Foundation under Grant No. 0410641. Any opinions, findings, and conclusions or recommendations expressed in this material are those of the author and do not necessarily reflect the views of the National Science Foundation.

This is an important feature of the notes, since the symbols assist the student and they also provide important cues for an instructor who wants to design a cohesive set of mandatory problems to be completed by the class.

The learning strategy of group discovery is based on the premise that most students learn well from working cooperatively with peers. Unlike many classes in which an expository lecture is the primary medium for conveying information, students in group discovery classes are actively involved in their own learning during class as well as outside the classroom. In the Bogart implementation of group discovery, most of the course grade comes from non-group work, including traditional in-class exams and the expectation that final solutions will be completed without consultation or collaboration.

During class time, the instructor must work to achieve a balance between giving students sufficient time to work successfully in groups and ensuring students are satisfactorily progressing through the material. Also, the notes are constructed so that an instructor can decide to guide some groups of students to push ahead and work on more difficult problems whose solution might even replace some required problems.

In the Bogart method, the instructor also guides by providing selective and thoughtful use of discussion which involves the whole class or a large portion of the class. This can range from a chance for students to discuss their understanding of a specific problem at the board to a review facilitated by the instructor. An interesting use of large-group discussion is to model successful group dynamics with no all-knowing participant. This can be especially effective when it is used at times when most of the students think they have hit a dead end on a problem sequence. As students are encouraged to explain what they and their group have tried or where they have trouble with a problem, the class can see that careful and patient listening to others can lead to progress.

Ken hoped guided discovery would strengthen the students' sense of their own responsibility for their learning, and that the instructor would foster an atmosphere where students themselves initiate discussions which involve a large percentage of the class. Careful use of large-group discussion is an integral part of the Bogart method, and can be an exemplary teaching method for the many future mathematics teachers who populate our classes.

3   The Adapted Notes

While using Ken's notes in Fall 2003, I was impressed by the student response to many of the problem sequences as well as by the amount of personal responsibility shown by most of the students. Some of this can be attributed to the type of problems, which are especially conducive to group discussion. This is true both because many of the problems are easily understood by students at different levels of mathematical maturity and because a slight change in wording can result in a major change in technique.

But we think most of the enthusiasm comes from the way the notes are constructed as well as the Bogart implementation of group discovery. As a check on our faithfulness to the original spirit, we asked Rosa Orellana to review our implementation from this point of view. Her report can be found at [6]. Rosa is a member of Dartmouth's faculty who has taught from Ken's notes and was also involved in their development.

3.1   Why adapt?

The prerequisites for Ken's course are: "comfort with sets, functions, and algebraic notation (including some summation notation); some experience with reading (and perhaps doing) proofs; and, ideally, a modest exposure to mathematical induction." A typical OSU math major takes this course concurrently with advanced calculus in the first term of their third year of study. As preparation, students have completed the calculus sequence and one course in matrix algebra. Since these are also service courses for engineering students, they provide little experience with mathematical proof and in general our students do not satisfy Ken's prerequisites when they begin this course.

Ken thought our class was the first in which his notes were used in a required course in which the usual enrollment[2] was at least twice as large as the maximum recommended by his original project. Ken encouraged us to apply for an NSF grant to adapt his notes and method for the larger classes which are more typical of required courses at state universities.

---

[2]The usual class size has been in the mid- to high-20s, and in Fall 2007 there are two sections with about 16 students each.

There are other differences between the course (and classes) at Oregon State and the assumptions made by Ken and his advisory board on the intended use of his notes and method. In addition to the smaller classes and the different level of mathematical preparation, Ken's prototype class was an elective class and the average student began the course with more interest in the material. Our more extensive syllabus of discrete mathematics also required some adjustment.

Even with these differences, the basic philosophy of the notes transfers well and also complements the more traditional course in advanced calculus which is usually taken concurrently by our majors.

3.2   Changes in the structure of the notes

Before giving an overview of some of the differences in mathematical content, we first describe some changes we have made in the basic structure of the notes.

Ken's original notes assumed a familiarity with basic topics, such as relations and mathematical induction, that we cannot expect from our students when they enter this course. Although Ken's appendices contained some review of this material, flipping back and forth from those sections to the main book interfered with the students' appreciation of what they were doing and how things fit together. In the adaptation, this basic material has been expanded and incorporated into the fabric of the early chapters. Three review appendices remain, but we find an average student doesn't need to do more than scan them outside of class.

Because the topic of Ken's course was enumerative combinatorics, some of his problem sequences cover material that is not basic to our more general syllabus of discrete mathematics. In our adaptation, much of that advanced material has been moved to optional sections which are not required in the mainstream problem sequences.

Ken's first chapter was very long, and we have spread that material over three chapters in the adaptation. The new presentation encourages students to review frequently and tries to facilitate that summarizing process by packaging the material in more obvious units. In addition, since many of our students don't have sufficient mathematical maturity to ferret out the definitions and theorems from the exercises, we have added more formal definitions, more statements of theorems, and an occasional example of a complete mathematical proof. Ken's advisory board and evaluator advocated the inclusion of more summarizing material, and we have designed separate summarizing handouts for each chapter. They are available as TEXdocuments as well as in PDF format so that instructors can tailor the summaries to their individual courses.

Since this course is required for math majors, we've found that some problems had to be re-worded to distinguish between whether a complete proof is expected or whether a less-formal explanation would suffice. Developing and improving this skill is an important goal for the course, and we cannot assume our students enter with much proof experience.

3.3   Changes in content

In this section we discuss three specific areas in which changes were made in the notes for our classes: functions, equivalence relations, and mathematical induction.

As the students progressed through Ken's notes, we found serious deficiencies in their concept of function. A basic technique in enumerative combinatorics is to count the size of a finite set by establishing a bijection between it and another set whose size has already been established.[3] In order to successfully apply this technique, the problem solver's understanding of function must have evolved from plugging values into a given function through determining whether or not a relation is a function. In fact, he or she should recognize how the notion of function is useful for the problem and then construct the proper function; that is, the student must have at least arrived at the object stage of conceptualization [1]. In the original project Ken and his advisory board assumed that students would enter with sufficient knowledge to understand something as being characterized as a function without having a specific expression. However, from our listening to group discussions and our reading and commenting on their written work, it soon became clear that many of our students were far from this stage. Later course instructors and Professor Orellana have analyzed the adaptation from the point of view of addressing this problem while maintaining the interest of students whose concept of function has evolved into the process stage.

---

[3] In the appendix to this article we give our modification of one of Ken's problem sequences using the Bijection Principle.

Another significant difference from the original notes is the adaptation's emphasis on equivalence relations. Many of Ken's problem sequences on the Quotient Principle and distributions have been changed to questions on equivalence relations. In addition to having wider application to equivalence classes of unequal size, using equivalence relations is an important skill to be obtained from a beginning course in discrete mathematics and is expected in the later required course in abstract algebra. The current edition has a full chapter on equivalence relations, absorbing much of the material from the original appendix as well as expanding problem sequences using the Quotient Principle. Ken supported this as a necessary and important change for a course in discrete mathematics.

Ken's notes introduce students to the Principle of Mathematical Induction through problem sequences which involve proof by contradiction. Specifically, problems ask students to prove a proposition indexed over all positive integers by contradicting the existence of a smallest counterexample. Once students have done problems using this technique, the commentary then introduces induction as the framework for a more natural direct proof. This was a successful new perspective for some advanced students in our classes, but others protested, saying that "this induction" looked nothing like the one they used in advanced calculus. These students were associating induction with formulaic exercises (such as finding a closed form for the sum of the first n positive integers) rather than as a method for proving a sequence of statements indexed by the positive integers. The problems in the notes were more varied than those they had encountered elsewhere, and we found that our students need more experience with problems that ask them to identify the underlying inductive process. Induction now occurs earlier in the adaptation, and more problems have been added, including a new sequence which builds student understanding of inductive processes. Ken and I were both intrigued by this from a pedagogical point of view as well as the more practical point of view of how it might be addressed.

4   The Adapted Method

The ongoing adaptation has been used in our discrete mathematics course since Fall 2004. The course is offered once a year, and three different instructors have used the adapted notes. In the longterm it is hoped that an instructor will teach the course for two consecutive years and then serve as a mentor for the next instructor (which could be in their second year if two sections are offered). The classes now meet twice a week in 75-minute sessions in order to provide sufficient immersion in the problems. For the first few years the classes met in 100-minute blocks, but we found it was more difficult for everyone to remain on task for that amount of time.

With our classes meeting on Mondays and Wednesdays, the usual format has been to assign a range of problem numbers on Monday, and this forms a minimal assignment to be completed by the beginning of the following Monday's class. Students spend class time working in four-person groups that have usually been formed by the instructor and whose composition changes about every other week.

Ken required students to solve 90% of the problems by the end of the term. In our ten-week courses, our classes regularly cover at most five of the seven chapters in the adaptation. It has been reasonable to expect the students to work about twenty problems a week for the first few weeks. As the term proceeds, problems become progressively harder, solutions often require making connections with earlier information, and completing about 12–15 problems is a reasonable expectation. Occasionally the list of mandatory problems is modified at the end of Wednesday's class, but in general it seems wise to set weekly expectations which the students strive to meet. Students are encouraged to work together outside class, but the final written draft of their problem solutions is expected to be done independently.

We have all found that quick and frequent feedback is essential, and our goal has been to collect written work on Monday and to return the graded work two days later in Wednesday's class. Ken's method allows for unlimited re-submission of problems, but in our classes both the number of re-submissions and the re-submittal window are restricted and varies according to the instructor.[4] All instructors have used our modification of Ken's 0—5—9/10 "triage" grading-scale, adding a possible grade of 7 as a compensation for fewer resubmits. Some instructors have assigned a separate grade which specifically assesses student writing.

Which problems are graded? There is a great deal of variability in this. Some instructors collect all assigned problems and grade a subset, while others give a short list of problems at the end of Wednesday's class. When I've taught

---

[4] I allowed for a pre-specified number (about 8–10) of resubmits during the the ten-week quarter and they must be handed in within two weeks of my grading. A problem could be submitted more than twice.

the course, I've purposely changed my expectations as the term progressed. For the first five weeks of the term, students turned in all assigned problems and I graded about five problems per week. Toward the end of the term, the students handed only in the smaller subset of problems which would be graded since by this time they should have learned what was expected for a complete mathematical solution. Although I encouraged the students to discuss their graded solutions with other members of their group before working on resubmits, time pressure from other courses made this an unrealistic expectation. The benefits of explaining work to peers and of analyzing others' work are many, enough that it might be worthwhile to encourage more of this type of collaboration. For instance, perhaps some peer review can be part of the grading scheme.

What about motivating students to do more? Students who understand the material more quickly can be encouraged to form groups which would do more problems, for instance ones in the optional sections. These sections contain problem sequences on material that is somewhat ancillary to our course (for instance Ramsey numbers) but can provide an intellectual challenge for these students. The second time I taught the course, three students formed a group which regularly worked on these sections. Two of those students were more experienced, and although sometimes the group work became close to a tutorial for the other student, all three students seemed satisfied with the arrangement. This group formed fairly early in the term and the midterm served as a check that they were learning the basic material. After the middle of the term we agreed that they would invest more time in harder problems with the understanding that they would be able to replace the final with take-home problems of greater difficulty. (And the less experienced student agreed to take the regular final.)

Ken and all of our instructors have agreed that in larger classes (especially those with fluid group membership) it is important for most of the students to be at about the same place most of the time: at a minimum, they should finish the chapters at about the same time. Students proceeded linearly through the chapters, and were regularly encouraged to return to problems which at first stumped them.

Although it is always a small percentage of class time, the amount and frequency of whole-class discussion is the feature with the most variability in the different classes. For instance, some instructors expect every student to present at least one problem to the class every term.

There are possible advantages to a larger class. For example, I found there were always students who were eager to report on their group's work or to voice their group's unresolved questions in an end-of-class wrap up. For harder problems, several different approaches would sometimes result in spirited arguments that continued after class, and the larger class size might have increased the chance of that happening. Because there were always five or six groups, each group knew that class could not be a private group tutorial with the instructor and many learned to make good use of the group time as well as any whole-class discussion time. In addition, most students seemed not at all fazed (and obliged) when asked to switch groups, and this might have been more difficult if there were fewer people in the room.

5 Concluding Remarks

In this project Ken Bogart's notes and method were adapted for larger class size, weaker mathematical background and motivation, and some difference in course material. A continuing challenge is maintaining discovery while expanding the interstitial discussion and encouragement, each of which seems to be necessary in a class with five or more groups.

These notes foster and develop important mathematical instincts in different ways from our other courses for math majors. Among these are: checking small cases first; formulating and testing conjectures; the importance of testing for counterexamples in combination with trying to construct a proof. By the end of each term, all students knew not to expect problems which were simply quick applications of cookbook algorithms and many had voiced at least some satisfaction with this feature of the course.

The current adaptation and supporting materials are available on the website [6]. More information for instructors will be added to the site in coming years. Please contact me if you'd like more information.

Appendix: An adapted problem sequence

The following is an example of our adaptation of a problem sequence which builds on the notion of function and was designed from an actual class experience.

At the start of a two-hour meeting of the Fall 2003 class, most of the groups were beginning Ken's problem sequence on the enumeration of labeled trees (Problems 104 to 116). By the end of the first hour, many groups saw a pattern emerging from counting the number for small vertex sets. Some students were ready to accept the conjecture without proof, while others wanted to find a general argument. After a short break, most of the students agreed to forge ahead and work on understanding Prüfer codes. A game evolved in which one group member would give a Prüfer code for their personal private tree and the rest of the group would then find the tree. By the end of the second hour, all of the groups had played the game and most students were convinced there would always exist a unique tree for every Prüfer code. Many of them left class trying to establish the necessary bijection by describing the sequences which were guaranteed to be Prüfer codes. Most of the students had progressed from barely knowing the definition of tree through making conjectures in the middle of class to leaving the class with some idea of how a proof might be constructed. The adapted sequence we give below uses this game.

Labeled Trees and Prufer Codes

Next you will explore the idea of *labeled* trees. Figure 1 gives all different labelings of a fixed tree with three vertices. Notice that the convention for labeling the vertices of trees is that the tree which has edges between vertices 1 and 2 and between vertices 2 and 3 is different from the tree that has edges between vertices 1 and 3 and between vertices 2 and 3.

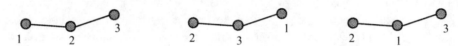

Figure 1. The three labeled trees on three vertices.

1. How many labeled trees are there on the vertex set $\{1, 2\}$? On the vertex set $\{1, 2, 3\}$? How many labeled trees are there on four vertices? How many labeled trees are there with five vertices? You don't have a lot of data to formulate a guess, but try to guess a formula for the number of labeled trees with vertex set $\{1, 2, \ldots, n\}$. When you get to four and especially five vertices, draw all the unlabeled trees you can think of, and then figure out in how many different ways you can put labels on the vertices.

The next problems will develop a method for proving the formula you just guessed in the last problem. In order to do this, an auxiliary sequence is defined.

Given a tree with $n \geq 2$ vertices which has been labeled in any way using the elements of $[n]$, define the auxiliary sequence $b_1, b_2, \ldots$ in the following inductive manner:

Step 1: If the tree has two vertices, the sequence consists of one term, the larger label, which means the sequence is $b_1 = 2$. Otherwise, let a_1 be the lowest-numbered vertex of degree 1 in the tree. (How do you know there is such a vertex?) Let b_1 be the label of the unique vertex in the tree adjacent to a_1 and write down b_1. (Why is b_1 unique?) For example, in the first graph in Figure 2, a_1 is 1 and b_1 is 2.

Step 2: Suppose a_1 through a_{i-1} have already been identified, and let a_i be the lowest-numbered vertex of degree 1 in the tree you get by deleting vertices a_1 through a_{i-1}. (How do you know the resulting graph is always a tree?) Let b_i be the unique vertex in this new tree adjacent to a_i. For example, in the first graph in Figure 2, $a_2 = 2$ and $b_2 = 3$. Then $a_3 = 5$ and $b_3 = 4$.

2. We use the letter B to stand for the sequence of b_is inductively obtained in this way. Use your earlier work to answer the questions posed in the above two-step algorithm.

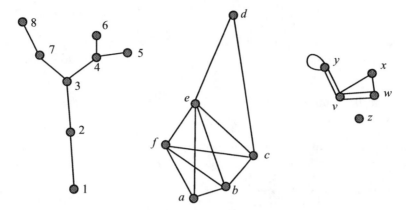

Figure 2. Three different graphs.

3. For the tree (the first graph) in Figure 2, the sequence B is 2344378. At this point, work with your group to draw some other labeled trees on eight vertices and construct the sequence B associated with each tree.

4. How long is the sequence B computed from a labeled tree with n vertices?

5. From your examples, decide if you can predict the last member of the sequence B. Explain.

6. Is it possible for a_1 to be in B? Can you tell from B what a_1 is?

For a labeled tree T, the associated sequence $P(T) := b_1, b_2, \ldots, b_{n-2}$ is called a **Prüfer coding** or **Prüfer code** of T. For instance, the Prüfer code for the labeled tree T of Figure 2 is $P(T) = 234437$. Notice that we do not include the last term of B in the Prüfer code because we know it is n.

Let S be the set of all labeled trees on nine vertices. For each tree $T \in S$, define $P(T)$ to be the Prüfer code for T.

7. Why is the relation $\{(T, P(T)) : T \in S\}$ a function with domain S? Find a co-domain for this function. (At this point you're not asked to find the *smallest* co-domain.)

8. Play the following game in your group: In turn, each of you should secretly write down a tree, determine its Prüfer code, and then share the code with the whole group. The other members of the group then should find all labeled trees that have your sequence as its Prüfer code. How many labeled trees are found? What does your answer say about the function P?

9. Now, as a group write down any sequence of seven integers from $\{1, 2, \ldots, 9\}$. Try to find a tree $T \in S$ for which your sequence is $P(T)$. Do this for several different sequences. Use this information to find the smallest co-domain for the function P?

10. Find a bijection between the set of labeled trees with n vertices and another set that you already now how to count.

11. Find the number of labeled trees with n vertices. Is this the formula you conjectured earlier in Problem 1?

(The idea of writing the last sequence of problems as a game originated with the Fall 2003 Math 399 class at Oregon State.)

In addition to providing a way to count labeled trees, there is a good bit of other interesting information encoded in the Prüfer code for a tree. You can begin to see this by working the next two problems and the problems in the optional section that follows them.

12. What can you say about the vertices of degree 1 from the Prüfer code for a tree labeled with the integers from 1 to n; that is, what vertex or vertices in the sequence $b_1, b_2, \ldots, b_{n-1}$ can have degree 1?

13. What can you say about the Prüfer code for a tree in which exactly two vertices have degree 1? Does this characterize such trees?

Bibliography

[1] Asiala, Mark, David J. DeVries, Ed Dubinsky, David Mathews, and Karen Thomas. "A framework for research and curriculum development in undergraduate mathematics education," *CBMS Issues in Mathematics Education* 6 (1996) 1–32.

[2] Bogart, Kenneth P. *Teaching Introductory Combinatorics by Guided Group Discovery*, www.math.dartmouth.edu/~kpbogart/.

[3] Davidson, Neil. "The small-group discovery method in secondary- and college-level mathematics," *Cooperative Learning in Mathematics*, Addison-Wesley, 1990.

[4] Dubinsky, Ed. "Teaching mathematical induction, I," *J. Math. Behavior* 6 (1987) 305–317.

[5] Dubinsky, Ed. "Teaching mathematical induction, II," *J. Math. Behavior* 8 (1989) 285–304.

[6] Flahive, Mary E. www.oregonstate.edu/~flahivem/DMnotes.htm.

[7] Hagelgans, N. L. *A practical guide to cooperative learning in collegiate mathematics*, The Mathematical Association of America, MAA Notes Number 37, 1995.

[8] Schoenfeld, Alan. *Mathematical Problem Solving*, Academic Press, 1985.

The Use of Logic in Teaching Proof

Susanna S. Epp
DePaul University

Introduction

Even rather simple proofs and disproofs are built atop a normally unexpressed substructure of great logical and linguistic complexity. For example, in [7] I described several of the many reasoning processes needed to establish the truth or falsity of the following statements: (1) The square of any rational number is rational; (2) For all real numbers a and b, if $a > b$ then $a^2 > b^2$; and (3) For all real numbers x, if x is irrational then $-x$ is irrational. The article cites evidence that a significant number of students taking college mathematics courses do not bring with them an intuitive feeling for the logic required to succeed in determining truth and falsity of such statements and argues that some explicit instruction in logical reasoning is needed in courses that require students to engage in proof writing. However, because proofs and disproofs of even elementary statements require a substantial base of understanding, a "clarifying" analysis for a proof may be so complex that if students could understand it, they would not need it in the first place. Figuring out how to present proof construction simply enough to be intelligible yet detailed enough to be effective is one of an instructor's greatest challenges.

The following sections contain ideas for helping students learn to construct simple proofs and disproofs. Most are approaches I have used myself and for which student reaction has been positive. Others are ideas for which colleagues have reported success.

For students who come to a course with reasonably good intuition for logical principles, merely seeing them stated and working a few examples can be a pleasure — like the delight of the Molière character who learned one day that he'd been speaking prose all his life. For many students, however, simple exposure to principles of logic is not sufficient to counteract deeply ingrained incorrect patterns of thought, and follow-up instruction is needed to illustrate the uses of the principles in mathematical contexts. Thus Section 1 contains not only suggestions for how to take advantage of having provided students with a brief introduction to basic principles of logical reasoning before requiring them to make serious attempts at mathematical proof, but also advice for how to help students develop a firmer and deeper grasp of reasoning principles as proof and disproof of various mathematical topics are discussed. Section 2 offers additional strategies to guide students through their initial proof efforts and lead them to see the desirability of expressing proofs with care, and Section 3 discusses additional ways to help students come to learn the need for proof.

1 Building on Initial Coverage of Logical Principles

Using Puzzle Problems

To make the transition from elementary logic to proof, it can be helpful to assign puzzle problems, such as Raymond Smullyan's knights and knaves [20]. These puzzles posit an island where each inhabitant is either a knight, who always tells the truth, or a knave, who always lies, but it is impossible to distinguish knights from knaves by their appearance. Each puzzle describes a situation in which certain inhabitants make certain statements, and the goal is to figure out who is lying and who is telling the truth. When solutions are discussed in class, quite a number of students make it clear that they do not have a natural feeling for the kind of indirect reasoning needed to solve most of the puzzles.

Nonetheless, almost all students seem to enjoy the puzzles, and working on them helps develop a basis of intuition for proof by contradiction. Discussing the solutions serves to illustrate how inference rules are used in practice and helps students develop a sense for the flow of deductive reasoning, which they will use later in mathematical proofs of all types.

Using Natural Deduction Proofs

John Barwise and John Etchemendy developed computer software called Tarski's World, named after the logician Alfred Tarski, to represent situations in a world that consists of a grid containing a number of geometric shapes in a variety of positions. The accompanying courseware [2, 3] shows students how, among other things, to produce natural deduction proofs of statements about the shapes in the world. Work of Lee and Stenning [11] supports the view that use of these materials improves students' ability to reason deductively. Another teacher who uses instruction in natural deduction to prepare students for reasoning in more general environments is Richard L. Morrow, a middle school mathematics coordinator with advanced training in logic [16]. When Morrow first taught geometry to a group of gifted eighth graders, his students finished the book a month before the end of the year. Thinking to fill in the extra time, Morrow began the course the next year with a few weeks study of formal logic, focusing on student construction of natural deduction proofs. While his students said they found the work difficult at first, they eventually all succeeded and, to his amazement, they then learned the geometry so much faster that they still finished the book a month early.

Using Disproof by Counterexample

In any course that asks students to write proofs, one can start by giving students statements to identify as true or false, asking them to justify a determination of true "as best as you can" and to support an answer of false by providing a counterexample. One reason for beginning in this way is that most students find it easier to understand and construct disproofs by counterexample than to understand and construct even simple direct proofs. A second reason is that the more experience students have in seeing that a single counterexample disproves a universal statement, the more likely they are to understand that a general argument is needed to show that no counterexample exists. Finally, offering students mathematical statements whose truth or falsity they have to determine themselves helps make the point that proof and counterexample are first and foremost problem-solving tools.

Direct Proof: Identifying the Starting Point and Conclusion to Be Shown

The most important initial point to communicate to beginning students about proving a universal statement is that they will have to move from something that is supposed to be true to something that must be shown to follow. It then becomes natural to

- identify *what* is supposed and *what* is to be shown, which determines the outer structure of the proof, and

- address the crucial question "How do I show that?" which determines the proof's inner structure and depends critically on the definitions of the terms in the statement.

The most common type of mathematical statement is universal and conditional, having the form

> For all elements x in a certain set, if [*hypothesis*] then [*conclusion*].

A direct proof of such a statement has the following outline:

- Suppose that x is a particular but arbitrarily chosen (or "generic") element of the set for which the hypothesis is true.

- We must show that x also makes the conclusion true.

The amazing thing about this proof technique is that merely by reasoning about a single element x, one deduces that the conclusion follows from the hypothesis for every element of the set — which is typically of infinite size. The validity of the reasoning is determined by the fact that x is arbitrarily chosen, or "generic," which means that it has all the characteristics and only those characteristics common to every other element of the set. Hence everything one deduces about it is equally true of every other element of the set, and thus a descriptive name for this type of reasoning is *generalizing from the generic particular*.

A dramatic way to emphasize the power of this proof method is to show how one can use it to structure proofs involving terms one does not even understand. For instance, given the statement "For all toths T, if T has a rath, then every wade of T is brillig," the starting point and conclusion to be shown for a proof would be "Suppose T is any toth that has a rath. We must show that every wade of T is brillig." This transformation may seem obvious to a mathematician, but it does not come naturally to many students. Yet as students venture further and further into realms of mathematical abstraction, instinctive ability to use the transformation becomes increasingly essential to their success.

Recognizing the "Suppose" and "To Show" in Proof by Contraposition and Proof by Contradiction

Once one has introduced proof by contraposition and proof by contradiction as well as direct proof, one can help students understand the differences among them by pointing out that while for each method there is something supposed and something to be shown, these "somethings" are dramatically different in each case. In a direct proof one supposes one has a particular but arbitrarily chosen object that satisfies the hypothesis, and one shows that this object satisfies the conclusion. In a proof by contraposition one supposes one has a particular but arbitrarily chosen object for which the conclusion is false, and one shows that for this object the hypothesis is also false. In a proof by contradiction one supposes that the entire statement to be proved is false, and one shows that this supposition leads to a contradiction.

Use of Definitions

Mathematically speaking, the most important part of a statement's proof is how one gets from the hypothesis to the conclusion. For most of the proofs undergraduate students are asked to construct, the majority of this task is achieved through a logico-linguistic analysis of definitions. The reason is that the inner structure of a straightforward, or routine, mathematical proof is largely determined by the meanings of the terms.

Note that, although they are frequently stated less formally, definitions are actually bidirectional. For instance, for n to be an even integer means that "n is even if, and only if, n equals twice some integer." Thus if we know that n is even, we can deduce that n equals twice some integer (from the "only if" part of the definition), and if we know that n equals twice some integer, we can deduce that n is even (from the "if" part of the definition).

To answer the question "How do I show that the conclusion follows from the hypothesis?" the prover needs an operational understanding of the "if" direction of the definitions of the mathematical terms in the conclusion. For example, to derive the conclusion that a certain quantity is rational, one needs to show that it can be expressed as a ratio of integers with a nonzero denominator. To derive the conclusion that one set is a subset of another, one needs to show that any element in the one set is an element in the other. To derive the conclusion that a function f is one-to-one, one needs to show that given any elements x_1 and x_2 in the domain for which $f(x_1) = f(x_2)$, one can conclude that $x_1 = x_2$. Similarly, to work forward from the hypothesis toward the conclusion, the prover needs an operational understanding of the "only if" direction of the mathematical terms in the hypothesis. Helping students translate the formal wording of a definition into such operational terms is one of the most important tasks facing a teacher in a course introducing students to proof.

One way to help students learn to use definitions is to try to induce them to see a definition as providing a test that has to be passed to decide whether something is the case. As soon as a new definition is introduced, one can introduce a range of examples, phrasing each as a question. For instance, immediately after defining rational, one can write "Is 0.873 rational?" and simultaneously ask the question out loud. To a student's answer of "yes," one can write "Yes, because" and look expectantly at the student. The student may be surprised that additional words seem to be called for but is generally able to supply the reason without difficulty (or other students may help out). One can move on to slightly more complicated examples (Is $-(5/3)$ rational? Is 0 rational? Is $0.25252525\ldots$ rational?), each time acting as if it is taken for granted that the student answering the question will give a reason. Soon students learn to give the reference to the definition without prompting, and gradually they come to understand the value of using the definition to answer such questions. Coming to see a definition as the ultimate test that determines whether or not a given object has a given property can help students accept certain facts, such as that 0 is an even number, which, surprisingly, they often disbelieve.

It is also useful to discuss alternative but logically equivalent ways to phrase definitions because it is often the case that the truth or falsity of a mathematical statement is more apparent if one uses one phrasing of a definition rather than another. Moore [15] gives several examples of student failure resulting from a lack of awareness of alternative versions of definitions. In an introductory course, an instructor needs to build in exploration of such alternative phrasings, reviewing the fact of their logical equivalence and showing how to operate with each version in the circumstances where it is superior to the others. Selden and Selden [19] argue that students' difficulty "packing and unpacking" the logic of mathematical definitions and theorems seriously undermines their ability to judge the correctness of mathematical arguments and to formulate arguments of their own. My experience supports this view. It is the main reason I give students practice in translating back and forth from formal mathematical statements to their many different informal versions. Because so many students find this difficult, I often continue to assign translation exercises throughout a large portion of the course.

Another reason to discuss alternative wordings for definitions is to compensate for the fact that quite a few students are still in the process of developing a more sophisticated concept of variable. For example, one way to state the definition of even is "n is even if, and only if, there is an integer k such that $n = 2k$," and in the usual development of many proofs it is important to be able to use this formulation. However, students with a naïve understanding of variables and quantification often make mistakes when they use it. For instance, to prove that the sum of any two even integers is even, they represent both as $2k$, thereby only considering the case where the integers are the same. To help them come to a more mature understanding of the definition, it is helpful (1) to restate it less formally without using an additional variable such as k (as in the second paragraph of this section), and (2) to write it several times using a variable but each time with a different symbol to represent it, pointing out that it is the existence of the integer k, not the symbol used for it, that is important.

In [21] Tall and Vinner introduced the notion of "concept image," which sheds considerable light on students' understanding of mathematical definitions. A concept image for a definition is "the total cognitive structure that is associated with the concept, which includes all the mental pictures and associated properties and processes." An overly narrow concept image leads to mistaken assumptions and may result in incorrect mathematical arguments. For students to develop concept images adequate to help them effectively evaluate abstract mathematical statements, they need experience with a broad range of examples for each newly defined term. They also need to become acquainted with the diagrams and other visual representations that mathematicians use in reasoning about the term. These might be arrow diagrams for relations and functions, the image of a nonspecific real number and its floor sitting on a number line for the study of the floor function, or a kind of blurry generic fraction with an indeterminate numerator and denominator for discussions about rational numbers.

2   Guiding Students' Fledgling Efforts

No matter how much one tries to prepare students for the process of writing proofs on their own, a certain number find it very difficult. It seems that some students cannot believe that an instructor is serious about demanding coherent expression, while others simply have difficulty putting all the pieces together in a way that makes sense. To learn as complex a skill as proof construction, most students need quite a bit of individual, back-and-forth interaction with an instructor. To the extent that one cannot act as a private tutor to every student, one can try to devise effective substitutes. For example, one can

- have students complete a few fill-in-the-blank proofs as homework to give them an out-of-class experience of participating in the development of a complete proof without making them responsible for its entire construction;

- supply a variety of model solutions for some of the homework problems to show students that their individual work is really supposed to resemble the kind of proofs that have been developed in class;

- suggest that students read their proofs out loud to test whether they are written in coherent sentences;

- discuss the kinds of errors often made in writing proofs.

Additional strategies are discussed in greater detail below.

Student Critiques of Proofs

A number of textbooks for "bridge" and discrete mathematics courses contain exercises asking students to determine whether a proposed proof for a given statement is valid or not. Campbell and Baker [4] developed activities that take these exercises one step further. Each activity "consists of a given statement and several different proposed proofs of that statement," some of which are valid and some of which are not. Students are divided into groups, and each group is given "one of the statement's proposed proofs, with directions... to determine if the proof is an acceptable argument," and, if so, to answer the following questions:

1) "What type of logical argument did the author use (direct, contradiction, contrapositive)?

2) How well written is the proof?

3) Was it easy to follow? Why or why not?

4) Can you think of some specific details which would make it clearer? If so, what are they?"

If students determine that the proposed proof is not an acceptable argument, they are asked to "identify all the major problems" they find with it. Each time a group finishes evaluating one proposed proof, it is given another, until each group has critiqued the entire collection. In the next class period, the students and the instructor discuss the various groups' critiques, "both on the level of identifying major issues, as well as minor problems such as style and clarity." Campbell reports that "having a variety of proposed proofs, all of the same statement, seems not only to help the students in recognizing certain logical errors, but also in developing a language of their own, recognizing that a statement can be correctly proven in a variety of ways, and learning the importance of reviewing one's work with a careful and objective eye." She also comments that students have benefited by becoming aware of the importance of format and of making proofs reader friendly.

Whole-Class Proofs

One technique for increasing student involvement in the proof-development process is for a teacher to do the writing on the board but have the students supply the individual steps. Richard L. Morrow [16] reported that when he uses this approach, he allows each student to give only one step so that as many students participate as possible. He wrote that "everyone gets to absorb the step, including its genesis or motivation, reason and role in the proof" and stated that the process makes it so that he "can

1) demonstrate how to go to the final steps and work backwards, when getting stuck approaching the proof from the beginning,

2) knowingly allow a proof to head off in the wrong direction and ask for suggestions on what to do when we get stuck — something which is sure to happen to many students when working alone,

3) demonstrate the value of marking up a diagram before writing down the steps,

4) show the value of getting a holistic view of the situation before putting down the series of steps — the right brain is especially useful in geometry proofs,

5) watch faces and judge how well the class or individuals are doing,

6) demonstrate that proofs do not need to be perfect or elegant to work,

7) let students know that everyone (or nearly so) is in the process of learning to do these things."

Identifying the Crux of a Proof

Many of the proofs one asks students to develop depend on a single central idea. Starting the proof-development process by trying to identify it accords with Leron's [12] suggestion to work down from a "top-level view of the proof." For other proofs, however, one may only come to realize the essential features after plowing mechanically through its details. Coming to see the crux of a proof in this way occurs, therefore, during the part of the problem-solving process Polya refers to as "looking back" [17]. A practiced mathematician can easily reconstruct a lengthy proof just by recollecting its essence, but students often have difficulty when told the main idea because they are still struggling to master the underlying logic of proof construction. Becoming aware that it is possible to reconstruct proofs from a few central ideas can help motivate them to develop facility with the more routine aspects of mathematical argumentation.

Using Informal Explanations

Hodgson and Morandi [10] report success following an idea of Mason, Burton, and Stacey [14] to have students first develop an informal explanation to convince a fellow student of the truth of a statement before trying to write a proof formally. Initially, the student verbalizes the explanation, using a tape recorder to refine it until a fellow student finds it convincing. Then the student writes up the explanation carefully. Only after completing these steps does the student rewrite the explanation, filling in any necessary details and using standard mathematical language. In their article, Hodgson and Morandi follow a student through the process as she develops a proof that for all integers n, the product $n(n + 1)(n + 2)$ is divisible by 6.

Student Presentations

Having students present proofs from homework assignments for the rest of the class at the board is especially effective if started in the very first class period after proofs have been assigned. It is important, however, to make sure to preserve the self-esteem of the presenters. One can thank them for being good sports when they volunteer and point out that to the extent that they make mistakes, discussion about them helps everyone in the class avoid similar errors in the future. If a student's proofs are good, the other students see that the demands made by their instructor can actually be met by one of their own kind. If a student's proofs contain mistakes or sections that are not well expressed, an instructor can ask for suggestions for improvement from the rest of the class. A ploy is to ask students to imagine they are a research team for a large company and that if they can collectively come up with a perfect answer they will get to share a sizeable bonus. After the class has finished its critique and some changes have been recorded, the instructor can take a turn, using the opportunity both to comment on significant errors that have gone undetected and also to show students the kinds of things the instructor will be looking for when grading students' work.

When I use this technique, I discuss small details as well as larger issues, but I try to put my criticisms in perspective, explaining frankly that certain corrections are more important than others, but that I also care about what might seem to be relatively minor points. For instance, if a student's proof states that a certain number, say n, is even because it equals $2k$, I would ask what was missing. Most likely, based on the emphasis I had previously placed on definitions, one of the other students would tell me to add "for some integer k." I would agree, pointing out that, for example, $1 = 2 \times (1/2)$ and yet 1 is not even and adding that it is not enough for n to be 2 times something — that *something* has to be an integer.

My primary reason for engaging in these kinds of critiques is to provide immediate feedback on students' proof writing, but an important secondary reason is to counteract student anxiety about how their proofs will be evaluated. Since there is more than one right way to construct any given proof and since different instructors may well have different standards of correctness, I feel obliged to try to give my students a sense of the range of proof styles I consider acceptable and to indicate which parts of a proof I consider most important. So when I critique student proofs, present my own, and write proofs at the board that have been developed collaboratively with members of the class, I discuss alternative ways of expressing the steps that I would consider acceptable. I also talk about conventions of mathematical writing, such as giving only part of the reason for a certain step, enough to indicate that the writer of the proof has considered and resolved the issue but not so much as to overload the proof with unnecessary verbiage. In addition, I point out that the amount of detail included in a proof varies considerably depending on the intended audience. In my courses I generally suggest that students address their proofs to a fellow student who has missed the last few classes.

At DePaul University some instructors have begun requiring students to present solutions to selected proof problems individually during office hours. Some require students to present one proof or disproof from each of the main types discussed in the course, while others offer students the possibility of raising their grade on a test by presenting a revised version of one of the problems they missed. Students are alerted that the instructor may stop to ask for clarification and base part of their grade on how effectively they respond, but because the presentations are private students do not need to worry about being embarrassed in front of their peers. In some cases the instructor's questions simply allow the student to demonstrate understanding of the reasons for certain steps; in other cases they raise more serious issues about the correctness of the argument or the incorrect use of terminology. Because several of the student's difficulties can be cleared up in the same session, such one-on-one student-instructor interaction can result in significant improvement in student understanding.

Rewriting Proofs

Requiring students to rewrite proofs until they are correct is a useful way to help students improve their proof-writing skills. In a large class it may be impossible for an instructor to find time to provide suggestions for improvement for the majority of assigned problems, but it may be possible to make sure that students rewrite at least one of each type of proof that is assigned. Nancy L. Hagelgans, Ursinus College, gave the following concrete suggestions [8].

1) Have students submit double-spaced, word-processed drafts electronically, except for first drafts taken from tests.

2) Write comments in pencil.

3) Comment on the appropriateness of the proof method or the lack of evident method.

4) If the method is appropriate, comment on the argument.

5) If the argument is valid, comment on the English composition.

6) Mention the good points: "A great first sentence!" "Clear organization!" "Good choice of method of proof!" "Excellent proof so far!"

7) Have conferences outside class with a few of the weakest students after several drafts.

Some variations she suggests are to have the whole class discuss selected first drafts that are projected on a screen, have student pairs discuss and write comments on each others' first drafts in class, and have students write comments on copies of selected first drafts. Another variation is to have students work on proofs in groups in class and go from group to group reviewing their work and offering hints on how to correct it.

Addressing Process Issues

To help students cope with the often frustrating enterprise of mathematical discovery, one can encourage class discussion about the psychological aspects of the process. For instance, if a few students have found a counterexample for a mathematical statement that stymied a majority of the class, one can ask the successful students to share the thoughts that went through their minds when the counterexample occurred to them. One can also point out that mathematical discovery may involve emotional ups and downs, that even the best mathematicians find mistakes in their arguments which force them to abandon one approach and seek another. For example, work of Schoenfeld [18] supports the view that while successful problem solvers are persistent, they readily change to new approaches when previous ones do not appear to be working, though they might eventually return to a previous approach if a new attempt seems unsuccessful.[1]

To assist students in structuring their time when they are trying to determine truth or falsity of a mathematical statement, one can suggest that they begin by imagining they actually have an object or objects satisfying the conditions described in the hypothesis. They can then ask themselves whether the conclusion must necessarily follow. If, after some effort, they do not see why this must be so, they can explore the possibility that the statement might be false by trying to think of elements that satisfy the hypothesis but not the conclusion. If this effort also fails, they can posit a situation where the hypothesis is true and the conclusion is false and try to derive a contradiction. If this method also seems to lead nowhere, the very process of having tried it and the other approaches may have resulted in insights that could lead to greater success when one of the previous approaches is tried again.

---

[1] The Nova program "The Proof," which describes Andrew Wiles' discovery of a proof for Fermat's Last Theorem, provides a powerful example for the effectiveness of this strategy.

3   Motivating the Need for Proof

A common use of proof is to affirm the general truth of properties that one has seen to be true in some cases, thereby coming to understand the essential reasons why the property always holds. While all introduction-to-proof courses try to convey this point, in courses where exploration and experimentation play a major role, it is the primary way the need for proof is introduced. For example, in the Mount Holyoke course Laboratory in Mathematical Experimentation [6] students work in groups on laboratory-style projects, most of which use the computer as an experimental tool to generate examples. Students are expected to come to see patterns and are then prompted to conjecture generalizations. Finally they are asked to support their conjectures with analytical arguments including, when possible, complete proofs. Projects are chosen from, among others, number theory, dynamical systems, and graph theory. The Franklin and Marshall College course Introduction to Higher Mathematics is structured in a similar way [13]. An initial "module" guides students through a sequence of increasingly pointed questions and activities to discover and verify basic properties of even and odd integers. Another module leads students to discover patterns related to the Fibonacci sequence by having them fill in values in a table for n, the nth Fibonacci number f_n, and the sum of the first n Fibonacci numbers $S(n)$. Later modules treat a variety of mathematical topics, such as polynomials and complex numbers, combinatorics and graph theory, difference equations and iteration, and number theory. The Foundations of Computing course at Butler University, developed by Peter Henderson [9], incorporates a once-a-week laboratory for exploration and experimentation alongside a more conventionally organized exposition of discrete mathematics that emphasizes logic and proof. In the lab sessions, students grapple with theoretical problems, which are often phrased engagingly as puzzles. These motivate the need for proof by requiring ingenuity and proof-like analysis (such as recursive thinking and identification of invariant properties) to solve.

When students are skeptical about the need for proof, a particularly effective way to motivate it is to have them evaluate statements about whose truth or falsity reasonable people might reasonably disagree. Fortunately, there are more such statements than one might think because what is obvious to a mathematician is not always obvious to a student. It is also possible to find relatively elementary statements upon which most people would need to reflect in order to reach a definitive answer. Such statements are especially effective when used for student presentations in class. For instance, consider the statement "For all integers a, b, and c, if a divides bc then a divides b or a divides c." If one assigns a homework problem asking students either to prove or provide a counterexample for this statement and then uses it for class discussion, it is common for one part of a class to claim it is false and another to say it is true. Once when two students from each group were chosen to go to the board to present their solutions, the result was one false proof, one partial "proof," one incorrect counterexample, and one correct counterexample. Such an outcome is a powerful argument for the importance of careful reasoning, especially if one points out that the ability to come up with correct answers to such mathematical questions provides the theoretical foundation to be able to engineer airplanes that do not crash, develop encryption systems to keep transmission of credit card information secure, and so forth.

This approach was developed as a formal teaching method, called "scientific debate," by a group of mathematics educators in France. In a first step, "the teacher initiates and organizes the production of scientific [mathematical] statements by the students. These are written on the blackboard without any immediate evaluation of their validity." In the second step, "the statements are put to the students for consideration and discussion. They come to a decision about their validity by taking a vote, with each opinion supported in some way, e.g., by scientific argument, by proof, by refutation, by counter-example, etc." In the third step, "the statements which can be validated by a full demonstration become theorems, whilst those which are established as incorrect are preserved as 'false-statements,' with a corresponding counter-example." [1]

The approach is taken even further in courses that use the "Moore method" or a "modified Moore method." In these courses students are given a list of definitions and an ordered set of statements proposed as possible theorems. They are given the job of proving the statements that are true and finding counterexamples for those that are false but are not allowed to consult textbooks or obtain solutions from an outside source. Classes consist primarily of presentations by students of their work, which is followed-up by questions and comments from members of the class. The method, originated by R. L. Moore for graduate courses at the University of Texas, has been modified by others to adapt it for use with a broader range of students and in less advanced courses. For instance, Chalice [5] includes elementary exercises on definitions to help the average student understand how they are applied to simple examples, and he encourages students to visit during office hours for hints on problems that give them difficulty. He also gives three exams during the semester, and when students are preparing for an exam he makes available to them careful proofs of the theorems that will be covered.

Conclusion

A few years ago I had an experience with one particular class that made a special impression on me. The class was unusually small, only twelve students, and was the second quarter of a sequence. The previous quarter had dealt with logic, an introduction to direct and indirect proof, mathematical induction, and elementary combinatorics, all interwoven with various computer science applications. The second quarter was to cover set theory, function properties, recursion, some analysis of algorithms, relations on sets, and an introduction to graph theory, also with an admixture of applications. The class met only once a week but in three-hour sessions.

The small size of the class and the length of the sessions gave me a chance to work with students more intimately than usual. I began each period by having students discuss in groups of three or four the homework they had prepared for that day and went from group to group talking with each at length. Overall the class atmosphere was excellent, and several students showed the kind of eager, enthusiastic intelligence that is a teacher's joy. What surprised me was that as the course moved from one topic to the next, almost all the students who had attained a relatively sophisticated level of achievement in dealing with a previous topic made it clear that they felt they had to struggle to succeed with the next. Yet as we worked through their questions and difficulties, they ultimately performed excellently with the new topic as well. Their understanding of general methodological principles clearly made it easier for them to learn the new material but it did not make it trivial for them.

This experience brought home to me more effectively than any before that abstract mathematical thinking is not something that either one is able to do or one is not able to do. Because of the experience I have become especially conscious of the need to respect my students and never to act surprised by their questions. Even when a student asks a question whose answer I have already discussed, I try to respond to it as if it were fresh. After all, nobody can concentrate 100% of the time when new ideas are coming in fast and furiously. In all likelihood the student was not mentally prepared to absorb the answer when I previously addressed the question. For the student to formulate the question means that they have thought about the issue, want to know the answer, and are probably ready to understand it. That is cause to celebrate. It may also be that clarifying the issue at this point in the course (if possible in a slightly different way from that presented earlier) will give the other students in the class greater insight also.

My main advice to those teaching courses whose goal is to develop students' mathematical reasoning powers is to play an activist role but recognize that achieving success is a long-term process. I have sometimes been surprised when students who in my view fell far short of achieving the levels of accomplishment I strive for tell me how valuable they found the course in helping them do better work in their other courses or (I am always pleased to hear) in their jobs.

The analogy I like to draw is of a child learning to walk. It takes months of daily effort for most children to take their first steps and several more months until they actually become steady on their feet. When a child is trying to move from one stage to the next in learning to walk and has failed several times, we don't say "Forget it." We remain calm, good humored, and encouraging. And when the child finally succeeds, we spare nothing in expressing our delight.

Bibliography

[1] Alibert, Daniel and Michael Thomas. "Research on Mathematical Proof," *Advanced Mathematical Thinking*, David Tall (editor), Kluwer Academic Publishers, 1991, 215–230.

[2] Barwise, Jon and John Etchemendy. *Hyperproof*, CSLI Publications, 1994.

[3] Barwise, Jon and John Etchemendy. *The Language of First-Order Logic*, CSLI Publications, 1999.

[4] Baker, Diane and Connie Campbell. "Fostering the Development of Mathematical Thinking: Observations from a Proofs Course," *PRIMUS* 14 (2004) 345–353.

[5] Chalice, D. "How to teach a class by the modified Moore method," *Amer. Math. Monthly* 102 (1995) 317–321.

[6] Cobb, George, et al. *Laboratories in Mathematical Experimentation: A Bridge to Higher Mathematics*, Key College Publishing, 1997.

[7] Epp, Susanna S. "The Role of Logic in Teaching Proof," *Amer. Math. Monthly* 110 (2003) 886–899.

[8] Hagelgans, Nancy. "Learning to Prove by Rewriting," presentation at the 2005 Joint Mathematics Meetings, Atlanta, GA.

[9] Henderson, Pete. *Nifty Examples for Discrete Mathematics*,
`http://blue.butler.edu/~phenders/sigcse2004/niftyworkshop`

[10] Hodgson, Ted and Pat Morandi. "Exploration, Explanation, Formalization: A Three-step Approach to Proof," *PRIMUS* 6 (1996) 49–57.

[11] Lee, John and Keith Stenning. "Cognitive Processes Involved in Learning Logic," *Proceedings of the International Symposium on Teaching Logic and Reasoning in an Illogical World*, Center for Discrete Mathematics and Theoretical Computer Science, Rutgers University.
`http://www.cs.cornell.edu/Info/People/gries/symposium/symp.htm.`

[12] Leron, Uri. "Heuristic Presentations: The Role of Structuring," *For the Learning of Mathematics* 5 (1985) 7–13.

[13] Levine, Alan and Ben Shanfelder. "The Transition to Advanced Mathematics," *PRIMUS* 10 (2000) 97–110.

[14] Mason, J., L. Burton, and K. Stacey. *Thinking Mathematically*, Addison-Wesley, 1982.

[15] Moore, Robert C. "Making the Transition to Formal Proof," *Educational Studies in Mathematics* 27 (1994) 249–266.

[16] Morrow, Richard L. Private communication, 2004.

[17] Polya, George. *How To Solve It*, Princeton University Press, 1945.

[18] Schoenfeld, Alan H. *Mathematical Problem Solving*, Academic Press, 1985.

[19] Selden, Annie and John Selden. "The Role of Logic in the Validation of Mathematical Proofs," *Proceedings of the International Symposium on Teaching Logic and Reasoning in an Illogical World*, Center for Discrete Mathematics and Theoretical Computer Science, Rutgers University.
`http://www.cs.cornell.edu/Info/People/gries/symposium/symp.htm.`

[20] Smullyan, Raymond. *What Is the Name of This Book?*, Prentice-Hall, 1978.

[21] Tall, David O. and Shlomo Vinner. "Concept Image and Concept Definition in Mathematics with Particular Reference to Limits and Continuity," *Educational Studies in Mathematics* 12 (1981) 151–169.

About the Editor

Brian Hopkins received his BS in mathematics and a BA in philosophy from the University of Texas in 1990, and earned his PhD in mathematics from the University of Washington in 1997. He teaches at Saint Peter's College, a Jesuit liberal arts institution in Jersey City, where he has led several undergraduate research projects and has been the recipient of the Varrichio Award for Teaching Excellence (awarded by the SPC Pi Mu Epsilon chapter) two times, in 2004 and 2007. Brian is the author, with Carl Swenson, of *Getting Started with the TI-92 in Calculus* (1998, John Wiley & Sons). Also, he has published several research articles in combinatorics and, with Robin J. Wilson, won the MAA's 2005 George Pólya Award for excellent expository writing in the *College Mathematics Journal* for "The Truth About Königsberg." Brian works with secondary school teachers in professional development projects with various organizations including the Institute for Advanced Study's Park City Mathematics Institute, the Northwest Math Interaction, the New Jersey Professional Development and Outreach group, the Institute for New Jersey Mathematics Teachers, and the Pikes Peak Math Teacher Circle Academy. He is a member of the Mathematical Association of America, the American Mathematical Society, the National Council of Teachers of Mathematics, and the National Association of Recording Arts and Sciences. Brian plays piano, sings with Cantori New York, and enjoys New York City with his partner Michael.